New Advances in Diagnostic Radiology of Ischemic Stroke

New Advances in Diagnostic Radiology of Ischemic Stroke

Editors

Gabriel Broocks
Lukas Meyer

Basel • Beijing • Wuhan • Barcelona • Belgrade • Novi Sad • Cluj • Manchester

Editors
Gabriel Broocks
MSH Medical School
Hamburg—University of
Applied Sciences and
Medical University
Hamburg
Germany

Lukas Meyer
University Medical Center
Hamburg-Eppendorf
Hamburg
Germany

Editorial Office
MDPI AG
Grosspeteranlage 5
4052 Basel, Switzerland

This is a reprint of articles from the Special Issue published online in the open access journal *Journal of Clinical Medicine* (ISSN 2077-0383) (available at: http://www.mdpi.com).

For citation purposes, cite each article independently as indicated on the article page online and as indicated below:

Lastname, A.A.; Lastname, B.B. Article Title. *Journal Name* **Year**, *Volume Number*, Page Range.

ISBN 978-3-7258-0069-8 (Hbk)
ISBN 978-3-7258-0070-4 (PDF)
doi.org/10.3390/books978-3-7258-0070-4

Contents

About the Editors

Gabriel Broocks

Gabriel Broocks is a professor of radiology and neuroradiology at the Medical School Hamburg (MSH) and the senior managing physician at the Helios Medical Center Schwerin, Germany. His research focuses on quantitative imaging biomarkers in ischemic stroke and the investigation of treatment effects in acute ischemia. To date, Dr. Broocks has currently published more than 160 PubMed-listed studies on ischemic stroke and coordinates several prospective scientific projects. He is the recipient of several awards, such as the Cornelius Dyke Award of the American Society of Neuroradiology and the Kurt Decker Award of the German Society of Neuroradiology (DGNR).

Lukas Meyer

Lukas Meyer is a radiology resident at the Department of Radiology and Neuroradiology at the University Medical Center Hamburg-Eppendorf, Germany. Dr. Meyer has published more than 100 papers on ischemic stroke, focusing on interventional treatment, and is the recipient of the Intervention Award of the European Society of Neuroradiology (ESNR).

Journal of
Clinical Medicine

New Advances in Diagnostic Radiology for Ischemic Stroke

Gabriel Broocks * and Lukas Meyer

Department of Diagnostic and Interventional Neuroradiology, University Medical Center Hamburg-Eppendorf, Martinistraße 52, 20246 Hamburg, Germany; lu.meyer@uke.de
* Correspondence: g.broocks@uke.de

Citation: Broocks, G.; Meyer, L. New Advances in Diagnostic Radiology for Ischemic Stroke. *J. Clin. Med.* **2023**, *12*, 6375. https://doi.org/10.3390/jcm12196375

Received: 7 September 2023
Accepted: 22 September 2023
Published: 6 October 2023

1. Introduction

Ischemic stroke, a leading cause of disability and mortality worldwide, occurs due to the sudden interruption of blood supply to a specific region of the brain. Timely and accurate diagnosis is crucial for providing effective treatment and improved patient outcomes. In recent years, diagnostic neuroimaging has witnessed significant advancements that revolutionized the way ischemic stroke is diagnosed, leading to improved patient care and outcomes. This editorial explores the recent advances in radiology techniques for diagnosing ischemic stroke and highlights their potential to enhance early detection, accurate characterization, and treatment decision making.

2. Advances in Stroke Neuroimaging and Treatment

2.1. Computed Tomography

Dual-energy CT (DECT), spectral CT, and photon-counting CT are innovative technologies that hold significant promise for enhancing the imaging triage of acute ischemic stroke. These advancements offer improved insights into stroke pathology, leading to more accurate and rapid diagnosis and treatment decisions. DECT employs two different X-ray energy levels, enabling better tissue differentiation and characterization. In stroke cases, this method can distinguish between ischemic and hemorrhagic strokes, resulting in swift intervention. Moreover, DECT can enhance vascular imaging, which is crucial for identifying vessel occlusions and determining treatment strategies like thrombectomy. Spectral CT goes beyond DECT by analyzing the energy-dependent attenuation of tissues. This technique allows for the enhanced visualization of subtle tissue changes associated with ischemic stroke, such as perfusion deficits and infarct expansion. Spectral CT's material decomposition capabilities offer detailed information on tissue composition, which can help to distinguish penumbra (potentially salvageable tissue) from core infarction [1]. Photon-counting CT, a cutting-edge technology, directly detects individual X-ray photons and enables energy-resolved imaging. This enables exquisite tissue differentiation, which is critical in stroke cases where tissue distinctions are subtle. It can improve the accuracy of perfusion imaging and provide insights into collateral circulation, which are both essential for making stroke management decisions.

In summary, these advanced CT technologies offer remarkable potential to transform the imaging triage of acute ischemic stroke. They enhance tissue differentiation, aid in stroke subtype classification, refine perfusion assessments, and ultimately facilitate quicker and more tailored interventions, leading to improved patient outcomes and reduced morbidity.

2.2. Perfusion Imaging and Automated Imaging Platforms

The assessment of cerebral hemodynamics has gained prominence in stroke diagnosis. Techniques like arterial spin labeling (ASL) perfusion MRI provide non-invasive measurements of cerebral blood flow, aiding in the identification of compromised perfusion territories. Dynamic susceptibility contrast (DSC) and dynamic contrast-enhanced (DCE) perfusion MRI offer insights into hemodynamic parameters, allowing for the differentiation between penumbra and infarct core.

Perfusion imaging and automated imaging platforms may enhance the accuracy of stroke diagnostics. Perfusion imaging provides dynamic information about the blood flow within the brain, enabling the identification of salvageable tissue (penumbra) and core infarction. This knowledge is crucial for guiding time-sensitive decisions like thrombolysis or thrombectomy, especially in subgroups with borderline indications for reperfusion therapy. Perfusion imaging also allows for clinicians to not only assess tissue viability and match treatment options to individual patients, but also to predict recovery potential more accurately. This information is important to anticipate the patients' courses and inform their family members. Automated platforms that are already used in daily clinical practice streamline the analysis of complex imaging data. These platforms rapidly process images to quantify perfusion deficits and penumbra volumes. This automation reduces inter-operator variability and time, allowing for clinicians to make timely decisions based on objectives and standardized data. Moreover, these platforms can be integrated into telestroke networks, facilitating expert consultation for remote hospitals that do not have stroke specialists onsite.

2.3. Machine Learning and Artificial Intelligence

Machine learning and artificial intelligence (AI) have brought about a paradigm shift in ischemic stroke diagnosis. These technologies enable the extraction of intricate patterns and insights from vast amounts of imaging and clinical data. AI algorithms can rapidly analyze images, identify subtle abnormalities, and quantify tissue characteristics, assisting radiologists in making accurate and timely diagnoses.

Deep learning models, when trained on large datasets, can detect small infarcts, assess lesion volume, and predict treatment responses. AI-driven software can also automatically segment brain structures and vessels, which saves time and reduces variability in radiological interpretations, leading to more standardized diagnoses.

2.4. Molecular Imaging and Biomarkers

Molecular imaging techniques offer a deeper understanding of ischemic stroke pathophysiology by visualizing molecular processes in real time. Positron emission tomography (PET) combined with radiotracers targeting specific molecular pathways can provide valuable information about inflammation, oxidative stress, and neurovascular remodeling in ischemic stroke cases.

CT densitometry-based methods used to directly quantify the net water uptake in the ischemic brain tissue have been increasingly applied in recent years [2]. There is potential for the net water uptake to be used as an imaging biomarker for the pathophysiology of infarcted lesions in clinical decision making. Artificial intelligence might help to address the lack of automation and standardization in the measurement of the net water uptake, which is an important current limitation.

2.5. Interventional Treatment

The use of thrombectomy in ischemic stroke cases has showcased remarkable advancements, expanding its application to patients with a large ischemic core, as supported by recent studies [3–5].

Currently running trials are investigating the effect of recanalization in distal vessel occlusions. Newer-generation stent retrievers and aspiration devices have demonstrated improved efficacy in reaching and removing clots lodged in smaller cerebral vessels. These innovations have widened the therapeutic window for endovascular treatment, enabling intervention in cases that were previously considered challenging.

Additionally, recent studies have investigated the efficacy of using thrombectomy to treat patients with a large ischemic core, where tissue damage was traditionally thought to be irreversible. Advanced imaging techniques, such as perfusion imaging, may help to identify salvageable tissue within the core, refining patient selection. Several trials have validated the benefit of thrombectomy even in these cases, showing improved functional outcomes when treatment is aligned with imaging-guided patient selection.

These breakthroughs underscore thrombectomy's transformative potential in ischemic stroke care. By addressing distal occlusions and extending treatment to patients with substantial core infarctions, thrombectomy continues to push the boundaries of stroke intervention.

3. Challenges and Future Directions

While the advances in diagnostic radiology for ischemic stroke are promising, challenges remain. Access to advanced imaging technologies and expertise may be limited in certain regions, hindering their widespread adoption. Additionally, integrating these advanced techniques into routine clinical practice requires training and familiarity for radiologists and clinicians.

Further research is needed to validate the clinical utility of emerging techniques and refine their applications. Prospective studies are essential to establish the long-term benefits of advanced imaging in guiding treatment decisions, predicting outcomes, and enhancing patient care.

4. Studies of the Research Topic—Hemorrhagic Stroke

Cao et al. [6] assessed the first available automated 3D segmentation tool for intracerebral hemorrhage (ICH) based on a 3D neural network before and after retraining. Their proposed model showed a decent generalization in an external validation cohort, and the authors concluded that external validation and retraining are significant steps that need to be assessed before applying deep learning models in novel clinical settings.

It is known that ICH is associated with a significant long-term morbidity and a high mortality and therefore has a significant health economic impact. In particular, outcomes are poor if the onset is not known, but there are no established imaging-based tools that can be used to estimate the onset. Rusche et al. [7] assessed whether the onset estimation of ICH patients using AI may be more accurate than human readers. Interestingly, the diagnostic accuracy of AI-based classifiers for the onset determination of patients with ICH was low, suggesting that accurate AI-based onset assessment for patients with ICH based on CT data may not currently be likely to alter decision making in daily clinical practice. It was concluded that in the future, multimodal AI-based approaches might improve the precision of ICH onset prediction.

The precise detection of cerebral microbleeds (CMBs) using susceptibility-weighted (SWI) magnetic resonance imaging (MRI) is important for the characterization of several neurological diseases. Rusche et al. [8] evaluated the diagnostic performance of whole-body low-field MRI for the detection of CMBs in a prospective cohort of suspected stroke patients compared to an established 1.5 T MRI and observed that low-field MRI at 0.55 T might have a similar accuracy compared to 1.5 T scanners for the detection of CMBs, and thus may have great potential as an interesting alternative for future studies. In a further study, Rusche et al. [9] performed a detailed comparative analysis of low-field MRI including 27 ischemic stroke patients who all received 1.5 T and 0.55 T imaging. The authors discussed that low-field MRI with 0.55 T might not be inferior to MR scanners with higher field strengths, emphasizing their potential as low-cost alternatives in stroke imaging triage. Some limitations should nevertheless be considered, particularly with regard to the detection of very small infarcts.

5. Studies of the Research Topic—Ischemic Stroke

Perfusion imaging is often utilized to identify hypoperfusion in acute situations, but lacks availability in many stroke centers. Bunker et al. [10] discussed FLAIR hyperintense vessels (FHVs) in various vascular regions as an alternative method for quantifying hypoperfusion. In conjunction with prior work, their results support the use of FLAIR imaging to estimate the amount and location of hypoperfusion when perfusion imaging is not available.

In this context, it is important to note that the capability to precisely detect ischemic stroke and to predict neurological recovery at an early stage is of great clinical value. The study by Guo et al. [11] aimed to test the diagnostic performance of whole-brain dynamic radiomics features (DRF) for the detection of ischemic stroke and the assessment of neurological impairment. This study discussed an interesting clinical tool with a high potential for use in clinical practice to improve diagnosis and early outcome estimation even before treatment.

Another technique utilized in ischemic stroke triage is the perfusion-based hypoperfusion intensity ratio (HIR), which is known to be associated with collateral status, reflecting impaired microperfusion of the ischemic brain tissue in stroke patients presenting with a large vessel occlusion (AIS-LVO). Van Horn et al. [12] investigated whether HIR is directly associated with an early edema progression rate (EPR), which was assessed using the ischemic net water uptake (NWU) in a multicenter observational study of AIS-LVO patients treated via mechanical thrombectomy. An established quantitative imaging biomarker was utilized, and the authors discussed that a favorable HIR was directly linked to a lower EPR as measured on baseline NCCT.

In the past, it was observed that the formation of ischemic edema, which is the imaging hallmark of progressive infarction, may be directly influenced by thrombolytic treatment and endovascular recanalization. Broocks et al. [13] hypothesized that intravenous therapy with alteplase (IVT) is linked to a worse clinical outcome and increased ischemic edema formation when administered to ischemic stroke patients who subsequently achieved complete recanalization (defined as TICI 3) after MT. In summary, it was observed that bridging IVT was associated with an increased edema volume and risk of symptomatic intracerebral hemorrhage as secondary injury volumes. The results of this study encourage direct MT approaches, particularly in patients with a higher likelihood of successful recanalization.

After stroke, angiogenesis and neuroplasticity are important recovery processes. Recent advances in neuroimaging techniques might be used to assess these processes and become quantifiable indicators, or they may to be used to assess treatment effects. Wlodarczyk et al. [14] discussed neuroimaging techniques as potential tools to assess angiogenesis and neuroplasticity processes after ischemic stroke and discussed the potential clinical implications for recovery prognosis and rehabilitation.

As a further current important research topic, the detailed analysis of thrombi retrieved from endovascular procedures may be of great importance in order to increase the success of interventional treatment. Viltuznik et al. [15] investigated the accuracy of cerebral thrombus characterization via computed tomography (CT), magnetic resonance (MR), and histology, and how several parameters attained using these methods are correlated with each other. Finally, these parameters' association with interventional procedures and clinical variables was investigated.

6. Conclusions

The field of diagnostic radiology has witnessed remarkable advances in the diagnosis of ischemic stroke, revolutionizing patient care and outcomes. Machine learning, advanced perfusion imaging, and quantitative imaging biomarkers have transformed our ability to detect and characterize ischemic stroke. These advances enable early intervention, accurate treatment decisions, and personalized care, ultimately leading to improved patient outcomes. As the momentum of innovation continues, collaborative efforts between clinicians, radiologists, and researchers will further drive the evolution of diagnostic radiology in the realm of ischemic stroke management.

Conflicts of Interest: The authors declare no conflict of interest.

References

1. Steffen, P.; Austein, F.; Lindner, T.; Meyer, L.; Bechstein, M.; Rumenapp, J.; Klintz, T.; Jansen, O.; Gellissen, S.; Hanning, U.; et al. Value of Dual-Energy Dual-Layer CT After Mechanical Recanalization for the Quantification of Ischemic Brain Edema. *Front. Neurol.* **2021**, *12*, 668030. [CrossRef] [PubMed]
2. Broocks, G.; Hanning, U.; Faizy, T.D.; Scheibel, A.; Nawabi, J.; Schon, G.; Forkert, N.D.; Langner, S.; Fiehler, J.; Gellissen, S.; et al. Ischemic lesion growth in acute stroke: Water uptake quantification distinguishes between edema and tissue infarct. *J. Cereb. Blood Flow. Metab.* **2020**, *40*, 823–832. [CrossRef] [PubMed]
3. Sarraj, A.; Hassan, A.E.; Abraham, M.G.; Ortega-Gutierrez, S.; Kasner, S.E.; Hussain, M.S.; Chen, M.; Blackburn, S.; Sitton, C.W.; Churilov, L.; et al. Trial of Endovascular Thrombectomy for Large Ischemic Strokes. *N. Engl. J. Med.* **2023**, *388*, 1259–1271. [CrossRef] [PubMed]
4. Huo, X.; Ma, G.; Tong, X.; Zhang, X.; Pan, Y.; Nguyen, T.N.; Yuan, G.; Han, H.; Chen, W.; Wei, M.; et al. Trial of Endovascular Therapy for Acute Ischemic Stroke with Large Infarct. *N. Engl. J. Med.* **2023**, *388*, 1272–1283. [CrossRef] [PubMed]
5. Yoshimura, S.; Sakai, N.; Yamagami, H.; Uchida, K.; Beppu, M.; Toyoda, K.; Matsumaru, Y.; Matsumoto, Y.; Kimura, K.; Takeuchi, M.; et al. Endovascular Therapy for Acute Stroke with a Large Ischemic Region. *N. Engl. J. Med.* **2022**, *386*, 1303–1313. [CrossRef] [PubMed]
6. Cao, H.; Morotti, A.; Mazzacane, F.; Desser, D.; Schlunk, F.; Guttler, C.; Kniep, H.; Penzkofer, T.; Fiehler, J.; Hanning, U.; et al. External Validation and Retraining of DeepBleed: The First Open-Source 3D Deep Learning Network for the Segmentation of Spontaneous Intracerebral and Intraventricular Hemorrhage. *J. Clin. Med.* **2023**, *12*, 4005. [CrossRef] [PubMed]
7. Rusche, T.; Wasserthal, J.; Breit, H.C.; Fischer, U.; Guzman, R.; Fiehler, J.; Psychogios, M.N.; Sporns, P.B. Machine Learning for Onset Prediction of Patients with Intracerebral Hemorrhage. *J. Clin. Med.* **2023**, *12*, 2631. [CrossRef] [PubMed]
8. Rusche, T.; Breit, H.C.; Bach, M.; Wasserthal, J.; Gehweiler, J.; Manneck, S.; Lieb, J.M.; De Marchis, G.M.; Psychogios, M.; Sporns, P.B. Prospective Assessment of Cerebral Microbleeds with Low-Field Magnetic Resonance Imaging (0.55 Tesla MRI). *J. Clin. Med.* **2023**, *12*, 1179. [CrossRef] [PubMed]
9. Rusche, T.; Breit, H.C.; Bach, M.; Wasserthal, J.; Gehweiler, J.; Manneck, S.; Lieb, J.M.; De Marchis, G.M.; Psychogios, M.N.; Sporns, P.B. Potential of Stroke Imaging Using a New Prototype of Low-Field MRI: A Prospective Direct 0.55 T/1.5 T Scanner Comparison. *J. Clin. Med.* **2022**, *11*, 2798. [CrossRef] [PubMed]
10. Bunker, L.D.; Hillis, A.E. Location of Hyperintense Vessels on FLAIR Associated with the Location of Perfusion Deficits in PWI. *J. Clin. Med.* **2023**, *12*, 1554. [CrossRef] [PubMed]
11. Guo, Y.; Yang, Y.; Cao, F.; Wang, M.; Luo, Y.; Guo, J.; Liu, Y.; Zeng, X.; Miu, X.; Zaman, A.; et al. A Focus on the Role of DSC-PWI Dynamic Radiomics Features in Diagnosis and Outcome Prediction of Ischemic Stroke. *J. Clin. Med.* **2022**, *11*, 5364. [CrossRef] [PubMed]
12. van Horn, N.; Broocks, G.; Kabiri, R.; Kraemer, M.C.; Christensen, S.; Mlynash, M.; Meyer, L.; Lansberg, M.G.; Albers, G.W.; Sporns, P.; et al. Cerebral Hypoperfusion Intensity Ratio Is Linked to Progressive Early Edema Formation. *J. Clin. Med.* **2022**, *11*, 2373. [CrossRef] [PubMed]
13. Broocks, G.; Meyer, L.; Ruppert, C.; Haupt, W.; Faizy, T.D.; Van Horn, N.; Bechstein, M.; Kniep, H.; Elsayed, S.; Kemmling, A.; et al. Effect of Intravenous Alteplase on Functional Outcome and Secondary Injury Volumes in Stroke Patients with Complete Endovascular Recanalization. *J. Clin. Med.* **2022**, *11*, 1565. [CrossRef] [PubMed]
14. Wlodarczyk, L.; Cichon, N.; Saluk-Bijak, J.; Bijak, M.; Majos, A.; Miller, E. Neuroimaging Techniques as Potential Tools for Assessment of Angiogenesis and Neuroplasticity Processes after Stroke and Their Clinical Implications for Rehabilitation and Stroke Recovery Prognosis. *J. Clin. Med.* **2022**, *11*, 2473. [CrossRef] [PubMed]
15. Viltuznik, R.; Bajd, F.; Milosevic, Z.; Kocijancic, I.; Jeromel, M.; Fabjan, A.; Kralj, E.; Vidmar, J.; Sersa, I. An Intermodal Correlation Study among Imaging, Histology, Procedural and Clinical Parameters in Cerebral Thrombi Retrieved from Anterior Circulation Ischemic Stroke Patients. *J. Clin. Med.* **2022**, *11*, 5976. [CrossRef] [PubMed]

Journal of
Clinical Medicine

Brief Report

Machine Learning for Onset Prediction of Patients with Intracerebral Hemorrhage

Thilo Rusche [1,*,†], Jakob Wasserthal [1,†], Hanns-Christian Breit [1], Urs Fischer [2], Raphael Guzman [3], Jens Fiehler [4], Marios-Nikos Psychogios [1] and Peter B. Sporns [1,4,5]

[1] Department of Neuroradiology, Clinic of Radiology & Nuclear Medicine, University Hospital Basel, 4031 Basel, Switzerland
[2] Department of Neurology, University Hospital Basel, 4031 Basel, Switzerland
[3] Department of Neurosurgery, University Hospital Basel, 4031 Basel, Switzerland
[4] Department of Diagnostic and Interventional Neuroradiology, University Medical Center Hamburg-Eppendorf, 55131 Hamburg, Germany
[5] Department of Radiology and Neuroradiology, Stadtspital Zürich, 8063 Zürich, Switzerland
* Correspondence: rusche.thilo@googlemail.com
† These authors contributed equally to this work.

Citation: Rusche, T.; Wasserthal, J.; Breit, H.-C.; Fischer, U.; Guzman, R.; Fiehler, J.; Psychogios, M.-N.; Sporns, P.B. Machine Learning for Onset Prediction of Patients with Intracerebral Hemorrhage. *J. Clin. Med.* **2023**, *12*, 2631. https://doi.org/10.3390/jcm12072631

Academic Editor: Mariarosaria Valente

Received: 16 February 2023
Revised: 13 March 2023
Accepted: 30 March 2023
Published: 31 March 2023

Abstract: Objective: Intracerebral hemorrhage (ICH) has a high mortality and long-term morbidity and thus has a significant overall health–economic impact. Outcomes are especially poor if the exact onset is unknown, but reliable imaging-based methods for onset estimation have not been established. We hypothesized that onset prediction of patients with ICH using artificial intelligence (AI) may be more accurate than human readers. Material and Methods: A total of 7421 computed tomography (CT) datasets between January 2007–July 2021 from the University Hospital Basel with confirmed ICH were extracted and an ICH-segmentation algorithm as well as two classifiers (one with radiomics, one with convolutional neural networks) for onset estimation were trained. The classifiers were trained based on the gold standard of 644 datasets with a known onset of >1 and <48 h. The results of the classifiers were compared to the ratings of two radiologists. Results: Both the AI-based classifiers and the radiologists had poor discrimination of the known onsets, with a mean absolute error (MAE) of 9.77 h (95% CI (confidence interval) = 8.52–11.03) for the convolutional neural network (CNN), 9.96 h (8.68–11.32) for the radiomics model, 13.38 h (11.21–15.74) for rater 1 and 11.21 h (9.61–12.90) for rater 2, respectively. The results of the CNN and radiomics model were both not significantly different to the mean of the known onsets ($p = 0.705$ and $p = 0.423$). Conclusions: In our study, the discriminatory power of AI-based classifiers and human readers for onset estimation of patients with ICH was poor. This indicates that accurate AI-based onset estimation of patients with ICH based only on CT-data may be unlikely to change clinical decision making in the near future. Perhaps multimodal AI-based approaches could improve ICH onset prediction and should be considered in future studies.

Keywords: artificial intelligence; onset prediction; intracerebral hemorrhage; machine learning

1. Introduction

Intracerebral hemorrhage (ICH) has an incidence of 10–30 per 100.000 people per year worldwide [1,2]. The by far most common form is spontaneous primary ICH, caused by rupture of small intraparenchymal arterial vessels and arterioles caused by preexisting vascular damage in the setting of chronic arterial hypertension or amyloid angiopathy [3]. Compared with ischemic strokes, hemorrhagic strokes have a higher morbidity, and an overall mortality of more than 30% in the first 30 days [3–5]. Moreover, half of ICH-associated mortality occurs during the first 24 h after the initial event [6], and a poor outcome with persistent physical impairment in 75% of all patients after 1 year [7]. Therefore, ICH results in a significant financial burden on the health care system due to the long-term treatment that is often required [8–10].

In many patients diagnosed with ICH by imaging, the exact onset is unknown [11]. However, therapeutic and triage decisions are based on factors such as the likelihood of rebleeding, brain edema, and herniation which are associated with the time since onset of the bleeding [12]. Thus, it has been shown that functional outcomes after 30 days are significantly worse in patients with unclear onset compared to patients with a known onset [13]. Possible treatment strategies [14] such as blood pressure reduction [15–17], neurosurgical evacuation [18], and application of tranexamic acid [19] may reduce the risk of hematoma expansion associated with neurological deterioration leading to poor outcomes and death [12,20]. Even though promising manual [12,20–22] and machine learning based algorithms have been developed and tested to predict the risk of hematoma growth and functional outcomes [23], to our knowledge there is no study that tested the reliability of machine-learning algorithms for onset estimation of patients with ICH based on imaging features. An AI-based approach is therefore particularly interesting and promising, as it can detect image information that may be hidden or primarily not obvious and use it for analysis processes. This opens new possibilities for data analysis. In recent years, AI-based applications have increasingly found their way into clinical routine, especially in the field of radiology [24,25]. For example, the automated detection of pulmonary artery emboli in CT scans of the thorax [26] or the detection and volumetry of ICH in CT scans of the head [27–29].

Therefore, we hypothesized that onset estimation in patients with ICH using machine learning may be feasible and may be more accurate than human readers.

2. Methods

2.1. Study Design

This study includes all consecutive patients diagnosed with ICH by computed tomography (CT) at the University Hospital Basel between January 2007–July 2021. Patients were identified using a self-programmed tool for internal Picture Archiving and Communication System (PACS)-based data retrieval (PACS crawler). Further inclusion criteria were being an age > 17 years at the timepoint of imaging and having a known onset of symptoms. Patients who had declined the use of personal data for research purposes were excluded.

2.2. Data Processing, Classifier Training and Image Assessment

CT datasets were exported to a research server (NORA Imaging Platform, Freiburg, Germany [10.1055/s-0037-1602977]). Manual segmentation of the hematoma in 1 mm thick axial slices (soft tissue window) was performed in a random subset of 319 CT datasets to train an algorithm for automatic ICH segmentation. All other CT datasets were segmented using the trained classifier (as described below).

Clinical data including time points of symptom onset were extracted from patients' clinical records.

For the automated segmentation of ICHs, we trained a nnU-Net on our manually annotated dataset [https://arxiv.org/abs/1809.10486, accessed on 10 September 2021] [10.1038/s41592-020-01008-z]. nnU-Net is a medical segmentation framework, which automatically configures the data preprocessing as well as the hyperparameters for training a U-Net. It can derive heuristics for optimally setting the data preprocessing parameters (e.g., normalization and resampling) as well as the U-Net configuration (e.g., number of layers and batch size) based on the characteristics of the input dataset. Furthermore, it performs extensive data augmentations (image rotation, blurring, etc.). On more than 20 public imaging segmentation challenges, this automatically configured segmentation pipeline was superior to other submissions. For this reason, the nnU-Net was used for the ICH segmentation.

For estimation of the onset of the patients with ICH, two different algorithms were tested: 1. Regression trees with radiomics features (Radiomics). 2. Convolutional neural network (CNN). The classifiers were trained based on the gold standard of 644 datasets with (a) a known onset of >1 h and <48 h and (b) a hematoma volume of >15 mm^3 (see flowchart in Figure 1). Criterion (a) was defined to include only acute to subacute ICHs in the training dataset as therapeutic and triage decisions are particularly relevant in this time frame. Criterion (b) was defined because a cut-off of >15 mm^3 provided reliable automated ICH-segmentation. A 5-fold cross-validation was used for evaluation. No extra validation set for hyperparameter optimization was used since only standard models and hyperparameters were used.

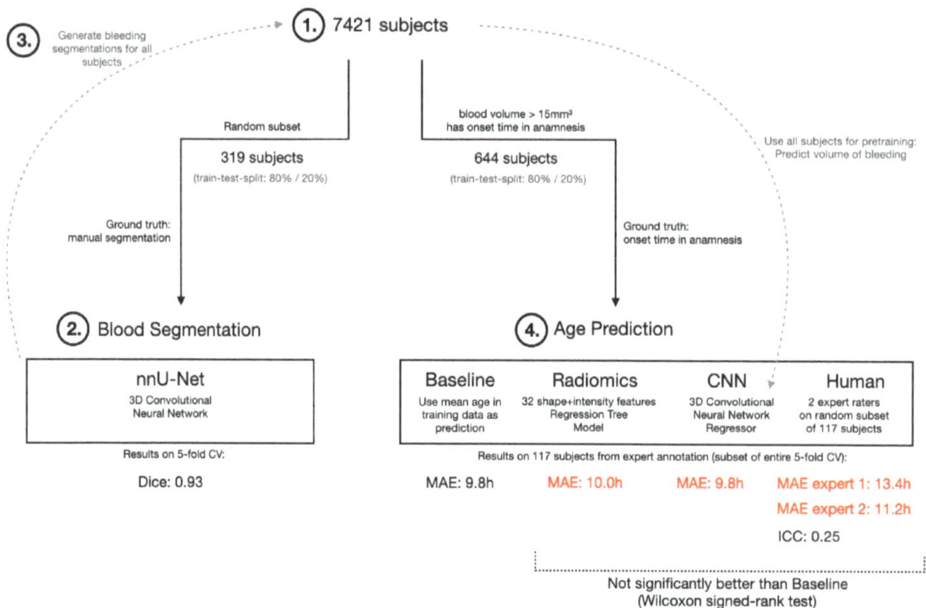

Figure 1. Flowchart of inclusion of patients. Overview study design: On a subset of the original study population (1.) a hematoma segmentation model is trained (2.) which is used to generate hematoma segmentations for the entire study population (3.) Then, the radiomics-, CNN-, and human-based age prediction is performed (4.) (MAE: Mean absolute error; CNN: Convolutional Neural Network, nnU-Net: Neural network for semantic segmentation).

The radiomics approach consisted of 32 radiomics features (14 shape features: elongation, flatness, least axis length, major axis length, max 2d diameter column, max 2d diameter row, max 2d diameter slice, max 3d diameter, mesh volume, minor axis length, sphericity, surface area, surface volume ratio, voxel volume; 18 first order intensity features: 10th percentile, 90th percentile, energy, entropy, interquartile range, kurtosis, maximum, mean absolute deviation, mean, median, minimum, range, robust mean absolute deviation, root mean squared, skewness, total energy, uniformity, variance), which were extracted from the CT images and the ICH segmentation using pyradiomics [10.1158/0008-5472.CAN-17-0339]. Those features were used to train a gradient boosted regression tree using XGBoost [https://doi.org/10.1145/2939672.2939785, accessed on 10 September 2021] (learning rate 0.01, number of trees 200, maximum depth 2. For the CNN approach, a 3d convolutional neural network regressor was trained. As architecture the CBR-tiny model from [https://papers.nips.cc/paper/2019/hash/eb1e78328c46506b46a4ac4a1e378 b91-Abstract.html, accessed on 10 September 2021] was used by replacing the 2D convolutions by 3D convolutions (network architecture: 4 sequential blocks of Convolution + BatchNorm + MaxPooling followed by adaptive average pooling and a fully connected layer). To improve the model performance, it was pretrained on a large dataset with an auxiliary task: on 7421 CT images containing ICHs the network was trained to predict the volume of the ICH (Figure 1). The volume was derived from the ICH segmentation generated by our segmentation model. After this first training, the model was finetuned on the estimation of onset task. For training, the following hyperparameters were used: batch size 8, learning rate 0.0005, number of epochs 15, learning rate decay to 0 over the entire training with cosine scheduler.

For both the segmentation as well as the onset prediction, a 5-fold cross-validation was used for evaluation.

In addition, a random subset of 117 CT datasets (Figure 1) was selected from the training dataset and analyzed by two radiologists (4 and 5 years of experience in ICH interpretation) regarding hemorrhage age based on their clinical–radiological experience and secondary factors (perifocal edema: relatively large edema = advanced bleeding age; Hounsfield Units (HU): acute bleeding approximately 60–70 HU, approximately 2 HU drop per 24 h). The hemorrhage age was given in hours. An interrater comparison and a comparison with the results of the AI-based age prediction were performed.

2.3. Statistics

For the calculation of confidence intervals, we used bootstrapping with 10,000 random permutations and for the comparison of human raters as well as of the AI algorithms with the baseline we used Wilcoxon signed-rank test since our data has no normal distribution as tested with a Kolmogorow–Smirnow test. Results with $p < 0.05$ were considered statistically significant. A paired test was used since different methods were compared on the same patients. All statistical analyses were performed with python 3.8.

3. Results

3.1. Study Cohort

The PACS crawler query for patients diagnosed with ICH by CT between January 2007–July 2021 yielded a total of 7733 subjects. Of these, 312 subjects did not have consent for the use of patient-related data for research purposes. Of the remaining 7421 subjects, 6121 had no or only an incomplete admission records with information on the symptom onset and thus no identifiable gold standard. Of the remaining 1300 datasets, 644 subjects fulfilled the final inclusion criteria (blood volume > 15 mm^3 and onset > 1 h and <48 h) and thus were used for the training model. The study workflow is summarized in Figure 1.

3.2. Onset Estimation of Classifiers and Human Raters

Both the AI-based classifiers and the radiologists had poor discrimination of the known onsets, with a mean absolute error (MAE) of 9.77 h (95% CI (confidence interval) = 8.52–11.03 h) for the convolutional neural network (CNN), 9.96 h (8.68–11.32 h) for the radiomics model, 13.38 h (11.21–15.74 h) for human rater 1 and 11.21 h (9.51–12.90 h) for rater 2, respectively (see Table 1).

Table 1. Results of human raters and machine learning classifiers for onset estimation of patients with intracerebral hemorrhage.

	Mean Absolute Error (MAE) in h	95% Confidence Interval (CI)
Rater 1	13.38	11.21, 15.74
Rater 2	11.21	9.51, 12.90
CNN	9.77	8.52, 11.03
Radiomics	9.96	8.68, 11.32
Mean of known onset in entire cohort	9.81	8.62, 11.06

MAE in hours and confidence interval (95%) for human raters and AI-based models (CNN and Radiomics)

Rater 1 and rater 2 were both significantly more inferior than the mean of the known onsets ($p = 0.006$ and $p = 0.045$, respectively). The results of the CNN and Radiomics model were both not significantly different to the mean of the known onsets ($p = 0.705$ and $p = 0.423$). The intraclass correlation between rater 1 and rater 2 was poor (intraclass correlation coefficient, ICC = 0.251).

4. Discussion

Our study shows that both human raters and machine learning algorithms only have poor discriminatory power to estimate the onset of patients with intracerebral hemorrhage based on CT imaging features.

This may be partially explained by the nature and course of the disease. It is known that acute ICH may show mixed densities on CT, representing blood of different ages [20–22,30,31]. Thus, onset estimation based on density in CT may be misleading. Another feature, perihematomal edema, is known to progress over time but also depends on other factors such as anticoagulation status [32]. Further, several other factors such as patient age [33], anticoagulation medication [34], and blood pressure [35] are also known factors that impact imaging appearance of ICH beyond time. Regarding the manual ratings available studies have shown promising interrater agreements for ICH shape and density features [36,37]. However, so far, no published studies have found a rating based or machine learning algorithm that could reliably determine the onset of ICH patients with unknown symptom onset.

Even though currently limited, therapeutic and triage decisions are based on factors such as the likelihood of rebleeding, brain edema and herniation which are associated with the time since onset of the bleeding [12]. Possible treatment strategies include blood pressure reduction [13,14], neurosurgical evacuation [16], and application of tranexamic acid [17,38], which may reduce the risk of hematoma expansion associated with neurological deterioration leading to poor outcomes and death [12,20]. These current approaches target hematoma expansion which is more likely to occur in the earlier time window [36]. In this context, prediction of the onset in patients with ICH and unknown onset may have had a great clinical relevance; however, the mean error of almost 10 h is probably too imprecise to be used for clinical decision making. Instead, other features that are likely to identify the risk of hematoma expansion will probably be used as an alternative. Also, for the prediction of hematoma growth and functional outcomes, machine learning algorithms have shown promising accuracy [23], most likely because several other features that can be extracted from imaging determine outcomes more than time since onset.

Our study has several limitations. First, the underlying ground truth (gold standard) of the timepoint of onset—on which the manual ratings and machine learning classifier are based—may be imprecise in some cases. Basically, the quality of the ground truth is a very important parameter for the performance and accuracy of an AI-based model. In other words, the AI model is only as good as the quality of the data on which it is based. Since the exact onset of ICH in our cohort could only be determined very imprecisely in part and depended on many factors (medical history, report by emergency physician, unclear clinical symptoms), the ground truth on which our AI-based model is based is also relatively imprecise. This was also supported by the fact that, considering the bleeding age of our total cohort, no decrease in density values (Hounsfield Units) could be reliably delineated over time (Figure 2). However, improving ground truth in general is problematic, as the exact onset of an ICH can never be determined with absolute certainty and will always be relatively inaccurate due to factors mentioned above.

In general, the acuity of clinical symptoms may be comparable to ischemic strokes, where reported symptom onsets are used for several studies that validated imaging-based approaches for onset estimation [39–43]. Second, it may be argued that a larger number of patients for training and validation of the classifier may have yielded more accurate results. Basically, the larger the dataset on which the algorithm is trained, the better AI-based methods perform. In other words, the data size of the cohort is a crucial factor. Some studies have shown that when the data size is very large, AI-based algorithms perform equally well or even better than the expert, for instance human rater [44]. In our training data set, only 644 subjects were included for the ICB onset estimation. It would therefore be interesting to see whether the results could be significantly changed or improved by including a significantly larger number of subjects. However, the current study already uses a comparably large dataset, and the results were still far from being clinically useful. Third, we must note that an average density (Hounsfeild Units) decrease of acute ICH would be expected over the time [45,46]. However, this is a crude measurement when used for shorter timespans of a few hours. Moreover, ICH often consists of blood of mixed densities which may bias the results even more [36]. Based on these factors and the above-mentioned results, it can be assumed that the information content of the CT scans in which ICH could be diagnosed does probably not contain or cover the information we are looking for (ICH onset). In our example, the human raters were also unable to read the bleeding age (onset time) from the CT dataset more accurately than approximately ±10 h. For this reason, this information probably does not exist sufficiently in the image. Conversely, if the human rater already cannot confidently read out the information, the AI-based approach probably cannot either. In this context, only the analysis of extremely large and perfectly annotated datasets would be promising (see discussion above). Fourth, we did not use multimodal data for our AI-based approach, but only CT-based data. However, for ICH onset prediction, other data such as perfusion indicators [47], hemodynamic metrics [48], morphology and anatomical differences of intracranial arterial vessels especially with regard to microvascular collateral circulation [49,50] or CT-derived secondary hemody-namic parameters, and laboratory test results could be used and even combined. Such a multimodal AI-based approach could again significantly improve ICH onset prediction and should be considered in future studies.

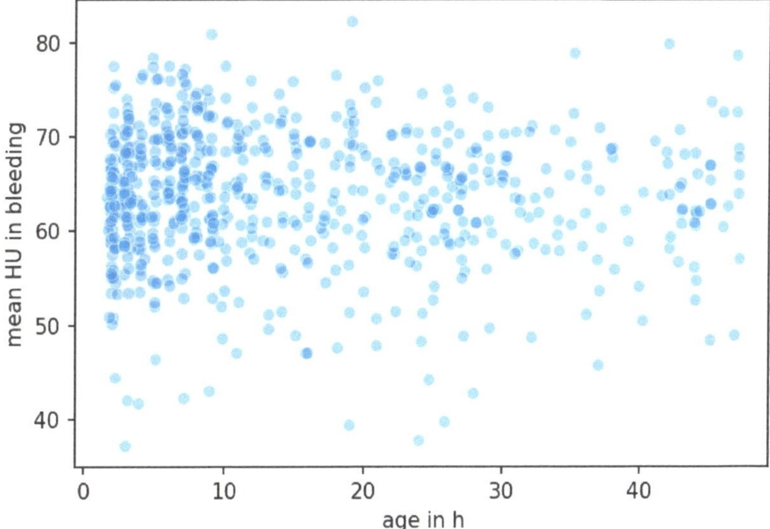

Figure 2. Mean density of intracerebral hemorrhage on admission imaging and association with symptom onset. HU = Hounsfield Units, age in h = time since onset in hours.

5. Conclusions

In our study, the discriminatory power of AI-based classifiers and human readers for onset estimation of patients with ICH was poor. This indicates that accurate AI-based onset estimation of patients with ICH based only on CT data may be unlikely to change clinical decision making in the near future. Perhaps multimodal AI-based approaches could improve ICH onset prediction and should be considered in future studies.

Author Contributions: Conceptualization, T.R., J.W. and P.B.S.; Methodology, T.R. and J.W.; Software, T.R. and J.W.; Validation, T.R., J.W. and H.-C.B.; Formal analysis, T.R., J.W., H.-C.B. and P.B.S.; Investigation, T.R.; Resources, T.R. and P.B.S.; Data curation, T.R. and J.W.; Writing—original draft, T.R., J.W. and P.B.S.; Writing—review & editing, T.R., J.W., U.F., R.G., J.F., M.-N.P. and P.B.S.; Visualization, T.R.; Supervision, T.R., M.-N.P. and P.B.S.; Project administration, T.R., M.-N.P. and P.B.S. All authors have read and agreed to the published version of the manuscript.

Funding: This research received no external funding.

Institutional Review Board Statement: The study was approved under the provisions of the local ethics committee (Swissethics Project ID: 2021-01831) in accordance with the Declaration of Helsinki.

Informed Consent Statement: Informed consent was obtained from all subjects involved in the study.

Data Availability Statement: The data presented in this study are available on request from the corresponding author.

Conflicts of Interest: The authors declare no conflict of interest.

Abbreviations

AI	Artificial intelligence
CI	Confidence interval
CT	Computed tomography
GCS	Glasgow coma scale
ICH	Intracerebral hemorrhage
ICC	Intraclass correlation coefficient
INR	International normalized ratio
IVH	Intraventricular hemorrhage
mRS	Modified ranking scale
MAE	Mean absolute error
MRI	Magnetic resonance imaging
OAC	Oral anticoagulants
PACS	Picture Archiving and Communication System
HU	Hounsfield Unit
RIS	Radiological Information System

References

1. Qureshi, A.I.; Mendelow, A.D.; Hanley, D.F. Intracerebral haemorrhage. *Lancet* **2009**, *373*, 1632–1644. [CrossRef] [PubMed]
2. Caceres, J.A.; Goldstein, J.N. Intracranial hemorrhage. *Emerg. Med. Clin. N. Am.* **2012**, *30*, 771–794. [CrossRef] [PubMed]
3. Elliott, J.; Smith, M. The acute management of intracerebral hemorrhage: A clinical review. *Anesth. Analg.* **2010**, *110*, 1419–1427. [CrossRef] [PubMed]
4. Broderick, J.P.; Brott, T.G.; Duldner, J.E.; Tomsick, T.; Huster, G. Volume of intracerebral hemorrhage. A powerful and easy-to-use predictor of 30-day mortality. *Stroke* **1993**, *24*, 987–993. [CrossRef]
5. Fogelholm, R.; Murros, K.; Rissanen, A.; Avikainen, S. Long term survival after primary intracerebral haemorrhage: A retrospective population based study. *J. Neurol. Neurosurg. Psychiatry* **2005**, *76*, 1534–1538. [CrossRef]
6. Hemphill, J.C., 3rd; Bonovich, D.C.; Besmertis, L.; Manley, G.T.; Johnston, S.C. The ICH score: A simple, reliable grading scale for intracerebral hemorrhage. *Stroke* **2001**, *32*, 891–897. [CrossRef]
7. van Asch, C.J.; Luitse, M.J.; Rinkel, G.J.; van der Tweel, I.; Algra, A.; Klijn, C.J. Incidence, case fatality, and functional outcome of intracerebral haemorrhage over time, according to age, sex, and ethnic origin: A systematic review and meta-analysis. *Lancet Neurol.* **2010**, *9*, 167–176. [CrossRef]
8. Russell, M.W.; Boulanger, L.; Joshi, A.V.; Neumann, P.J.; Menzin, J. The economic burden of intracerebral hemorrhage: Evidence from managed care. *Manag. Care Interface* **2006**, *19*, 24–28, 34.
9. Smeds, M.; Skrifvars, M.B.; Reinikainen, M.; Bendel, S.; Hoppu, S.; Laitio, R.; Ala-Kokko, T.; Curtze, S.; Sibolt, G.; Martinez-Majander, N.; et al. One-year healthcare costs of patients with spontaneous intracerebral hemorrhage treated in the intensive care unit. *Eur. Stroke J.* **2022**, *7*, 267–279. [CrossRef]
10. Thomas, S.M.; Reindorp, Y.; Christophe, B.R.; Connolly, E.S., Jr. Systematic Review of Resource Use and Costs in the Hospital Management of Intracerebral Hemorrhage. *World Neurosurg.* **2022**, *164*, 41–63. [CrossRef]
11. Soyland, M.H.; Tveiten, A.; Eltoft, A.; Oygarden, H.; Varmdal, T.; Indredavik, B.; Mathiesen, E.B. Wake-up stroke and unknown-onset stroke; occurrence and characteristics from the nationwide Norwegian Stroke Register. *Eur. Stroke J.* **2022**, *7*, 143–150. [CrossRef] [PubMed]

12. Sporns, P.B.; Schwake, M.; Schmidt, R.; Kemmling, A.; Minnerup, J.; Schwindt, W.; Cnyrim, C.; Zoubi, T.; Heindel, W.; Niederstadt, T.; et al. Computed Tomographic Blend Sign Is Associated With Computed Tomographic Angiography Spot Sign and Predicts Secondary Neurological Deterioration After Intracerebral Hemorrhage. *Stroke* **2017**, *48*, 131–135. [CrossRef] [PubMed]

13. Inoue, Y.; Miyashita, F.; Koga, M.; Minematsu, K.; Toyoda, K. Unclear-onset intracerebral hemorrhage: Clinical characteristics, hematoma features, and outcomes. *Int. J. Stroke* **2017**, *12*, 961–968. [CrossRef] [PubMed]

14. Kim, J.Y.; Bae, H.J. Spontaneous Intracerebral Hemorrhage: Management. *J. Stroke* **2017**, *19*, 28–39. [CrossRef] [PubMed]

15. Qureshi, A.I. Antihypertensive Treatment of Acute Cerebral Hemorrhage (ATACH): Rationale and design. *Neurocrit. Care* **2007**, *6*, 56–66. [CrossRef] [PubMed]

16. Qureshi, A.I.; Palesch, Y.Y. Antihypertensive Treatment of Acute Cerebral Hemorrhage (ATACH) II: Design, methods, and rationale. *Neurocrit. Care* **2011**, *15*, 559–576. [CrossRef]

17. Anderson, C.S.; Heeley, E.; Huang, Y.; Wang, J.; Stapf, C.; Delcourt, C.; Lindley, R.; Robinson, T.; Lavados, P.; Neal, B.; et al. Rapid blood-pressure lowering in patients with acute intracerebral hemorrhage. *N. Engl. J. Med.* **2013**, *368*, 2355–2365. [CrossRef]

18. Luzzi, S.; Elia, A.; Del Maestro, M.; Morotti, A.; Elbabaa, S.K.; Cavallini, A.; Galzio, R. Indication, Timing, and Surgical Treatment of Spontaneous Intracerebral Hemorrhage: Systematic Review and Proposal of a Management Algorithm. *World Neurosurg.* **2019**, *124*, e769–e778. [CrossRef]

19. Sprigg, N.; Flaherty, K.; Appleton, J.P.; Al-Shahi Salman, R.; Bereczki, D.; Beridze, M.; Christensen, H.; Ciccone, A.; Collins, R.; Czlonkowska, A.; et al. Tranexamic acid for hyperacute primary IntraCerebral Haemorrhage (TICH-2): An international randomised, placebo-controlled, phase 3 superiority trial. *Lancet* **2018**, *391*, 2107–2115. [CrossRef]

20. Sporns, P.B.; Schwake, M.; Kemmling, A.; Minnerup, J.; Schwindt, W.; Niederstadt, T.; Schmidt, R.; Hanning, U. Comparison of Spot Sign, Blend Sign and Black Hole Sign for Outcome Prediction in Patients with Intracerebral Hemorrhage. *J. Stroke* **2017**, *19*, 333–339. [CrossRef]

21. Sporns, P.B.; Kemmling, A.; Minnerup, J.; Hanning, U.; Heindel, W. Imaging-based outcome prediction in patients with intracerebral hemorrhage. *Acta Neurochir.* **2018**, *160*, 1663–1670. [CrossRef] [PubMed]

22. Sporns, P.B.; Kemmling, A.; Schwake, M.; Minnerup, J.; Nawabi, J.; Broocks, G.; Wildgruber, M.; Fiehler, J.; Heindel, W.; Hanning, U. Triage of 5 Noncontrast Computed Tomography Markers and Spot Sign for Outcome Prediction After Intracerebral Hemorrhage. *Stroke* **2018**, *49*, 2317–2322. [CrossRef] [PubMed]

23. Nawabi, J.; Kniep, H.; Elsayed, S.; Friedrich, C.; Sporns, P.; Rusche, T.; Bohmer, M.; Morotti, A.; Schlunk, F.; Duhrsen, L.; et al. Imaging-Based Outcome Prediction of Acute Intracerebral Hemorrhage. *Transl. Stroke Res.* **2021**, *12*, 958–967. [CrossRef] [PubMed]

24. Reyes, M.; Meier, R.; Pereira, S.; Silva, C.A.; Dahlweid, F.M.; von Tengg-Kobligk, H.; Summers, R.M.; Wiest, R. On the Interpretability of Artificial Intelligence in Radiology: Challenges and Opportunities. *Radiol. Artif. Intell.* **2020**, *2*, e190043. [CrossRef]

25. Richardson, M.L.; Garwood, E.R.; Lee, Y.; Li, M.D.; Lo, H.S.; Nagaraju, A.; Nguyen, X.V.; Probyn, L.; Rajiah, P.; Sin, J.; et al. Noninterpretive Uses of Artificial Intelligence in Radiology. *Acad. Radiol.* **2021**, *28*, 1225–1235. [CrossRef]

26. Weikert, T.; Winkel, D.J.; Bremerich, J.; Stieltjes, B.; Parmar, V.; Sauter, A.W.; Sommer, G. Automated detection of pulmonary embolism in CT pulmonary angiograms using an AI-powered algorithm. *Eur. Radiol.* **2020**, *30*, 6545–6553. [CrossRef]

27. Rao, B.; Zohrabian, V.; Cedeno, P.; Saha, A.; Pahade, J.; Davis, M.A. Utility of Artificial Intelligence Tool as a Prospective Radiology Peer Reviewer—Detection of Unreported Intracranial Hemorrhage. *Acad. Radiol.* **2021**, *28*, 85–93. [CrossRef]

28. Rava, R.A.; Seymour, S.E.; LaQue, M.E.; Peterson, B.A.; Snyder, K.V.; Mokin, M.; Waqas, M.; Hoi, Y.; Davies, J.M.; Levy, E.I.; et al. Assessment of an Artificial Intelligence Algorithm for Detection of Intracranial Hemorrhage. *World Neurosurg.* **2021**, *150*, e209–e217. [CrossRef]

29. Voter, A.F.; Meram, E.; Garrett, J.W.; Yu, J.J. Diagnostic Accuracy and Failure Mode Analysis of a Deep Learning Algorithm for the Detection of Intracranial Hemorrhage. *J. Am. Coll. Radiol.* **2021**, *18*, 1143–1152. [CrossRef]

30. Sporns, P.B.; Psychogios, M.N.; Boulouis, G.; Charidimou, A.; Li, Q.; Fainardi, E.; Dowlatshahi, D.; Goldstein, J.N.; Morotti, A. Neuroimaging of Acute Intracerebral Hemorrhage. *J. Clin. Med.* **2021**, *10*, 1086. [CrossRef]

31. Barras, C.D.; Tress, B.M.; Christensen, S.; MacGregor, L.; Collins, M.; Desmond, P.M.; Skolnick, B.E.; Mayer, S.A.; Broderick, J.P.; Diringer, M.N.; et al. Density and shape as CT predictors of intracerebral hemorrhage growth. *Stroke* **2009**, *40*, 1325–1331. [CrossRef] [PubMed]

32. Nawabi, J.; Elsayed, S.; Morotti, A.; Speth, A.; Liu, M.; Kniep, H.; McDonough, R.; Broocks, G.; Faizy, T.; Can, E.; et al. Perihematomal Edema and Clinical Outcome in Intracerebral Hemorrhage Related to Different Oral Anticoagulants. *J. Clin. Med.* **2021**, *10*, 2234. [CrossRef] [PubMed]

33. Sporns, P.B.; Psychogios, M.N.; Fullerton, H.J.; Lee, S.; Naggara, O.; Boulouis, G. Neuroimaging of Pediatric Intracerebral Hemorrhage. *J. Clin. Med.* **2020**, *9*, 1518. [CrossRef] [PubMed]

34. Zimmer, S.; Meier, J.; Minnerup, J.; Wildgruber, M.; Broocks, G.; Nawabi, J.; Morotti, A.; Kemmling, A.; Psychogios, M.; Hanning, U.; et al. Prognostic Value of Non-Contrast CT Markers and Spot Sign for Outcome Prediction in Patients with Intracerebral Hemorrhage under Oral Anticoagulation. *J. Clin. Med.* **2020**, *9*, 1077. [CrossRef] [PubMed]

35. Morotti, A.; Nawabi, J.; Schlunk, F.; Poli, L.; Costa, P.; Mazzacane, F.; Busto, G.; Scola, E.; Arba, F.; Brancaleoni, L.; et al. Characteristics of Early Presenters after Intracerebral Hemorrhage. *J. Stroke* **2022**, *24*, 425–428. [CrossRef]

36. Morotti, A.; Boulouis, G.; Dowlatshahi, D.; Li, Q.; Barras, C.D.; Delcourt, C.; Yu, Z.; Zheng, J.; Zhou, Z.; Aviv, R.I.; et al. Standards for Detecting, Interpreting, and Reporting Noncontrast Computed Tomographic Markers of Intracerebral Hemorrhage Expansion. *Ann. Neurol.* **2019**, *86*, 480–492. [CrossRef]

37. Nawabi, J.; Elsayed, S.; Kniep, H.; Sporns, P.; Schlunk, F.; McDonough, R.; Broocks, G.; Duhrsen, L.; Schon, G.; Gotz, T.; et al. Inter- and Intrarater Agreement of Spot Sign and Noncontrast CT Markers for Early Intracerebral Hemorrhage Expansion. *J. Clin. Med.* **2020**, *9*, 1020. [CrossRef]

38. Guo, Y.; Guo, X.M.; Li, R.L.; Zhao, K.; Bao, Q.J.; Yang, J.C.; Zhang, Q.; Yang, M.F. Tranexamic Acid for Acute Spontaneous Intracerebral Hemorrhage: A Meta-Analysis of Randomized Controlled Trials. *Front. Neurol.* **2021**, *12*, 761185. [CrossRef]

39. Broocks, G.; Leischner, H.; Hanning, U.; Flottmann, F.; Faizy, T.D.; Schon, G.; Sporns, P.; Thomalla, G.; Kamalian, S.; Lev, M.H.; et al. Lesion Age Imaging in Acute Stroke: Water Uptake in CT Versus DWI-FLAIR Mismatch. *Ann. Neurol.* **2020**, *88*, 1144–1152. [CrossRef]

40. Nawabi, J.; Flottmann, F.; Kemmling, A.; Kniep, H.; Leischner, H.; Sporns, P.; Schon, G.; Hanning, U.; Thomalla, G.; Fiehler, J.; et al. Elevated early lesion water uptake in acute stroke predicts poor outcome despite successful recanalization—When "tissue clock" and "time clock" are desynchronized. *Int. J. Stroke* **2021**, *16*, 863–872. [CrossRef]

41. Sporns, P.B.; Hohne, M.; Meyer, L.; Krogias, C.; Puetz, V.; Thierfelder, K.M.; Duering, M.; Kaiser, D.; Langner, S.; Brehm, A.; et al. Simplified Assessment of Lesion Water Uptake for Identification of Patients within 4.5 Hours of Stroke Onset: An Analysis of the MissPerfeCT Study. *J. Stroke* **2022**, *24*, 390–395. [CrossRef] [PubMed]

42. Sporns, P.B.; Kemmling, A.; Minnerup, J.; Meyer, L.; Krogias, C.; Puetz, V.; Thierfelder, K.; Duering, M.; Kaiser, D.; Langner, S.; et al. CT Hypoperfusion-Hypodensity Mismatch to Identify Patients With Acute Ischemic Stroke Within 4.5 Hours of Symptom Onset. *Neurology* **2021**, *97*, e2088–e2095. [CrossRef] [PubMed]

43. Thomalla, G.; Simonsen, C.Z.; Boutitie, F.; Andersen, G.; Berthezene, Y.; Cheng, B.; Cheripelli, B.; Cho, T.H.; Fazekas, F.; Fiehler, J.; et al. MRI-Guided Thrombolysis for Stroke with Unknown Time of Onset. *N. Engl. J. Med.* **2018**, *379*, 611–622. [CrossRef] [PubMed]

44. Esteva, A.; Kuprel, B.; Novoa, R.A.; Ko, J.; Swetter, S.M.; Blau, H.M.; Thrun, S. Dermatologist-level classification of skin cancer with deep neural networks. *Nature* **2017**, *542*, 115–118. [CrossRef]

45. Kornbluth, J.; Nekoovaght-Tak, S.; Ullman, N.; Carhuapoma, J.R.; Hanley, D.F.; Ziai, W. Early Quantification of Hematoma Hounsfield Units on Noncontrast CT in Acute Intraventricular Hemorrhage Predicts Ventricular Clearance after Intraventricular Thrombolysis. *AJNR Am. J. Neuroradiol.* **2015**, *36*, 1609–1615. [CrossRef]

46. Ziya, A. Determination of bleeding time by hounsfield unit values in computed tomography scans of patients diagnosed with intracranial hemorrhage: Evaluation results of computed tomography scans of 666 patients. *Clin. Neurol. Neurosurg.* **2022**, *217*, 107258. [CrossRef]

47. Xu, J.; Dai, F.; Wang, B.; Wang, Y.; Li, J.; Pan, L.; Liu, J.; Liu, H.; He, S. Predictive Value of CT Perfusion in Hemorrhagic Transformation after Acute Ischemic Stroke: A Systematic Review and Meta-Analysis. *Brain Sci.* **2023**, *13*, 156. [CrossRef]

48. Tian, X.; Fang, H.; Lan, L.; Ip, H.L.; Abrigo, J.; Liu, H.; Zheng, L.; Fan, F.S.Y.; Ma, S.H.; Ip, B.; et al. Risk stratification in symptomatic intracranial atherosclerotic disease with conventional vascular risk factors and cerebral haemodynamics. *Stroke Vasc. Neurol.* **2023**, *8*, 77–85. [CrossRef]

49. Lee, R.M. Morphology of cerebral arteries. *Pharmacol. Ther.* **1995**, *66*, 149–173. [CrossRef]

50. Niederberger, E.; Gauvrit, J.Y.; Morandi, X.; Carsin-Nicol, B.; Gauthier, T.; Ferre, J.C. Anatomic variants of the anterior part of the cerebral arterial circle at multidetector computed tomography angiography. *J. Neuroradiol.* **2010**, *37*, 139–147. [CrossRef]

Journal of
Clinical Medicine

Review

The Assessment of Endovascular Therapies in Ischemic Stroke: Management, Problems and Future Approaches

Tadeusz J. Popiela [1,*], Wirginia Krzyściak [2,*], Fabio Pilato [3], Anna Ligęzka [2,4,5], Beata Bystrowska [6], Karolina Bukowska-Strakova [7], Paweł Brzegowy [1], Karthik Muthusamy [4] and Tamas Kozicz [4,5]

[1] Department of Radiology, Medical College, Jagiellonian University, 31-501 Kraków, Poland; p.brzegowy@uj.edu.pl
[2] Department of Medical Diagnostics, Medical College, Jagiellonian University, 31-688 Kraków, Poland; ligezka.anna@mayo.edu
[3] Neurology, Neurophysiology and Neurobiology Unit, Department of Medicine, Università Campus Bio-Medico di Roma, 00128 Rome, Italy; f.pilato@policlinicocampus.it
[4] Department of Clinical Genomics, Mayo Clinic, Rochester, MN 55902, USA; muthusamy.karthik@mayo.edu (K.M.); kozicz.tamas@mayo.edu (T.K.)
[5] Center for Individualized Medicine, Mayo Clinic, Rochester, MN 55902, USA
[6] Department of Toxicology, Medical College, Jagiellonian University, Medyczna 9, 30-688 Kraków, Poland; beata.bystrowska@uj.edu.pl
[7] Department of Clinical Immunology and Transplantology, Institute of Pediatrics, Medical College, Jagiellonian University, 30-663 Kraków, Poland; k.bukowska-strakova@uj.edu.pl
[*] Correspondence: tadeusz.popiela@uj.edu.pl (T.J.P.); wirginiakrzysciak@cm-uj.krakow.pl (W.K.)

Citation: Popiela, T.J.; Krzyściak, W.; Pilato, F.; Ligezka, A.; Bystrowska, B.; Bukowska-Strakova, K.; Brzegowy, P.; Muthusamy, K.; Kozicz, T. The Assessment of Endovascular Therapies in Ischemic Stroke: Management, Problems and Future Approaches. *J. Clin. Med.* 2022, 11, 1864. https://doi.org/10.3390/jcm11071864

Academic Editor: Georgios Tsivgoulis

Received: 28 February 2022
Accepted: 25 March 2022
Published: 28 March 2022

Publisher's Note: MDPI stays neutral with regard to jurisdictional claims in published maps and institutional affiliations.

Abstract: Ischemic stroke accounts for over 80% of all strokes and is one of the leading causes of mortality and permanent disability worldwide. Intravenous administration of recombinant tissue plasminogen activator (rt-PA) is an approved treatment strategy for acute ischemic stroke of large arteries within 4.5 h of onset, and mechanical thrombectomy can be used for large arteries occlusion up to 24 h after onset. Improving diagnostic work up for acute treatment, reducing onset-to-needle time and urgent radiological access angiographic CT images (angioCT) and Magnetic Resonance Imaging (MRI) are real problems for many healthcare systems, which limits the number of patients with good prognosis in real world compared to the results of randomized controlled trials. The applied endovascular procedures demonstrated high efficacy, but some cellular mechanisms, following reperfusion, are still unknown. Changes in the morphology and function of mitochondria associated with reperfusion and ischemia-reperfusion neuronal death are still understudied research fields. Moreover, future research is needed to elucidate the relationship between continuously refined imaging techniques and the variable structure or physical properties of the clot along with vascular permeability and the pleiotropism of ischemic reperfusion lesions in the penumbra, in order to define targeted preventive procedures promoting long-term health benefits.

Keywords: ischemic stroke; thrombolysis; rt-PA; endovascular therapies; mechanical thrombectomy; clots; mitochondria

1. Introduction

Ischemic stroke is caused by interruption of the blood supply, most often by a blood clot [1]. The histological structure of the thrombus may determine the subtype of stroke, depending on its etiology: cardioembolic, atherothrombotic, or stroke of unknown cause [2], and it may influence reperfusion outcome. Fibrin and platelets dominate in cardioembolic thrombi with a small amount of neutrophil extracellular traps (NETs) which impair tissue plasminogen activator tPA-mediated thrombolysis [3]. In atherothrombosis, thrombosis, red blood cells and fibrin predominate, while the relative proportion of each component is undefined for cryptogenic strokes [4]. The knowledge of clot histological structure, percentage composition and architecture of the clot components, as well as the patient's clinical features, and the molecular changes in the area of cerebral ischemia seem to be crucial for understanding the results of the current acute stroke treatments and long-term clinical benefits of patients with ischemic stroke [5–7].

Ischemic stroke annually affects 17 million people worldwide, of which as many as 6 million die, and a significant proportion are permanently burdened with disability [8–10]. Ischemic stroke accounts for approximately 80% of all stroke cases (0.2% according to Béjot et al. refers to the world population and applies to all stroke cases in 2016; 0.2% is estimated as 7.9 billion people lived in the world in 2021) [11–13]. Estimates indicate that the increase in stroke incidence in the general population will reach 25% between 2010 and 2030; in men aged 65 to 74, it accounts for approximately 70 cases per 100,000, while in the general population over 74, women are more likely to experience acute stroke with more severe consequences [14,15]. The projected number of strokes in all EU countries, Iceland, Norway, and Switzerland, will increase by 27% (from 1.1 million in 2000 to 1.5 million in 2025) [16–19].

Studying data collected by Global Burden of Disease, Injuries, and Risk Factors Study 2017 (GBD 2017), including separate estimates of the global burden and trends for each type of stroke, a 2-fold increase was observed in the absolute number of people who had a new stroke, died, survived, or remained disabled after their stroke within the 12 million stroke-related incidents. Most of the stroke burden occurred in low- and middle-income countries, 65% of those stroke cases were ischemic stroke, 26% was primary intracerebral haemorrhage (PICH), and 9% were represented by subarachnoid haemorrhages [20–22].

The above-mentioned epidemiological data related to the incidence of stroke varies from country to country and fluctuates according to the burden in high-, low-, and middle-income countries. Individual records from hospital stroke units are incomplete and their availability is limited [23,24]. The increase in the number of registered cases of people suffering from ischemic stroke will be found mainly in the populations of developing countries [14,25].

One of the reasons for the increase in morbidity in the world is the change in the structure of the population—the high level of aging compared to the number of births. Moreover, in the last two years, COVID-19 pandemic deeply affected stroke management [15] and, probably, even more effects will be evident in the future, especially in younger people [16–19]. Based on the meta-analysis of patients with COVID-19 developing acute cerebrovascular diseases, it was observed that among the elderly, stroke episodes occurred more frequently in patients with comorbidities, i.e., diabetes, coronary artery disease, and severe infections, as compared to patients who had a stroke without infection. Younger people with registered stroke cases had higher NIHSS, higher incidence of large vessel obstruction, and higher in-hospital mortality [26]. It may be due to the inflammatory response in endothelial cells, characterized by increased cytokine secretion and expression of adhesion molecules [27].

In Poland, in 2050, it has been estimated that the society will consist of 11% of people in the pre-working age, 57% in working age, and 32.7% over 65 years [28]. Ischemic stroke is associated with almost twice as many cases of dementia compared to the general population. The dramatic increase in the incidence of stroke in the population prompts epidemiologists to describe it as a form of a pandemic [29]. This together means that all efforts to improve epidemiological data take a high priority of healthcare systems, both in terms of primary prevention of stroke occurrence and secondary prevention of subsequent episodes [30]. Stroke is not only one of the most common causes of hospitalization, but also comes at a high economic cost. Moreover, stroke is a main problem not only in older ages but also in young adult causing increased long-lasting health-related costs [31]. Today, about 34% of total healthcare expenditure in the world is spent on this disease. The reason for the increased economic burden of stroke treatment is the need for long hospitalization and rehabilitation services, which are the basis of conservative treatment [32].

2. Qualification for the Causal Treatment of Ischemic Stroke

Currently approved treatments for acute ischemic stroke (AIS) are mechanical thrombectomy (MT), which accounted for 1.9% of treatments in 2019 in Poland (TM was performed in 1413 patients out of 75,213 of all patients hospitalized due to ischemic stroke) and 3.5% of treatments in 2020 year (TM was performed in 2526 patients out of 70,926 of all patients hospitalized due to ischemic stroke) [33] while in the United States, MTs was performed (3.1% out of examined acute ischemic stroke population, i.e., approximately 424,330; data from 2016) [22]. On the other hand, the intravenous thrombolysis (IVT) with the administration of the recombinant tissue plasminogen activator (rt-PA) accounts for 7.3% treatments of AIS patients in Europe [34,35] and 3.6–6.5% treatments of AIS patients in the USA [36–38].

Current stroke guidelines report that a fast diagnostic protocol, in qualifying stroke patients for acute treatments should be performed [39]. First step is the exclusion of a hemorrhagic stroke. For this purpose, a computed tomography (CT) of the head or a magnetic resonance imaging (MRI) is performed. Due to the limited availability of the latter, a CT of the head is more used and faster than MRI in acute setting. After the exclusion of cerebral hemorrhage and the features of a completed ischemic stroke in the CT scan, intravenous thrombolytic therapy is started with the administration of rt-PA within a 4.5 h time window from the onset of the first clinical symptoms of ischemic stroke. When the stroke etiology is a large vessel occlusion (LVO) in anterior circulation, MT is the standard of care for acute stroke [39]. It should be noted that in patients with an anterior cerebral ischemic stroke, the results of MT started within 4.5 h alone compared with the combined IVT, measured on day 90 on the modified Rankin scale (mRS) were comparable. These were the results of a multicenter randomized non-inferiority clinical trial conducted at 33 stroke centers in China. Only the clinical evaluation related to functional improvement after 90 days (0–2 in mRS) was taken into account as the endpoint [40]. Moreover, MT is the treatment of choice when rt-PA is contraindicated, i.e., in patients treated with anticoagulants. Depending on time of stroke onset, the qualification is based on additional specialist examinations, such as CT angiography (angioCT) and CT perfusion (pCT). In the case of occlusion of large vessels in the anterior cerebral circulation, the DAWN and DEFUSE criteria apply in patients within 6 to 24 h after symptom onset [41,42] and depending on availability, the CT scan may be replaced with appropriate MR sequences.

2.1. A Standard Operating Procedure of Imaging Methods Which Is Employed in Ischemic Stroke in NSSU Diagnostic Imaging Unit

A 56-year-old patient with symptoms of ischemic stroke of the right hemisphere which appeared 2 h before admission to the hospital emergency department was demonstrated.

Two years ago, the patient had his first ischemic stroke of the left hemisphere of the brain, which left him with only slight neurological deficits—mRS 1. At the present admission to the hospital, he was confused and had full left limb paresis—NIHSS 16. The CT scan was started 2 h 20 min after the onset of the symptoms of the stroke. The film (https://drive.google.com/file/d/1xRIdjTL7EYZZz0cSCqWGEr49Y2AbPnj9/view?usp=sharing, accessed on 27 February 2022) presents the imaging techniques used in sequence:

1. CT scan without contrast assisted by RAPID software. The scans automatically analyzed (RAPID software) on the ASPECT scale show only old ischemic changes in the left hemisphere of the brain. In the right hemisphere of the brain there are no signs of the presence of hypodense areas that would indicate new areas of stroke.
2. Perfusion CT scan assisted by RAPID software. The reference levels in the RAPID analysis are set according to the criteria developed in the DEFUSE 3 study. The levels of 2 parameters are investigated: CBF (cerebral blood flow) less than 30% compared to the opposite hemisphere indicates necrosis—purple color area, Tmax value greater than 6 sec signifies the area of the penumbra. Additionally, a quantitative analysis is performed. The calculated mismatch is at the level of 1.6, which according to the adopted DEFUSE 3 criteria, should disqualify the patient from mechanical thrombectomy. On the other hand, European recommendations suggest that up to 6 h

after the appearance of the first symptoms of the disease, we do not always have to follow the pCT results, but rather rely on the CT scan.

3. CT angiography of the cerebral and intracerebral arteries taken from the level of the aortic arch. The patient moved during the examination, because of this, the head is bent to the right and slightly upwards. Full occlusion of the right internal carotid artery is visible in its proximal part.

4. MR Diffusion Weighted Imaging (DWI) and corresponding apparent diffusion coefficient (ADC) sequences. A. In the DWI sequence a large area with a hyperintense signal corresponding to cerebral ischemia is visible in the right hemisphere. B. In the ADC sequence the same area has a hypointense signal which may correspond, including DWI images, to the acute nature of ischemia.

5. MR Fluid-attenuated inversion recovery (FLAIR) and DWI sequences. DWI-FLAIR-mismatch. No marked parenchymal hyperintensity is detected on fluid attenuated inversion recovery (FLAIR) images on right hemisphere (C), while acute ischemic lesion is clearly visible on DWI (D), indicating DWI-FLAIR-mismatch. Old post-stroke lesions are visible in the left hemisphere of the brain.

6. Digital Subtraction Angiography—DSA. The distal end of the aspiration catheter through which a contrast agent is administered is inserted into the proximal segment of the right internal carotid artery. The place of the occlusion is visible through which the contrast agent does not flow.

7. Rotary Digital Subtraction Angiography—3D DSA. Control examination performed directly after successful mechanical thrombectomy—mTICI 3. Visible flow of contrast blood both through the main trunk of the internal carotid artery and its branches.

2.2. Prehospital Triage in Ischemic Stroke: Problems and Needs in Poland

Substantial uncertainty exists on the benefit of organizational paradigms in stroke networks and several studies compared functional outcome between the mothership and the drip and ship models [43] and probably they mainly depend on local infrastructures and organization.

As a result of the VII Symposium on Acute Brain Stroke and the experiences of Polish leaders of the Pilot Mechanical Thrombectomy in Poland on 2–4 December 2021 [44], attention was drawn to the difficulties associated with the already applied endovascular treatment in ischemic stroke in Poland, i.e., difficult fast transport (from call to realization) for endovascular treatment (EVT) of patients with acute ischemic stroke due to intracranial large vessel occlusion (LVO) with appropriate final qualification for causal treatment (rt-PA and TM) and access to urgent brain scan (angioCT and MRI). The criteria for qualifying for the performance of MT (including, e.g., angioCT) in coordinated care due to ischemic stroke are a real problem that can be solved systematically by introducing, inter alia, training for teams of neuro-interventionalists in causal treatment. One of the elements of the training would be the detection of LVO, i.e., middle cerebral artery (M1), the proximal second segment of the middle cerebral artery (M2), and the internal carotid artery (ICA) in CT, preceded by cross-validation on pre-processed images (which are the standard in image learning methods) as support for neurointervention centers of acute stroke that do not have radiologists on-site. This is based on unsupervised analysis of convolutional neural networks (CNNs) for the detection of LVO in angioCT (used in training in Nashville, TN, USA) [45]. The presented approach would allow for the elimination of costs related to unjustified transport to the superior intervention center, guaranteeing the continuity of the hospital's operation.

Another problem is the identification of high-risk patients (with intracranial occlusion) on the basis of the observed clinical symptoms or individual assessment of the prognosis after endovascular procedures.

Creating a pilot network with referring patients to a superior center (depending on the occlusion of large vessels detected in angioCT) as a reference center would allow avoiding local problems related to a shortage of specialist local staff, and facilitate a personalized approach, guaranteeing the comfort of work of all persons participating in the pilot project as it was done in Poland, like as in China [46].

Incomplete formalized reimbursement services for the mechanical thrombectomy (related to the lack of dedicated procedures) are also a real problem, which, in the case of randomized trials, show that the use of EVT as an option for the treatment of acute ischemic stroke eliminates the costs associated with patient care during the first year of stroke and generates savings in the next years of the patient's life [47,48]. In the case of staffing problems and the growing number of stroke patients due to the aging society, the solution could be employment standards analogous to those in intensive care units: 1:1 or 1:2.

The presented problems related to delays in EVT concern many healthcare systems [48], thus limiting the number of patients with a good prognosis compared to the results of randomized trials.

In a randomized clinical trial to evaluate rapid EVT of ischemic stroke, quality control was highlighted as a guideline for ultimate success in the treatment of causative agents with weekly monitoring of imaging speed with feedback to central centers via teleconference [49]. Training in fast and effective EVT and imaging methods has helped meet these requirements. The time criterion was met with a validated time from non-contrast CT scan to groin puncture of up to 60 min and from non-contrast CT scan to first reperfusion (with first median cerebral artery flow) of <90 min. Meeting the strict time criteria goals ensured the introduction of an effective EVT with much better prognosis for patients with acute ischemic stroke. A clear limitation in the use of endovascular procedures was: difficult accessibility due to the tortuosity of the cerebral vessels or the unavailability of the neuro-interventional team, which coincides with the problems of the non-randomized studies presented above.

Ongoing efforts should aim to precisely define the occurrence of intracranial occlusion (LVO) in the transmitted available imaging test results (angioCT) with a precise estimation of the time needed to influence pre-hospital triage decisions in order to eliminate delays in causal treatment of individual patients. Successive efforts should focus on identifying real obstacles and problems guaranteeing the improvement of pre-hospital triage and the availability of pre- and post-processing IT systems, combined with timely monitoring of causal treatment of ischemic stroke and long-term care.

Future efforts should also counter the importance of telemedicine facilitating the pre-registration process using devices with GPS navigation applications, which will further facilitate the proper selection of data based on real-time information with the possibility of verifying times: while driving with synchronization of the work time of the neuro-intervention team that will review current regimens and provide long-term health benefits in patient care for ischemic stroke induced by intracranial LVO.

3. Causes of Reperfusion Procedures Failure and Potentials Risks Related to Reperfusion

Intravenous thrombolysis with rt-PA and endovascular thrombectomy are the current standard of care for acute ischemic stroke [50,51]. Their beneficial effect is mainly due to reperfusion of cerebral vessels occlusion. However, there is a proportion of patients who do not achieve clinical improvement despite successful recanalization of the occluded artery and reperfusion of the ischemic area. Despite successful recanalization by endovascular procedures, some neurological impairments may be unrecovered and some patients may develop early complications of ischemic stroke, including early neurological deterioration, symptomatic hemorrhagic transformation (so-called secondary hemorrhage) and cerebral edema [52–54], but the causes of reperfusion injuries are still debated.

The pathophysiology of reperfusion injuries include: damage to the blood–brain barrier (BBB) [55], structural remodeling of endothelial cells associated with breaking tight connections, a consistently non-specific inflammatory response induced by oxidative, nitrosative (overproduction of reactive oxygen and nitrogen species) [56], and metabolic stress (associated with insufficient oxygen supply and hypoglycemia in neuronal mitochondria).

In the event of reduced blood flow to the brain, apart from structural degeneration, there is also a functional dysfunction of blood vessels based on reduced Na^+/K^+-ATPase activity, overload of cells with sodium, calcium [57], partial depolarization, and loss of membrane potential, resulting in an increased inflow of sodium ions. This, in turn, leads to the penetration of chlorine ions and water inside the cell, causing swelling of neurons [58,59]. Subsequently, a cascade of biochemical processes related to the release of the main excitatory neurotransmitters, i.e., glutamic acid, produced from presynaptic terminals by depolarization of synaptosomes and hindered uptake by hypoxic astrocytes takes place [60]. Ischemia-induced (with blood flow to the brain below 20 mL/100 g/min), persistently elevated glutamate levels occur in all regions of the brain, leading to neurological activation associated with glutamate excitotoxicity in ischemic neurons [61]. As a consequence, continuous glutamatergic stimulation impairs the basic functions of mitochondria [62,63] and damages the BBB. It comes to an increased concentration of lactates, carbonate and glutamate, and a decreased level of alanine, citrate, glycine, tyrosine, methionine, and tryptophan in the peripheral blood [64], which emphasizes the key role of these amino acids in bio-energetic homeostasis in both ischemic brain areas and in the peripheral blood. Tyrosine, lactate, and tryptophan have been identified as potential biomarkers of acute ischemic stroke, which are associated with enhanced glycolysis and inhibition of the tricarboxylic acid cycle (TAC) in patients with acute ischemic stroke (AIS). Lactate as an indicator of the severity of anaerobic metabolism associated with ischemia and subsequent hypoxia is elevated in the cerebrospinal fluid, brain tissues, and blood serum of AIS patients. Additionally, the concentration of lactate reflects the level of neuronal necrosis and the prognosis after ischemic stroke [65].

Single results of clinical data in humans are inconclusive due to the occurrence of perifocal edema and leakage of the BBB by labeled metabolites, which complicate the interpretation of effectiveness of reperfusion procedures [66,67]. In addition, progressive neurocognitive disorders in more than half of stroke patients are still a severe and non-decreasing burden on healthcare systems around the world [68–70].

The effectiveness of the reperfusion procedures used in ischemic stroke may be limited by the secondary ischemia-reperfusion injury of the brain, which may increase the volume of ischemia and aggravate the cerebral infarction [71–73]. The results of both animal and human clinical trials show that despite early reperfusion after stroke, which protects healthy brain tissue from further ischemia, selective loss of peri-infarction neurons or emerging micro infarcts may be adverse effects resulting in long-term sensorimotor deficits after stroke [74]. An additional obstacle in the effectiveness of the performed reperfusion procedures is the lack of actual evaluation of the histopathological structure of the embolic material in real time or thrombus perviousness evaluated by neuroradiological techniques, which means that mechanical thrombus removal during thrombectomy may be quite difficult and not bring the expected results [75]. This is due to the fact that some clots require multiple removal attempts in order to achieve successful recanalization (with the preferred recanalization rate, classification of modified treatment in cerebral ischemia, mTICI), while other thrombi are effectively removed in the first attempt, resulting in better final results [76]. There are also situations where, despite a preliminary assessment of the seemingly known consistency of the clot, it turns out that the structure is so diverse that it may cause unexpected fragmentation of the clot during mechanical thrombectomy with its dispersion to the distal branches of the cerebral arteries extending the infarct zone [77]. The risk of clot defragmentation strictly depends on its composition, namely the dissection resistance increases significantly with an increase in the content of fibrin or neutrophilic networks (NETs, consisting of: DNA, histones, and proteolytic enzymes produced by activated neutrophils in various mechanisms) [78], while red blood clots are more prone to rupture during emerging stresses [79] but it also may depend on the patient's features, i.e., antithrombotic therapy [7]. In the presence of NETs and related extracellular DNA and histones, the structure of fibrin changes, becoming more resistant to both mechanical destruction using EVT devices, thus increasing the difficulty of effectively removing the thrombus during mechanical thrombectomy and enzymatic degradation using rt-PA. Knowledge of the content and amount of NETs in the composition of the embolic material could, therefore, be a potential goal of assessing the effectiveness of mechanical thrombectomy in terms of successful recanalization [80]. This approach supports the concept of multi-target combined therapy (apart from the components, targeted to the percentage and architecture of the individual components of the clot structure) in the causal treatment of ischemic stroke of large vessels of the cerebral arteries [81].

With regard to the reperfusion procedures, the current efforts are focused on achieving complete reperfusion already during the first thrombectomy attempt as a favorable end result of EVT using mechanical thrombectomy and/or a strategy combining thrombectomy with thromboaspiration [82]. Nevertheless, this approach does not solve the problem of the lack of knowledge of the structure of the embolic material prior to mechanical thrombectomy.

In the literature on the subject, attention was paid to the relationship between the appearance of the symptom of the so-called hyperdense middle cerebral artery sign of the brain shown in the CT scan in the acute phase of stroke with the size of Hounsfield units (HU), which are associated with the number of red blood cells and their parameters, i.e., hemoglobin and hematocrit [83]. According to the accepted theory, the presence of hemoglobin determines the weakening of the clot, and the susceptibility to lysis seems to increase the hematocrit level. Thus, red (cardioembolic) thrombi containing erythrocytes and fibrin result in a higher HU count and greater susceptibility to fibrinolytic agents, while the so-called white clots (atherosclerotic) consisting of varying amounts of platelets, atherosclerotic debris, and cellular debris with low numbers of red blood cells, results in lower HU values and greater resistance to fibrinolytic agents. An additional problem that makes it impossible to generalize the assumptions are clots with a heterogeneous histological structure at different lengths, containing extracellular neutrophil traps and microcalcifications only in specific areas of the thrombus, which together causes resistance to mechanical thrombectomy [76], and, at the same time, their structure cannot be determined in advance with the available diagnostic imaging methods [83].

In addition to thrombolytic activity in blood vessels, rt-PA seems to have a neuroprotective effect consisting in the suppression of oxidative stress during reperfusion, which is associated with increasing phosphorylation of 5′adenosine monophosphate-activated protein kinase, increasing glucose uptake in neurons and promoting mitochondrial ATP production [84].

The knowledge of the physical properties of the clot, i.e., its density or permeability, may influence its successful removal with MT [85,86]. In addition to the above-mentioned clot composition and permeability, clot length and volume are also important for successful recanalization and favorable clinical outcomes in AIS [87,88]. The likelihood of clot dissolution increases, the greater the surface area of the clot (in contact with the blood flow) as determined by cirHU (circle HU) units for the range of blood flow around the clot. CirHU is a marker of good collateral and/or residual flow, which helps in the effective rate of reperfusion, deciding on the accuracy of the adopted indicators [89]. This is especially important in those patients who are less prone to recanalization after intravenous thrombolysis in the initial evaluation [90]. Multiphase angioCT is also used to predict the clinical outcome of AIS EVT, which is used as part of the diagnostic standard to assess the existence (or absence) of collateral circulation to the ischemic area of the brain. Collateral circulation is an important prognostic factor for the volume of the so-called penumbra, i.e., tissue that is ischemic but salvageable after successful recanalization [91–94]. In addition, it turns out that the patency of the arteries may be genetically determined; in the case of patients with a lower family burden, a greater flow is observed at cerebral arteries [95].

4. The Role of Mitochondria in the Pathophysiology of Ischemic Stroke and Recanalization

Cellular mechanisms developing in neurons after reperfusion are still debated. The recanalization therapies in ischemic stroke, apart from their undoubtedly beneficial influence on the restoration of normal cerebral circulation, might also start cellular pathways causing ischemic-reperfusion neuronal death along with changes in the morphology and function of mitochondria with hemorrhage transformation of the tissue necrosis zone after ischemia [96] and the comprehension of these mechanisms may even more improve a patient's outcome. Impairment of mitochondrial function results from calcium flowing into the cell, which increases its concentration in the mitochondria as early as 24 h after reperfusion procedures [97]. Intra-mitochondrial Ca^{2+} ingress in excess of the buffer capacity leads to osmotic edema of the mitochondrial crest, release of NADH, cytochrome c and cytosolic double-stranded DNA (dsDNA) into the cytoplasm of astrocytes. This, in turn, triggers a cascade of events that lead to increased expression of various pro- and anti-apoptotic proteins by initiating inflammatory responses, GMP-AMP synthase (cGAMP, cGAS), sequentially to cell apoptosis in the penumbra [98,99].

The cycle of events entails the opening of large conduction channels in the inner mitochondrial membrane, known as the mitochondrial permeability transitional pores (MPTP). This, in turn, contributes to the unrestricted penetration of small solutes, disturbance of ion gradients and alteration of the mitochondrial membrane potential, what leads to impairment of oxidative–antioxidant pathways [100]. The activity of proteins related to glycolysis, pyruvate dehydrogenase complex, tricarboxylic acid cycle (TCA), and oxidative phosphorylation are inhibited, which cause metabolic imbalance between the cytosol and mitochondria and lead to breakdowns of energy metabolism [101]. One of the key molecular mechanisms of stroke is believed to be oxidative stress caused by the overproduction of reactive oxygen species (ROS) in the tissues of the brain after the restoration of blood flow or reperfusion. Reactive oxygen species and reactive nitrogen species (including neurotoxic NO) oxidize mitochondrial lipids, protein sulfhydryl groups, and iron–sulfur complexes essential for maintaining the function of respiratory oxidative phosphorylation enzymes. In the case of excess of NO, apoptotic cell death is initiated by activating the p53-dependent pathway, caspase activation, chromatin condensation, and DNA fragmentation [102] As a result of these events, ATP synthesis is inhibited and the p53-dependent mechanism is induced, which is caused by the reaction of NO with O_2^- creating a much stronger oxidant $ONOO^-$, which is highly reactive and is mainly responsible for the toxicity of NO in hypoxic brain tissues [103].

In the literature, the oxidative stress induced by damage to the brain tissue was indirectly related to the markers and characteristic metabolic abnormalities, however, the actual dynamics of ROS has not been recorded in vivo so far.

The resulting oxidative and nitrosative stress induce iron-dependent ferroptosis, a non-apoptotic form of cell death consequently leading to mitochondrial contraction and switching their functions off in the area of neurovascular units (Figure 1) [104,105]. In the course of revascularization procedures, the concentration of cytosolic Ca^{2+} increases secondary to glutamate excitotoxicity, which further inactivates contracted astrocyte mitochondria located near glutamate [106], leading to the death of neurons. Reperfusion is essential for saving ischemic tissue, but, paradoxically, it can also exacerbate neuronal damage by generating mitochondrial damage. Therefore, the timely elimination of dysfunctional mitochondria is crucial for maintaining a healthy mitochondrial network during revascularization procedures [107].

Research is currently under way to precisely define the role of mitochondria in response to the functioning of neurons, astrocytes, and glial cells in ischemic brain damage [100]. The diverse functions of mitochondria in response to ischemia affect all nerve cells, including cortical astrocytes, and are observed with respect to the delivery of glucose-derived ATP energy, mitochondrial membrane potential, cytosolic calcium release, and generation of mitochondrial permeability transitional pore opening (MPTP) in response to pro-apoptotic factors and levels of ketoglutarate dehydrogenase. All the indicated mechanisms related to mitochondrial bioenergetics determine the resistance of astrocytes to ischemia, as well as control the local processes responsible for the survival and death of nerve cells [108,109].

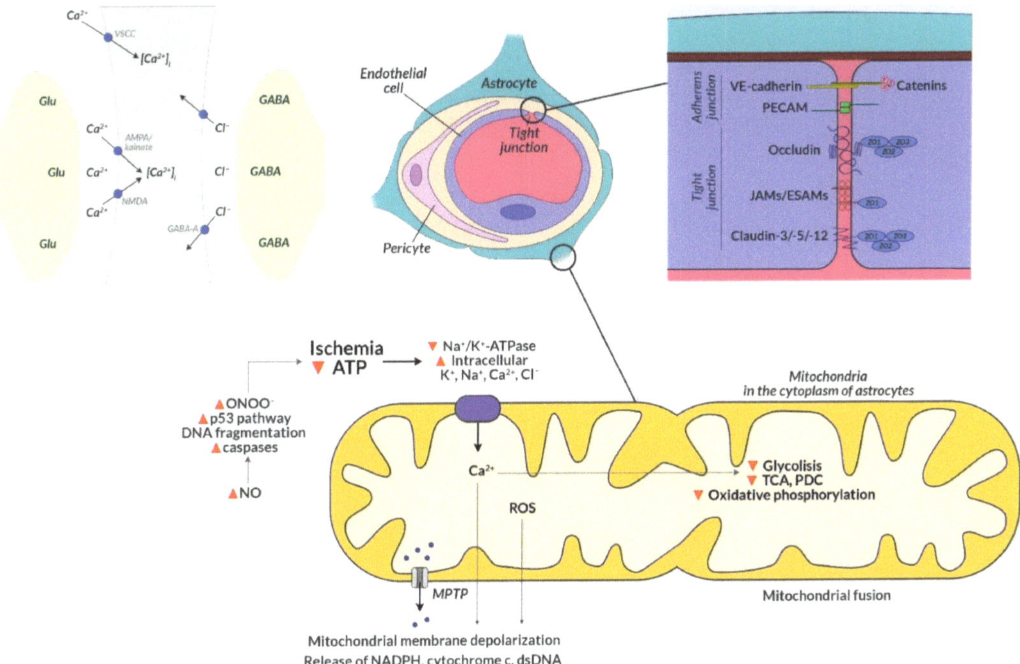

Figure 1. The response of astrocytic mitochondria to hypoxia in the course of ischemic stroke. AMPA: α-amino-3-hydroxy-5-methyl-4-isoxazole propionate; dsDNA: double-stranded DNA; ESAM: endothelial cell-selective adhesion molecule; GABA: γ-aminobutyric acid; Glu: glutamate; JAM: junctional adhesion molecule; MPTP: mitochondrial permeability transitional pore; NADPH: nicotinamide adenine dinucleotide phosphate; NMDA: N-methyl-d-aspartate; NO: nitric oxide; PDC: pyruvate dehydrogenase complex; PECAM: platelet-endothelial cell adhesion molecule; ROS: reactive oxygen species; TCA: tricarboxylic acid; VE: vascular epithelium; and VSCC: L-type voltage-sensitive calcium channels, calcium influx, and overload upon hypoxia.

The role of mitochondria in astrocytes is now extensively studied, gaining importance in the emerging field of astroneurology, where small astrocytic mitochondria exhibit neuro-adaptive abilities in an oxygen- and glucose-deprived environment. Thus, astrocyte mitochondria guarantee the maintenance of balance by phagocytosing synapses, fragments of axonal and neuronal mitochondria, and damaged proteins [110]. During hypoxia and lack of glucose, such as ischemic stroke or reperfusion, astrocyte mitochondria are depolarized early, switching glycolysis-related oxygen metabolism to lactate energy supply (a mechanism known as lactate transfer from astrocytes to neurons, astrocyte–neuron lactate shuttle ANLS, associated with simultaneous lactate import and export) [111], released into the extracellular space and taken up by neurons via monocarboxylic acid transporters (MCT), thus preventing neuronal death [112].

Glutamate excitotoxicity related to, inter alia, changes in transport activity as well as altered expression of glutamate transporters, plays a key role in this process, as research shows that calcium activation of glutamate transport by astrocytes seems to be a causative factor affecting the inhibition of mitochondrial function [113,114]. Microglial activation occurs first after cerebral ischemia, and is activated by the M1 and M2 pathways dependent on angiogenic functions and enhancing BBB integrity. This is mainly due to increase of the expression of tight junction proteins (TJP) and the release of matrix metalloproteinases (MMPs) within the neurovascular units (NVU) [115] by inhibiting their functioning and, thus, leading to a metabolic disconnection between the neurons and the proximal blood flow. The change in the architecture of the mitochondria-forming networks is related to the adaptive role of microglia, which determines the survival of surrounding neurons, limiting the extent of losses caused by ischemia-reperfusion damage to the brain [116].

Thanks to these abilities, it is possible to regenerate these organelles and survive in conditions of exposure to environmental stress of ischemic areas of the brain, which is a promising target of therapy improving the clinical condition of patients after stroke and reperfusion procedures.

5. The Role of Inflammation in the Pathophysiology of Ischemic Stroke and in the Assessment of the Prognosis of Reperfusion Procedures

Currently conducted clinical trials and animal models indicate that stroke is a network of interactions within the central nervous system (CNS) [117]. In the course of these events, the brain calls for help by releasing a number of factors, such as hypoxia-induced factor 1a (HIF-1a), protein S100b, ATP, which activate the central and peripheral immune systems. By mobilization of the cells of the immune system, i.e., white blood cells, neutrophils, connective tissue cells (mast cells), microglia cells, to the ischemic site, toxic substances of their metabolism are released and involved in the inflammatory response, i.e., pro-inflammatory cytokines, and a number of others, including reactive species of oxygen, nitrogen, sulfur (ROS, RNS), or matrix metalloproteinases, for example MMP-9 [118]. MMP-9 plays an important role in rt-PA-related bleeding complications, while ROS increase the effect of rt-PA on MMP activation in the mechanism of loss of caveolin-1 (cav-1), a protein encoded in the *cav-1* gene, which is a critical determinant of unsealing BBB during reperfusion [119]. Due to the release of toxic components of the blood plasma, apart from the unsealing of the BBB, the blood flow in the area of the cerebral microcirculation is stopped. As one of the released factors of the immune system, cytokines simultaneously stimulate a cellular and/or humoral response. Cells of the peripheral immune system, i.e., monocytes, neutrophils, T-lymphocytes, platelets, and macrophages, get through the damaged BBB from the systemic circulation to the area of cerebral ischemia, contributing to further irreversible ischemic brain damage by penetrating the brain parenchyma and exacerbating the ongoing the inflammatory process [120]. Calprotectin (a heterodimer composed of two cytosolic proteins: S100A8 and S100A9) released by these cells is one of the main proteins in the course of inflammation in patients after ischemic stroke and is a potential predictor of resistance to reperfusion and, thus, determines the effectiveness of reperfusion treatments [121]. In the case of neutrophils, the matter is complicated by their diverse phenotype influencing changes in the structure of these cells depending on the degree of maturity (or related to the presence of TLR4) and differentiation under the influence of regulatory signals released under ischemia [122,123].

The assessment of the amount of released neutrophils after a stroke is a prognostic factor for the severity of a stroke or the degree of bleeding complications. The assessment of an increased neutrophil/lymphocyte ratio is associated with poor neurological improvement after ischemic stroke, which is attributed to the neurotoxic role of neutrophils in the post-ischemic brain region [124].

In general, clinical, imaging, and laboratory biomarkers are important in the assessment of the prognosis of AIS or reperfusion procedures. In predicting annual mortality after ischemia, a worse prognosis is for those with higher NIHSS values and higher carotid intimal and middle membrane thickness (cIMT), lower coagulation parameters, i.e., antithrombin levels, lower platelet count, protein C, and albumin concentration, lower HDL cholesterol, higher concentration of factor VIII, von Willebrand factor (vWF), higher absolute white blood cell count, higher concentrations of tumor necrosis factor α (TNF-α), interleukin 10, high sensitivity C-reactive protein (hsCRP), vascular cell adhesion molecule 1 (VCAM-1), apoB, LDL cholesterol, and triglycerides [125,126]. In the construction of ROC curves of complications and annual survival after ischemic stroke or applied revascularization, a multimodal approach is used that combines a number of biomarkers, i.e., variables using NIHSS, cIMT, age, IL-6, TNF-α, hsCRP, HDL, protein C, protein S, vWF, and platelet endothelial cell adhesion molecule 1 (PECAM-1), which have a larger area under the curve (AUC/ROC), i.e., 0.975 (accuracy ca. 93%, 100% sensitivity, and 85.7% specificity) than either indicators separately [127]. The assessment of monocyte-to-cholesterol high-density lipoprotein (MHR) and monocyte-to-lymphocyte (MLR) ratios as a combined approach is also a better predictor of ischemic stroke/reperfusion than analyzing these parameters separately.

The severity of inflammation associated with damage to the BBB after reperfusion procedures is associated with the risk of hemorrhagic transformation after ischemic stroke [128], being a key factor in its pathophysiology [129]. It is the mitochondria that regulate a number of cellular mechanisms, such as autophagy, apoptosis, energy production, and expression of genes related to mitochondrial biogenesis, thus influencing the direct inflammatory response [108,130] of neurons and, indirectly, of microglial cells [131].

The brain and the immune system constitute functional neuro–humoral connectivity, therefore, the death of nerve cells in the course of acute stroke leads to the release of a number of humoral response factors causing local inflammation in the damaged brain [132]. These factors lead to the expression of receptors on microglial cells and astrocytes, which, in turn, recruit cells of the peripheral immune system to the infarcted brain tissue, consequently contributing to exacerbation of neurological changes. Recruited T lymphocytes and natural killer (NK) cells mediate the impairment of cerebral microcirculation through the adhesion of leukocytes to the cerebral vessel walls and the initiation of secondary micro-thrombosis. Inflammation and/or infection may promote autoimmune responses against brain antigens in stroke patients [133]. Hypoglycemia and insufficient energy to maintain the membrane potential of the nerve cells may degrade the functional assessment of ischemic areas of the brain due to a diminished hormonal response and greater systemic effects [134] independent of the humoral response. The role of the vascular endothelium is crucial in maintaining perfusion and patency of cerebral microcirculation vessels. Brain ischemia causes endothelial damage and a cascade of positive feedback loop events between blood flow and brain tissue along with damage to the BBB, transcytosis, death of endothelial cells, and recruitment of immune cells [135,136].

6. Examples of Neuroprotection in the Treatment of Ischemic Stroke

A particular interesting field about acute stroke management is related to neuroprotective therapies. Some reports suggest a potential role of these therapies [137], although underlying mechanisms are still unknown and evidence of their usefulness in clinical setting is lacking [47]. An example of endogenous neuroprotection of damaged neurons after a stroke stimulating the remodeling of the mitochondrial network architecture are the steps of gradual cooling down to 33 °C, which facilitates the transfer of mitochondria from astrocytes to damaged neurons and endothelial cells, while increasing the content of intracellular ATP, mitochondrial membrane potential (MMP), and cell viability during hypoxia [138,139]. Depolarization of the mitochondrial membrane potential of primary cortical neurons is inhibited after excitotoxic glutamate stimulation, which results in NAD^+ administration, ultimately inhibiting apoptotic neuronal death [140].

Hopes related to neuroprotection are also created by hyperbaric oxygen (HBO) therapy, whose mechanism is associated with the inhibition of mitochondrial apoptosis and disturbances in energy metabolism of these structures [141]. The administration of 100% oxygen at approximately three times higher atmospheric pressure allows for higher arterial oxygen pressure and greater oxygen supply by increasing dissolved oxygen in the plasma, which stimulates cellular respiration and supports ATP synthesis in ischemic areas of the brain [142]. Oxidative stress and the expression of proteins related to apoptosis is alleviated; while the decrease in ATP levels, the activity of enzymes of the mitochondrial complex and the activity of Na^+/K^+-ATPase, which maintain the energy metabolism at a constant level, are inhibited [143].

The excessive release of glutamate has become the target of a therapy using checkpoints related to protein kinases regulating the signaling pathways of the N-methyl-D-aspartate (NMDA) receptor [144]. These include mainly antagonists of the glutamate-binding site to the NMDA receptor, e.g., amino acids and their derivatives containing basic acid and phosphate groups with strong hydrophilicity, which makes most of these compounds highly polar and prevents their penetration through the BBB [145]. An equally promising strategy influencing glutamate excitotoxicity is the inhibition of microRNA-29b by suppressing oxidative stress and apoptosis [146]. MicroRNA (miRNA) as single-stranded, relatively short RNA molecules regulate the expression of proteins, thus influencing both physiological processes and diseases of the nervous system. Stroke-specific miRNAs include, inter alia, miR-223, miR-181, miR-125a, miR-125b, miR-1000, miR-132, and miR-124a, which affect glutamate receptors and the related regulation of neuronal and astrocytic proteins after stroke [147–149].

Preventing neuronal death and, thus, reducing neurological damage are complex tasks that cannot be successfully solved by targeting individual mechanisms.

One of the treatment strategies discussed in the literature would be the inhibition of TAK1 (transforming growth factor-β-activated kinase 1) in microglial cells in order to protect before stroke by inhibiting the death of neurons, e.g., through the pro-apoptotic JNK/c-Jun pathway. As demonstrated in the mouse MCAO/reperfusion model, TAK1 does not show any neuroprotective activity. Moreover, TAK1 activation was observed in post-ischemic neurons, linking its presence with the death of neuronal cells through many signaling pathways, e.g., JNK/c-Jun, p38 and NF-κB [150]. To date, the molecular mechanism underlying the pathological effects of TAK1 activation in the post-ischemic brain remains unclear. TAK1 inhibition, which promotes gray and white matter integrity, may be a promising therapeutic strategy after ischemic stroke [151].

Post-stroke regenerative therapies currently focus on improving neural plasticity to reverse the disturbances in the nervous structure and improve the functioning of the neural network both during recovery and in the long term.

Stem cell-based therapies are of interest because of their neuro-regenerative potential to promote neurogenesis and protect surviving neurons. Bone marrow stem cells (BMSC) and mesenchymal stem cells (MSC) promote neurogenesis in preclinical models of intracerebral transplantation with immunomodulatory properties capable of suppressing inflammation following stroke and, potentially, improving recovery [152].

Cell therapies based on protective cell phenotypes are also an interesting approach of improvement after reperfusion therapy in ischemic stroke patients. Polarized microglial cells or peripheral blood mononuclear cells are promising therapeutic strategies because of their pleiotropic effects, which, depending on the microenvironment and adaptation properties, determine the variable phenotypes of cells in response to brain injuries [153,154]. Microglial activation has significant effects on spontaneous regeneration after stroke, including structural and functional restoration of neurovascular networks, neurogenesis, axonal remodeling, and blood vessel regeneration. Polarized cell therapies are gaining increasing attention in the treatment of strokes and neurological diseases [155].

An enhanced immune system response during ischemic stroke is the target of advanced immunotherapies of ischemic inflammatory cascade while extending the therapeutic time window [156] over conventional rt-PA treatment or revascularization therapy. An example is natalizumab, a monoclonal antibody that shows therapeutic effects in both preclinical and clinical trials related to stroke [157]. The mechanism of action of the used biological therapy is that the antibody interacts with its specific target, be it a ligand or a receptor, inhibiting the cytotoxic signaling cascade. Thus, immunotherapy improves cell viability or delays cell death [158], therefore ultimately reducing inflammatory injuries and, at the same time, stimulating peripheral immunity by acting neuroprotectively. Natalizumab prevents systemic migration of brain leukocytes or inhibits the neurotoxic production of inflammatory mediators, excluding the incidence of bacterial infections during stroke changes. It should be noted, however, that stem cell therapies bring better results mainly among people who are prone to stroke.

Targeting receptors or proteins of channels and enzymes (CD68, Iba-1, GFAP) involved in microglia-dependent neuritis is a promising goal of therapy after recanalization procedures, especially in the treatment of secondary damage caused by cerebral ischemia (where many types of cells are involved, including microglia, astrocytes, oligodendrocytes, and peripheral T lymphocytes) [159]. This is due to a different phenotype of neurodegenerative lesions resulting from secondary thalamic trauma compared to the primary cortical injury following stroke. Signaling changes from initiated nervous system inflammation appear to represent a key canonical signaling pathway between primary and secondary changes in both brain regions following stroke [160].

7. Potential Goals of Ischemic Stroke Therapy and the Evaluation of the Effectiveness of the Applied Treatment

Selection of endpoints is a significant problem in assessing the effectiveness of the applied recanalization procedures both in clinical and pre-clinical evaluation. This applies in particular to the confirmed scientific evidence on the effectiveness of new drugs, which would be classified in terms of not only survival or functional improvement expressed according to the modified Rankin Scale (mRS) as the main endpoint, but also other measurable scales for the obtained results, e.g., independent measurements of imaging or laboratory tests [161]. Examples include the determination of the perfusion volume, the volume of the ischemic intact part of the brain, or real-time histomorphometric measurements of the embolic material during reperfusion procedures.

The search for a possible treatment of ischemic stroke has evolved from finding a relationship between selective molecular events controlling the known mechanism of action of selected individual stages of the ischemic cascade, such as the use of glutamate receptor-specific antagonists (for which clinical trials have failed) to synergistic and pleiotropic therapies combining glutamate antagonists and GABA agonists [162], which provided scientific evidence of their effectiveness. Similarly, the consideration of single peripheral markers for assessing recanalization efficacy and long-term functional benefit after stroke was more predictive of the concept of combining multiple biochemical biomarkers than considering each of the indicators individually.

Based on our previous experience related to the search for objective biomarkers of schizophrenia, which is characterized by the complexity of clinical symptoms and endophenotypic differentiation, we can draw conclusions that one must look at a new perspective in the search for the prognosis and treatment of strokes related to the cascade of events [163]. In the constructed statistical model, the results of clinical, imaging, and laboratory tests will be taken into account in a number of variables, which will allow to achieve long-term success only when taken together.

The ongoing efforts focused on a number of variables covering all cells of the nervous system, i.e., neurons, astrocytes, and cells that form the blood–brain barrier, are focused around translational medicine. From the laboratory table and in vitro research, it will take the changes at the level of preclinical and clinical research involving the human brain, setting a new horizon of thought presumably capable of predicting complex neurological functions.

A limitation of this approach is the complex nature of the neurovascular units (NVU) that form networks of connections throughout the brain that cannot be linked to unicellular in vitro models. Studies show the isolated functions of individual types of NVU cells, ultimately causing many problems of an interpretative nature in relation to the functions performed in vivo [164,165], in particular in the area of the ischemic penumbra, where the observed residual blood flow supporting cell survival cannot be reproduced in vitro.

Single-cell models, despite a number of limitations in relation to in vivo models, are simple and repeatable, thus, it becomes possible to control potential candidates for neuroprotective drugs and to assess their protective effects against glial cells and the functioning of the blood–brain barrier.

Studies on the circulation of mitochondria between astrocytes and neurons during ischemia/reperfusion provide evidence of potential new targets for stroke therapy to address mitochondrial-related energy failure. Initial preclinical studies related to the rescue of neurons through astrocytic mitochondria have shown the first spectacular results in this area [166].

Despite the triumphant recanalization therapies, thrombolysis and thrombectomy in the treatment of ischemic stroke, currently perceived treatment will require a redefinition of basic concepts to recognize the differential susceptibility associated with the complexity of the neurovascular units that make up the brain networks, requiring a pleiotropic approach to work through many different mechanisms of action.

Author Contributions: Conceptualization, T.J.P. and W.K.; writing—original draft preparation, W.K., F.P., B.B., A.L., K.B.-S. and P.B.; writing—review and editing, W.K., T.J.P., F.P., K.M. and T.K.; visualization, W.K.; supervision, T.J.P. and W.K.; project administration, T.J.P., F.P. and K.M.; funding acquisition, T.J.P., B.B. and W.K. All authors have read and agreed to the published version of the manuscript.

Funding: This research was supported by the grants from the Jagiellonian University Medical College, Poland.

Institutional Review Board Statement: Not applicable.

Informed Consent Statement: Not applicable.

Conflicts of Interest: The authors declare no conflict of interest.

References

1. Mandalaneni, K.; Rayi, A.; Jillella, D.V. *Stroke Reperfusion Injury*; StatPearls Publishing: Treasure Island, FL, USA, 2021.
2. Boeckh-Behrens, T.; Kleine, J.F.; Zimmer, C.; Neff, F.; Scheipl, F.; Pelisek, J.; Schirmer, L.; Nguyen, K.; Karatas, D.; Poppert, H. Thrombus Histology Suggests Cardioembolic Cause in Cryptogenic Stroke. *Stroke* **2016**, *47*, 1864–1871. [CrossRef] [PubMed]
3. Desilles, J.; Nomenjanahary, M.S.; Consoli, A.; Ollivier, V.; Faille, D.; Bourrienne, M.; Hamdani, M.; Dupont, S.; Di Meglio, L.; Escalard, S.; et al. Impact of COVID-19 on thrombus composition and response to thrombolysis: Insights from a monocentric cohort population of COVID-19 patients with acute ischemic stroke. *J. Thromb. Haemost.* **2022**, *20*, 919–928. [CrossRef] [PubMed]
4. Alkarithi, G.; Duval, C.; Shi, Y.; Macrae, F.L.; Ariëns, R.A. Thrombus Structural Composition in Cardiovascular Disease. *Arter. Thromb. Vasc. Biol.* **2021**, *41*, 2370–2383. [CrossRef] [PubMed]
5. Saver, J.L. Time is brain-quantified. *Stroke* **2006**, *37*, 263–266. [CrossRef]
6. Da Silva-Candal, A.; Dopico-López, A.; Pérez-Mato, M.; Rodríguez-Yáñez, M.; Pumar, J.; Ávila-Gómez, P.; Castillo, J.; Sobrino, T.; Campos, F.; Hervella, P.; et al. Characterization of a Temporal Profile of Biomarkers as an Index for Ischemic Stroke Onset Definition. *J. Clin. Med.* **2021**, *10*, 3136. [CrossRef]
7. Pilato, F.; Valente, I.; Calandrelli, R.; Alexandre, A.; Arena, V.; Dell'Aquila, M.; Broccolini, A.; Della Marca, G.; Morosetti, R.; Frisullo, G.; et al. Clot evaluation and distal embolization risk during mechanical thrombectomy in anterior circulation stroke. *J. Neurol. Sci.* **2022**, *432*, 120087. [CrossRef]
8. Badwaik, D.G.; Badwaik, P. Influence of Psychological Disorders on the Functional Outcomes in the Survivors of Ischemic Stroke. *J. Stroke Cerebrovasc. Dis.* **2021**, *30*, 105486. [CrossRef]
9. Zhang, S.R.; Phan, T.G.; Sobey, C.G. Targeting the Immune System for Ischemic Stroke. *Trends Pharmacol. Sci.* **2021**, *42*, 96–105. [CrossRef]
10. Feigin, V.; Krishnamurthi, R. *Oxford Textbook of Stroke and Cerebrovascular Disease*; Oxford University Press: Oxford, UK, 2014.
11. Béjot, Y.; Daubail, B.; Giroud, M. Epidemiology of stroke and transient ischemic attacks: Current knowledge and perspectives. *Rev. Neurol.* **2016**, *172*, 59–68. [CrossRef]
12. Worldometers. Current World Population. Available online: https://www.worldometers.info/world-population/ (accessed on 27 December 2021).
13. United States Census Bureau. U.S. and World Population Clock. Available online: https://www.census.gov/popclock/ (accessed on 27 December 2021).
14. Roy-O'Reilly, M.; McCullough, L.D. Age and Sex Are Critical Factors in Ischemic Stroke Pathology. *Endocrinology* **2018**, *159*, 3120–3131. [CrossRef]
15. Howard, V.J.; Madsen, T.E.; Kleindorfer, D.O.; Judd, S.E.; Rhodes, J.D.; Soliman, E.Z.; Kissela, B.M.; Safford, M.M.; Moy, C.S.; McClure, L.A.; et al. Sex and Race Differences in the Association of Incident Ischemic Stroke with Risk Factors. *JAMA Neurol.* **2019**, *76*, 179–186. [CrossRef] [PubMed]
16. Béjot, Y.; Bailly, H.; Durier, J.; Giroud, M. Epidemiology of stroke in Europe and trends for the 21st century. *Presse Med.* **2016**, *45*, e391–e398. [CrossRef] [PubMed]
17. Truelsen, T.; Piechowski-Jozwiak, B.; Bonita, R.; Mathers, C.; Bogousslavsky, J.; Boysen, G. Stroke incidence and prevalence in Europe: A review of available data. *Eur. J. Neurol.* **2006**, *13*, 581–598. [CrossRef] [PubMed]

18. Truelsen, T.; Begg, S.; Mathers, C.D.; Satoh, T. Global burden of cerebrovascular disease in the year 2000. In *GBD 2000 Working Paper*; WHO: Geneva, Switzerland, 2002.

19. Wafa, H.A.; Wolfe, C.D.; Emmett, E.; Roth, G.A.; Johnson, C.O.; Wang, Y. Burden of Stroke in Europe: Thirty-Year Projections of Incidence, Prevalence, Deaths, and Disability-Adjusted Life Years. *Stroke* **2020**, *51*, 2418–2427. [CrossRef]

20. Krishnamurthi, R.V.; Ikeda, T.; Feigin, V.L. Global, Regional and Country-Specific Burden of Ischaemic Stroke, Intracerebral Haemorrhage and Subarachnoid Haemorrhage: A Systematic Analysis of the Global Burden of Disease Study 2017. *Neuroepidemiology* **2020**, *54*, 171–179. [CrossRef]

21. Lozano, R.; Naghavi, M.; Foreman, K.; Lim, S.; Shibuya, K.; Aboyans, V.; Abraham, J.; Adair, T.; Aggarwal, R.; Ahn, S.Y.; et al. Global and regional mortality from 235 causes of death for 20 age groups in 1990 and 2010: A systematic analysis for the Global Burden of Disease Study 2010. *Lancet* **2012**, *380*, 2095–2128. [CrossRef]

22. The GBD 2016 Lifetime Risk of Stroke Collaborators; Feigin, V.L.; Nguyen, G.; Cercy, K.; Johnson, C.O.; Alam, T.; Parmar, P.G.; Abajobir, A.A.; Abate, K.H.; Abd-Allah, F.; et al. Global, regional, and country-specific lifetime risks of stroke, 1990 and 2016. *N. Engl. J. Med.* **2018**, *379*, 2429–2437. [CrossRef]

23. Kim, J.; Thayabaranathan, T.; Donnan, G.A.; Howard, G.; Howard, V.J.; Rothwell, P.M.; Feigin, V.; Norrving, B.; Owolabi, M.; Pandian, J.; et al. Global Stroke Statistics 2019. *Int. J. Stroke* **2020**, *15*, 819–838. [CrossRef]

24. Thrift, A.G.; Thayabaranathan, T.; Howard, G.; Howard, V.J.; Rothwell, P.M.; Feigin, V.L.; Norrving, B.; Donnan, G.A.; Cadilhac, D. Global stroke statistics. *Int. J. Stroke* **2017**, *12*, 13–32. [CrossRef]

25. Muratova, T.; Khramtsov, D.; Stoyanov, A.; Vorokhta, Y. Clinical Epidemiology of Ischemic Stroke: Global Trends and Regional Differences. *Georgian Med. News* **2020**, *2*, 83–86.

26. Nannoni, S.; de Groot, R.; Bell, S.; Markus, H.S. Stroke in COVID-19: A systematic review and meta-analysis. *Int. J. Stroke* **2021**, *16*, 137–149. [CrossRef] [PubMed]

27. Yaghi, S.; Ishida, K.; Torres, J.; Mac Grory, B.; Raz, E.; Humbert, K.; Henninger, N.; Trivedi, T.; Lillemoe, K.; Alam, S.; et al. SARS-CoV-2 and Stroke in a New York Healthcare System. *Stroke* **2020**, *51*, 2002–2011, Erratum in *Stroke* **2020**, *51*, e179. [CrossRef] [PubMed]

28. Kwiatkowska, A. Contemporary Problems of Managing Human Resources and Corporate Finance. In *Knowledge–Economy–Society*; Kraków University of Economics: Kraków, Poland, 2017; p. 17.

29. Underlying Cause of Death, 1999–2000. Available online: https://wonder.cdc.gov/wonder/help/ucd.html (accessed on 27 February 2022).

30. MacKenzie, I.E.; Moeini-Naghani, I.; Sigounas, D. Trends in Endovascular Mechanical Thrombectomy in Treatment of Acute Ischemic Stroke in the United States. *World Neurosurg.* **2020**, *138*, e839–e846. [CrossRef] [PubMed]

31. Renna, R.; Pilato, F.; Profice, P.; Della Marca, G.; Broccolini, A.; Morosetti, R.; Frisullo, G.; Rossi, E.; De Stefano, V.; Di Lazzaro, V. Risk Factor and Etiology Analysis of Ischemic Stroke in Young Adult Patients. *J. Stroke Cerebrovasc. Dis.* **2014**, *23*, e221–e227. [CrossRef]

32. Rochmah, T.; Rahmawati, I.; Dahlui, M.; Budiarto, W.; Bilqis, N. Economic Burden of Stroke Disease: A Systematic Review. *Int. J. Environ. Res. Public Health* **2021**, *18*, 7552. [CrossRef]

33. Priorities in Healthcare. Warsaw, Session: Directions for Further Optimization of Stroke Care in Poland. Presentation by Prof. A. Słowik under the Title "Access to Causal Treatment of Ischemic Stroke in Poland 2019–2021". Available online: https://www.termedia.pl/Konferencja-PRIORYTETY-W-OCHRONIE-ZDROWIA-2022-OBEJRZYJ-NAGRANIA,1624,18667.html (accessed on 26 January 2022).

34. Psychogios, K.; Tsivgoulis, G. Intravenous thrombolysis for acute ischemic stroke: Why not? *Curr. Opin. Neurol.* **2022**, *35*, 10–17. [CrossRef]

35. Aguiar de Sousa, D.; von Martial, R.; Abilleira, S.; Gattringer, T.; Kobayashi, A.; Gallofré, M.; Szikora, I.; Feigin, V.; Caso, V.; Fischer, U.; et al. Access to and delivery of acute ischaemic stroke treatments: A survey of national scientific societies and stroke experts in 44 European countries. *Eur. Stroke J.* **2019**, *4*, 13–28. [CrossRef]

36. Hirsch, J.A.; Yoo, A.J.; Nogueira, R.G.; Verduzco, L.A.; Schwamm, L.H.; Pryor, J.C.; Rabinov, J.D.; Gonzalez, R.G. Case volumes of intra-arterial and intravenous treatment of ischemic stroke in the USA. *J. NeuroInterv. Surg.* **2009**, *1*, 27–31. [CrossRef]

37. Joo, H.; Wang, G.; George, M.G. Use of intravenous tissue plasminogen activato and hospital costs for patients with acute ischaemic stroke aged 18–64 years in the USA. *Stroke Vasc. Neurol.* **2016**, *1*, 8–15. [CrossRef]

38. Zeng, X.Y.; Li, Y.C.; Liu, J.M.; Liu, Y.N.; Liu, S.W.; Qi, J.L.; Zhou, M.G. [Estimation of the impact of risk factors control on non-communicable diseases mortality, life expectancy and the labor force lost in China in 2030]. *Zhonghua Yu Fang Yi Xue Za Zhi* **2017**, *51*, 1079–1085.

39. Powers, W.J.; Rabinstein, A.A.; Ackerson, T.; Adeoye, O.M.; Bambakidis, N.C.; Becker, K.; Biller, J.; Brown, M.; Demaerschalk, B.M.; Hoh, B.; et al. Guidelines for the Early Management of Patients with Acute Ischemic Stroke: 2019 Update to the 2018 Guidelines for the Early Management of Acute Ischemic Stroke: A Guideline for Healthcare Professionals from the American Heart Association/American Stroke Association. *Stroke* **2019**, *50*, e344–e418; Correction in *Stroke* **2019**, *50*, e440–e441. [CrossRef] [PubMed]

40. Zi, W.; Qiu, Z.; Li, F.; Sang, H.; Wu, D.; Luo, W.; Liu, S.; Yuan, J.; Song, J.; Shi, Z.; et al. Effect of Endovascular Treatment Alone vs Intravenous Alteplase Plus Endovascular Treatment on Functional Independence in Patients with Acute Ischemic Stroke: The DEVT Randomized Clinical Trial. *JAMA* **2021**, *325*, 234–243. [CrossRef] [PubMed]

41. Turc, G.; Bhogal, P.; Fischer, U.; Khatri, P.; Lobotesis, K.; Mazighi, M.; Schellinger, P.D.; Toni, D.; de Vries, J.; White, P.; et al. European Stroke Organisation (ESO)–European Society for Minimally Invasive Neurological Therapy (ESMINT) Guidelines on Mechanical Thrombectomy in Acute Ischaemic StrokeEndorsed by Stroke Alliance for Europe (SAFE). *Eur. Stroke J.* **2019**, *4*, 6–12. [CrossRef] [PubMed]

42. Berkhemer, O.A.; Fransen, P.S.S.; Beumer, D.; Berg, L.A.V.D.; Lingsma, H.F.; Yoo, A.J.; Schonewille, W.J.; Vos, J.A.; Nederkoorn, P.J.; Wermer, M.J.H.; et al. A Randomized Trial of Intraarterial Treatment for Acute Ischemic Stroke. *N. Engl. J. Med.* **2015**, *372*, 11–20. [CrossRef]

43. Romoli, M.; Paciaroni, M.; Tsivgoulis, G.; Agostoni, E.C.; Vidale, S. Mothership versus Drip-and-Ship Model for Mechanical Thrombectomy in Acute Stroke: A Systematic Review and Meta-Analysis for Clinical and Radiological Outcomes. *J. Stroke* **2020**, *22*, 317–323. [CrossRef]

44. Udar Mózgu–Problem Interdyscyplinarny. Available online: https://vimeo.com/652844074/ae3fbf2c0c (accessed on 4 December 2021).

45. Remedios, L.W.; Lingam, S.; Remedios, S.W.; Gao, R.; Clark, S.W.; Davis, L.T.; Landman, B.A. Comparison of convolutional neural networks for detecting large vessel occlusion on computed tomography angiography. *Med. Phys.* **2021**, *48*, 6060–6068. [CrossRef]

46. Jin, H.; Qu, Y.; Guo, Z.-N.; Yan, X.-L.; Sun, X.; Yang, Y. Impact of Jilin Province Stroke Emergency Maps on Acute Stroke Care Improvement in Northeast China. *Front. Neurol.* **2020**, *11*, 734. [CrossRef]

47. Sevick, L.K.; Demchuk, A.M.; Shuaib, A.; Smith, E.E.; Rempel, J.L.; Butcher, K.; Menon, B.K.; Jeerakathil, T.; Kamal, N.; Thornton, J.; et al. A Prospective Economic Evaluation of Rapid Endovascular Therapy for Acute Ischemic Stroke. *Can. J. Neurol. Sci./J. Can. Sci. Neurol.* **2021**, *48*, 791–798. [CrossRef]

48. Venema, E.; Burke, J.F.; Roozenbeek, B.; Nelson, J.; Lingsma, H.F.; Dippel, D.W.; Kent, D.M. Prehospital Triage Strategies for the Transportation of Suspected Stroke Patients in the United States. *Stroke* **2020**, *51*, 3310–3319. [CrossRef]

49. Goyal, M.; Demchuk, A.M.; Menon, B.K.; Eesa, M.; Rempel, J.L.; Thornton, J.; Roy, D.; Jovin, T.G.; Willinsky, R.A.; Sapkota, B.L.; et al. Randomized Assessment of Rapid Endovascular Treatment of Ischemic Stroke. *N. Engl. J. Med.* **2015**, *372*, 1019–1030. [CrossRef]

50. Katyal, A.; Bhaskar, S. CTP-guided reperfusion therapy in acute ischemic stroke: A meta-analysis. *Acta Neurol. Scand.* **2021**, *143*, 355–366. [CrossRef] [PubMed]

51. Luan, D.; Zhang, Y.; Yang, Q.; Zhou, Z.; Huang, X.; Zhao, S.; Yuan, L. Efficacy and Safety of Intravenous Thrombolysis in Patients with Unknown Onset Stroke: A Meta-Analysis. *Behav. Neurol.* **2019**, *2019*, 5406923. [CrossRef] [PubMed]

52. Beyeler, M.; Weber, L.; Kurmann, C.C.; Piechowiak, E.I.I.; Mosimann, P.J.; Zibold, F.; Meinel, T.R.; Branca, M.; Goeldlin, M.; Pilgram-Pastor, S.M.; et al. Association of reperfusion success and emboli in new territories with long term mortality after mechanical thrombectomy. *J. NeuroInterv. Surg.* **2022**, *14*, 326–332. [CrossRef] [PubMed]

53. Nowak, K.; Derbisz, J.; Pęksa, J.; Łasocha, B.; Brzegowy, P.; Slowik, J.; Wrona, P.; Pulyk, R.; Popiela, T.; Slowik, A. Post-stroke infection in acute ischemic stroke patients treated with mechanical thrombectomy does not affect long-term outcome. *Adv. Interv. Cardiol.* **2020**, *16*, 452–459. [CrossRef] [PubMed]

54. Luchowski, P.; Szmygin, M.; Wojczal, J.; Prus, K.; Sojka, M.; Luchowska, E.; Rejdak, K. Stroke patients from rural areas have lower chances for long-term good clinical outcome after mechanical thrombectomy. *Clin. Neurol. Neurosurg.* **2021**, *206*, 106687. [CrossRef]

55. Bernardo-Castro, S.; Sousa, J.A.; Brás, A.; Cecília, C.; Rodrigues, B.; Almendra, L.; Machado, C.; Santo, G.; Silva, F.; Ferreira, L.; et al. Pathophysiology of Blood–Brain Barrier Permeability Throughout the Different Stages of Ischemic Stroke and Its Implication on Hemorrhagic Transformation and Recovery. *Front. Neurol.* **2020**, *11*, 1605. [CrossRef]

56. Sun, M.-S.; Jin, H.; Sun, X.; Huang, S.; Zhang, F.-L.; Guo, Z.-N.; Yang, Y. Free Radical Damage in Ischemia-Reperfusion Injury: An Obstacle in Acute Ischemic Stroke after Revascularization Therapy. *Oxidative Med. Cell. Longev.* **2018**, *2018*, 3804979. [CrossRef]

57. Alluri, H.; Shaji, C.A.; Davis, M.L.; Tharakan, B. Oxygen-Glucose Deprivation and Reoxygenation as an In Vitro Ischemia-Reperfusion Injury Model for Studying Blood-Brain Barrier Dysfunction. *J. Vis. Exp.* **2015**, *99*, e52699. [CrossRef]

58. Kawase, H.; Takeda, Y.; Mizoue, R.; Sato, S.; Fushimi, M.; Murai, S.; Morimatsu, H. Extracellular Glutamate Concentration Increases Linearly in Proportion to Decreases in Residual Cerebral Blood Flow After the Loss of Membrane Potential in a Rat Model of Ischemia. *J. Neurosurg. Anesthesiol.* **2021**, *33*, 356–362. [CrossRef]

59. Zaitoun, I.S.; Shahi, P.K.; Suscha, A.; Chan, K.; McLellan, G.J.; Pattnaik, B.R.; Sorenson, C.M.; Sheibani, N. Hypoxic–ischemic injury causes functional and structural neurovascular degeneration in the juvenile mouse retina. *Sci. Rep.* **2021**, *11*, 12670. [CrossRef]

60. Chamorro, Á.; Dirnagl, U.; Urra, X.; Planas, A.M. Neuroprotection in acute stroke: Targeting excitotoxicity, oxidative and nitrosative stress, and inflammation. *Lancet Neurol.* **2016**, *15*, 869–881. [CrossRef]

61. Shimada, N.; Graf, R.; Rosner, G.; Wakayama, A.; George, C.P.; Heiss, W.-D. Ischemic Flow Threshold for Extracellular Glutamate Increase in Cat Cortex. *J. Cereb. Blood Flow Metab.* **1989**, *9*, 603–606. [CrossRef] [PubMed]

62. Kostandy, B.B. The role of glutamate in neuronal ischemic injury: The role of spark in fire. *Neurol. Sci.* **2012**, *33*, 223–237. [CrossRef] [PubMed]

63. Lin, C.-H.; Chen, H.-Y.; Wei, K.-C. Role of HMGB1/TLR4 Axis in Ischemia/Reperfusion-Impaired Extracellular Glutamate Clearance in Primary Astrocytes. *Cells* **2020**, *9*, 2585. [CrossRef] [PubMed]

64. Wang, D.; Kong, J.; Wu, J.; Wang, X.; Lai, M. GC–MS-based metabolomics identifies an amino acid signature of acute ischemic stroke. *Neurosci. Lett.* **2017**, *642*, 7–13. [CrossRef]
65. Cvoro, V.; Wardlaw, J.M.; Marshall, I.; Armitage, P.A.; Rivers, C.S.; Bastin, M.E.; Carpenter, T.K.; Wartolowska, K.; Farrall, A.J.; Dennis, M.S. Associations Between Diffusion and Perfusion Parameters, N -Acetyl Aspartate, and Lactate in Acute Ischemic Stroke. *Stroke* **2009**, *40*, 767–772. [CrossRef]
66. Guadagno, J.V.; Jones, P.S.; Aigbirhio, F.I.; Wang, D.; Fryer, T.D.; Day, D.J.; Antoun, N.; Nimmo-Smith, I.; Warburton, E.A.; Baron, J.C. Selective neuronal loss in rescued penumbra relates to initial hypoperfusion. *Brain* **2008**, *131*, 2666–2678. [CrossRef]
67. Tijssen, M.P.M.; Hofman, P.A.M.; Stadler, A.A.R.; Van Zwam, W.; De Graaf, R.; Van Oostenbrugge, R.J.; Klotz, E.; Wildberger, J.E.; Postma, A.A. The role of dual energy CT in differentiating between brain haemorrhage and contrast medium after mechanical revascularisation in acute ischaemic stroke. *Eur. Radiol.* **2014**, *24*, 834–840. [CrossRef]
68. Barbay, M.; Diouf, M.; Roussel, M.; Godefroy, O.; GRECOGVASC Study Group. Systematic Review and Meta-Analysis of Prevalence in Post-Stroke Neurocognitive Disorders in Hospital-Based Studies. *Dement. Geriatr. Cogn. Disord.* **2018**, *46*, 322–334. [CrossRef]
69. Pendlebury, S.T.; Rothwell, P.M. Prevalence, incidence, and factors associated with pre-stroke and post-stroke dementia: A systematic review and meta-analysis. *Lancet Neurol.* **2009**, *8*, 1006–1018. [CrossRef]
70. Mitchell, A.J.; Sheth, B.; Gill, J.; Yadegarfar, M.; Stubbs, B.; Yadegarfar, M.; Meader, N. Prevalence and predictors of post-stroke mood disorders: A meta-analysis and meta-regression of depression, anxiety and adjustment disorder. *Gen. Hosp. Psychiatry* **2017**, *47*, 48–60. [CrossRef] [PubMed]
71. Hosoo, H.; Marushima, A.; Nagasaki, Y.; Hirayama, A.; Ito, H.; Puentes, S.; Mujagic, A.; Tsurushima, H.; Tsuruta, W.; Suzuki, K.; et al. Neurovascular Unit Protection from Cerebral Ischemia–Reperfusion Injury by Radical-Containing Nanoparticles in Mice. *Stroke* **2017**, *48*, 2238–2247. [CrossRef] [PubMed]
72. Nawabi, J.; Flottmann, F.; Hanning, U.; Bechstein, M.; Schön, G.; Kemmling, A.; Fiehler, J.; Broocks, G. Futile Recanalization with Poor Clinical Outcome Is Associated with Increased Edema Volume After Ischemic Stroke. *Investig. Radiol.* **2019**, *54*, 282–287. [CrossRef] [PubMed]
73. Gauberti, M.; Lapergue, B.; de Lizarrondo, S.M.; Vivien, D.; Richard, S.; Bracard, S.; Piotin, M.; Gory, B. Ischemia-Reperfusion Injury After Endovascular Thrombectomy for Ischemic Stroke. *Stroke* **2018**, *49*, 3071–3074. [CrossRef] [PubMed]
74. Emmrich, J.V.; Ejaz, S.; Williamson, D.J.; Hong, Y.T.; Sitnikov, S.; Fryer, T.D.; Aigbirhio, F.I.; Wulff, H.; Baron, J.-C. Assessing the Effects of Cytoprotectants on Selective Neuronal Loss, Sensorimotor Deficit and Microglial Activation after Temporary Middle Cerebral Occlusion. *Brain Sci.* **2019**, *9*, 287. [CrossRef]
75. Staessens, S.; François, O.; Desender, L.; Vanacker, P.; Dewaele, T.; Sciot, R.; Vanhoorelbeke, K.; Andersson, T.; De Meyer, S.F. Detailed histological analysis of a thrombectomy-resistant ischemic stroke thrombus: A case report. *Thromb. J.* **2021**, *19*, 11. [CrossRef] [PubMed]
76. Bai, X.; Zhang, X.; Yang, W.; Zhang, Y.; Wang, T.; Xu, R.; Wang, Y.; Li, L.; Feng, Y.; Yang, K.; et al. Influence of first-pass effect on recanalization outcomes in the era of mechanical thrombectomy: A systemic review and meta-analysis. *Neuroradiology* **2021**, *63*, 795–807. [CrossRef]
77. Weyland, C.S.; Neuberger, U.; Potreck, A.; Pfaff, J.A.R.; Nagel, S.; Schönenberger, S.; Bendszus, M.; Möhlenbruch, M.A. Reasons for Failed Mechanical Thrombectomy in Posterior Circulation Ischemic Stroke Patients. *Clin. Neuroradiol.* **2021**, *31*, 745–752. [CrossRef]
78. Ducroux, C.; Di Meglio, L.; Loyau, S.; Delbosc, S.; Boisseau, W.; Deschildre, C.; Ben Maacha, M.; Blanc, R.; Redjem, H.; Ciccio, G.; et al. Thrombus Neutrophil Extracellular Traps Content Impair tPA-Induced Thrombolysis in Acute Ischemic Stroke. *Stroke* **2018**, *49*, 754–757. [CrossRef]
79. Fereidoonnezhad, B.; Dwivedi, A.; Johnson, S.; McCarthy, R.; McGarry, P. Blood clot fracture properties are dependent on red blood cell and fibrin content. *Acta Biomater.* **2021**, *127*, 213–228. [CrossRef]
80. Abbasi, M.; Larco, J.A.; Mereuta, M.O.; Liu, Y.; Fitzgerald, S.; Dai, D.; Kadirvel, R.; Savastano, L.; Kallmes, D.F.; Brinjikji, W. Diverse thrombus composition in thrombectomy stroke patients with longer time to recanalization. *Thromb. Res.* **2022**, *209*, 99–104. [CrossRef] [PubMed]
81. Vallés, J.; Lago, A.; Santos, M.T.; Latorre, A.M.; Tembl, J.I.; Salom, J.B.; Nieves, C.; Moscardó, A. Neutrophil extracellular traps are increased in patients with acute ischemic stroke: Prognostic significance. *Thromb. Haemost.* **2017**, *117*, 1919–1929. [CrossRef] [PubMed]
82. Di Maria, F.; Kyheng, M.; Consoli, A.; Desilles, J.-P.; Gory, B.; Richard, S.; Rodesch, G.; Labreuche, J.; Girot, J.-B.; Dargazanli, C.; et al. Identifying the predictors of first-pass effect and its influence on clinical outcome in the setting of endovascular thrombectomy for acute ischemic stroke: Results from a multicentric prospective registry. *Int. J. Stroke* **2021**, *16*, 20–28. [CrossRef] [PubMed]
83. Mokin, M.; Morr, S.; Natarajan, S.K.; Lin, N.; Snyder, K.V.; Hopkins, L.N.; Siddiqui, A.H.; Levy, E.I. Thrombus density predicts successful recanalization with Solitaire stent retriever thrombectomy in acute ischemic stroke. *J. Neurointerv. Surg.* **2015**, *7*, 104–107. [CrossRef] [PubMed]
84. Cai, Y.; Yang, E.; Yao, X.; Zhang, X.; Wang, Q.; Wang, Y.; Liu, J.; Fan, W.; Yi, K.; Kang, C.; et al. FUNDC1-dependent mitophagy induced by tPA protects neurons against cerebral ischemia-reperfusion injury. *Redox Biol.* **2021**, *38*, 101792. [CrossRef] [PubMed]

85. Froehler, M.T.; Tateshima, S.; Duckwiler, G.; Jahan, R.; Gonzalez, N.; Vinuela, F.; Liebeskind, D.; Saver, J.; Villablanca, J.P.; for the UCLA Stroke Investigators. The hyperdense vessel sign on CT predicts successful recanalization with the Merci device in acute ischemic stroke. *J. NeuroInterv. Surg.* **2013**, *5*, 289–293. [CrossRef]

86. Patel, T.R.; Fricano, S.; Waqas, M.; Tso, M.; Dmytriw, A.A.; Mokin, M.; Kolega, J.; Tomaszewski, J.; Levy, E.I.; Davies, J.M.; et al. Increased Perviousness on CT for Acute Ischemic Stroke is Associated with Fibrin/Platelet-Rich Clots. *Am. J. Neuroradiol.* **2020**, *42*, 57–64. [CrossRef]

87. Sivan-Hoffmann, R.; Gory, B.; Rabilloud, M.; Gherasim, D.N.; Armoiry, X.; Riva, R.; Labeyrie, P.-E.; Gonike-Sadeh, U.; Eldesouky, I.; Turjman, F. Patient Outcomes with Stent-Retriever Thrombectomy for Anterior Circulation Stroke: A Meta-Analysis and Review of the Literature. *Isr. Med Assoc. J.* **2016**, *18*, 561–566.

88. Sporns, P.B.; Krähling, H.; Psychogios, M.N.; Jeibmann, A.; Minnerup, J.; Broocks, G.; Meyer, L.; Brehm, A.; Wildgruber, M.; Fiehler, J.; et al. Small thrombus size, thrombus composition, and poor collaterals predict pre-interventional thrombus migration. *J. Neurointerv. Surg.* **2021**, *13*, 409–414. [CrossRef]

89. Mishra, S.M.; Dykeman, J.; Sajobi, T.; Trivedi, A.; Almekhlafi, M.; Sohn, S.I.; Bal, S.; Qazi, E.; Calleja, A.; Eesa, M.; et al. Early Reperfusion Rates with IV tPA Are Determined by CTA Clot Characteristics. *Am. J. Neuroradiol.* **2014**, *35*, 2265–2272. [CrossRef]

90. Riedel, C.H.; Zimmermann, P.; Jensen-Kondering, U.; Stingele, R.; Deuschl, G.; Jansen, O. The importance of size: Successful recanalization by intravenous thrombolysis in acute anterior stroke depends on thrombus length. *Stroke* **2011**, *42*, 1775–1777. [CrossRef] [PubMed]

91. Qiu, W.; Kuang, H.; Ospel, J.M.; Hill, M.D.; Demchuk, A.M.; Goyal, M.; Menon, B.K. Automated Prediction of Ischemic Brain Tissue Fate from Multiphase Computed Tomographic Angiography in Patients with Acute Ischemic Stroke Using Machine Learning. *J. Stroke* **2021**, *23*, 234–243. [CrossRef] [PubMed]

92. Wang, Z.; Xie, J.; Tang, T.-Y.; Zeng, C.-H.; Zhang, Y.; Zhao, Z.; Zhao, D.-L.; Geng, L.-Y.; Deng, G.; Zhang, Z.-J.; et al. Collateral Status at Single-Phase and Multiphase CT Angiography versus CT Perfusion for Outcome Prediction in Anterior Circulation Acute Ischemic Stroke. *Radiology* **2020**, *296*, 393–400. [CrossRef] [PubMed]

93. Kuang, H.; Qiu, W.; Boers, A.M.; Brown, S.; Muir, K.; Majoie, C.B.; Dippel, D.W.; White, P.; Epstein, J.; Mitchell, P.J.; et al. Computed Tomography Perfusion–Based Machine Learning Model Better Predicts Follow-Up Infarction in Patients with Acute Ischemic Stroke. *Stroke* **2021**, *52*, 223–231. [CrossRef] [PubMed]

94. Brugnara, G.; Neuberger, U.; Mahmutoglu, M.A.; Foltyn, M.; Herweh, C.; Nagel, S.; Schönenberger, S.; Heiland, S.; Ulfert, C.; Ringleb, P.A.; et al. Multimodal Predictive Modeling of Endovascular Treatment Outcome for Acute Ischemic Stroke Using Machine-Learning. *Stroke* **2020**, *51*, 3541–3551. [CrossRef]

95. Ledru, F.; Blanchard, D.; Battaglia, S.; Jeunemaitre, X.; Courbon, D.; Guize, L.; Guermonprez, J.-L.; Ducimetière, P.; Diébold, B. Relation between severity of coronary artery disease, left ventricular function, and myocardial infarction, and influence of the ACE I/D gene polymorphism. *Am. J. Cardiol.* **1998**, *82*, 160–165. [CrossRef]

96. Lai, Y.; Lin, P.; Chen, M.; Zhang, Y.; Chen, J.; Zheng, M.; Liu, J.; Du, H.; Chen, R.; Pan, X.; et al. Restoration of L-OPA1 alleviates acute ischemic stroke injury in rats via inhibiting neuronal apoptosis and preserving mitochondrial function. *Redox Biol.* **2020**, *34*, 101503. [CrossRef] [PubMed]

97. Zaidan, E.; Sims, N. Alterations in the glutathione content of mitochondria following short-term forebrain ischemia in rats. *Neurosci. Lett.* **1996**, *218*, 75–78. [CrossRef]

98. Uzdensky, A.B. Apoptosis regulation in the penumbra after ischemic stroke: Expression of pro- and antiapoptotic proteins. *Apoptosis* **2019**, *24*, 687–702. [CrossRef]

99. Li, Q.; Cao, Y.; Dang, C.; Han, B.; Han, R.; Ma, H.; Hao, J.; Wang, L. Inhibition of double-strand DNA-sensing cGAS ameliorates brain injury after ischemic stroke. *EMBO Mol. Med.* **2020**, *12*, e11002. [CrossRef] [PubMed]

100. Shih, E.K.; Robinson, M.B. Role of Astrocytic Mitochondria in Limiting Ischemic Brain Injury? *Physiology* **2018**, *33*, 99–112. [CrossRef] [PubMed]

101. Datta, A.; Akatsu, H.; Heese, K.; Sze, S.K. Quantitative clinical proteomic study of autopsied human infarcted brain specimens to elucidate the deregulated pathways in ischemic stroke pathology. *J. Proteom.* **2013**, *91*, 556–568. [CrossRef] [PubMed]

102. Lim, W.; Kim, J.-H.; Gook, E.; Kim, J.; Ko, Y.; Kim, I.; Kwon, H.; Lim, H.; Jung, B.; Yang, K.; et al. Inhibition of mitochondria-dependent apoptosis by 635-nm irradiation in sodium nitroprusside-treated SH-SY5Y cells. *Free Radic. Biol. Med.* **2009**, *47*, 850–857. [CrossRef] [PubMed]

103. Bladowski, M.; Gawrys, J.; Gajecki, D.; Szahidewicz-Krupska, E.; Sawicz-Bladowska, A.; Doroszko, A. Role of the Platelets and Nitric Oxide Biotransformation in Ischemic Stroke: A Translative Review from Bench to Bedside. *Oxidative Med. Cell. Longev.* **2020**, *2020*, 2979260. [CrossRef]

104. Carinci, M.; Vezzani, B.; Patergnani, S.; Ludewig, P.; Lessmann, K.; Magnus, T.; Casetta, I.; Pugliatti, M.; Pinton, P.; Giorgi, C. Different Roles of Mitochondria in Cell Death and Inflammation: Focusing on Mitochondrial Quality Control in Ischemic Stroke and Reperfusion. *Biomedicines* **2021**, *9*, 169. [CrossRef]

105. Weiland, A.; Wang, Y.; Wu, W.; Lan, X.; Han, X.; Li, Q.; Wang, J. Ferroptosis and Its Role in Diverse Brain Diseases. *Mol. Neurobiol.* **2019**, *56*, 4880–4893. [CrossRef]

106. Jackson, J.; O'Donnell, J.; Takano, H.; Coulter, D.; Robinson, M.B. Neuronal Activity and Glutamate Uptake Decrease Mitochondrial Mobility in Astrocytes and Position Mitochondria Near Glutamate Transporters. *J. Neurosci.* **2014**, *34*, 1613–1624. [CrossRef]

107. Anzell, A.R.; Maizy, R.; Przyklenk, K.; Sanderson, T.H. Mitochondrial Quality Control and Disease: Insights into Ischemia-Reperfusion Injury. *Mol. Neurobiol.* **2018**, *55*, 2547–2564. [CrossRef]

108. Yang, J.-L.; Mukda, S.; Chen, S.-D. Diverse roles of mitochondria in ischemic stroke. *Redox Biol.* **2018**, *16*, 263–275. [CrossRef]

109. Schousboe, A. Astrocytic Metabolism Focusing on Glutamate Homeostasis: A Short Review Dedicated to Vittorio Gallo. *Neurochem. Res.* **2020**, *45*, 522–525. [CrossRef]

110. Lee, S.Y.; Chung, W.-S. The roles of astrocytic phagocytosis in maintaining homeostasis of brains. *J. Pharmacol. Sci.* **2021**, *145*, 223–227. [CrossRef] [PubMed]

111. Roosterman, D.; Cottrell, G.S. Astrocytes and neurons communicate via a monocarboxylic acid shuttle. *AIMS Neurosci.* **2020**, *7*, 94–106. [CrossRef] [PubMed]

112. Povysheva, N.; Nigam, A.; Brisbin, A.K.; Johnson, J.W.; Barrionuevo, G. Oxygen–Glucose Deprivation Differentially Affects Neocortical Pyramidal Neurons and Parvalbumin-Positive Interneurons. *Neuroscience* **2019**, *412*, 72–82. [CrossRef] [PubMed]

113. Ludhiadch, A.; Sharma, R.; Muriki, A.; Munshi, A. Role of Calcium Homeostasis in Ischemic Stroke: A Review. *CNS Neurol. Disord. Drug Targets* **2022**, *21*, 52–61. [CrossRef]

114. Morken, T.S.; Brekke, E.; Håberg, A.; Widerøe, M.; Brubakk, A.-M.; Sonnewald, U. Altered Astrocyte–Neuronal Interactions After Hypoxia-Ischemia in the Neonatal Brain in Female and Male Rats. *Stroke* **2014**, *45*, 2777–2785. [CrossRef] [PubMed]

115. Hernández, I.H.; Villa-González, M.; Martín, G.; Soto, M.; Pérez-Álvarez, M.J. Glial Cells as Therapeutic Approaches in Brain Ischemia-Reperfusion Injury. *Cells* **2021**, *10*, 1639. [CrossRef] [PubMed]

116. Qin, C.; Zhou, L.-Q.; Ma, X.-T.; Hu, Z.-W.; Yang, S.; Chen, M.; Bosco, D.B.; Wu, L.-J.; Tian, D.-S. Dual Functions of Microglia in Ischemic Stroke. *Neurosci. Bull.* **2019**, *35*, 921–933. [CrossRef]

117. Cai, W.; Liu, S.; Hu, M.; Huang, F.; Zhu, Q.; Qiu, W.; Hu, X.; Colello, J.; Zheng, S.G.; Lu, Z. Functional Dynamics of Neutrophils After Ischemic Stroke. *Transl. Stroke Res.* **2020**, *11*, 108–121. [CrossRef]

118. Zhong, C.; Yang, J.; Xu, T.; Peng, Y.; Wang, A.; Wang, J.; Peng, H.; Li, Q.; Ju, Z.; Geng, D.; et al. Serum matrix metalloproteinase-9 levels and prognosis of acute ischemic stroke. *Neurology* **2017**, *89*, 805–812. [CrossRef]

119. Foster, A.J.; Smyth, M.; Lakhani, A.; Jung, B.; Brant, R.F.; Jacobson, K. Consecutive fecal calprotectin measurements for predicting relapse in pediatric Crohn's disease patients. *World J. Gastroenterol.* **2019**, *25*, 1266–1277. [CrossRef]

120. Pluta, R.; Januszewski, S.; Czuczwar, S. Neuroinflammation in Post-Ischemic Neurodegeneration of the Brain: Friend, Foe, or Both? *Int. J. Mol. Sci.* **2021**, *22*, 4405. [CrossRef]

121. Marta-Enguita, J.; Navarro-Oviedo, M.; Rubio-Baines, I.; Aymerich, N.; Herrera, M.; Zandio, B.; Mayor, S.; Rodriguez, J.-A.; Páramo, J.-A.; Toledo, E.; et al. Association of calprotectin with other inflammatory parameters in the prediction of mortality for ischemic stroke. *J. Neuroinflamm.* **2021**, *18*, 3. [CrossRef]

122. Chen, C.; Huang, T.; Zhai, X.; Ma, Y.; Xie, L.; Lu, B.; Zhang, Y.; Li, Y.; Chen, Z.; Yin, J.; et al. Targeting neutrophils as a novel therapeutic strategy after stroke. *J. Cereb. Blood Flow Metab.* **2021**, *41*, 2150–2161. [CrossRef]

123. García-Culebras, A.; Durán-Laforet, V.; Peña-Martínez, C.; Moraga, A.; Ballesteros, I.; Cuartero, M.I.; de la Parra, J.; Palma-Tortosa, S.; Hidalgo, A.; Corbí, A.L.; et al. Role of TLR4 (Toll-Like Receptor 4) in N1/N2 Neutrophil Programming After Stroke. *Stroke* **2019**, *50*, 2922–2932. [CrossRef]

124. Silvestre-Roig, C.; Braster, Q.; Ortega-Gomez, A.; Soehnlein, O. Neutrophils as regulators of cardiovascular inflammation. *Nat. Rev. Cardiol.* **2020**, *17*, 327–340. [CrossRef]

125. Lehmann, A.L.C.F.; Alfieri, D.F.; de Araújo, M.C.M.; Trevisani, E.R.; Nagao, M.R.; Pesente, F.S.; Gelinski, J.R.; de Freitas, L.B.; Flauzino, T.; Lehmann, M.F.; et al. Immune-inflammatory, coagulation, adhesion, and imaging biomarkers combined in machine learning models improve the prediction of death 1 year after ischemic stroke. *Clin. Exp. Med.* **2021**, *22*, 111–123. [CrossRef]

126. Yuan, S.; Tang, B.; Zheng, J.; Larsson, S.C. Circulating Lipoprotein Lipids, Apolipoproteins and Ischemic Stroke. *Ann. Neurol.* **2020**, *88*, 1229–1236. [CrossRef]

127. Plubell, D.; Fenton, A.M.; Rosario, S.; Bergstrom, P.; Wilmarth, P.A.; Clark, W.M.; Zakai, N.A.; Quinn, J.F.; Minnier, J.; Alkayed, N.J.; et al. High-Density Lipoprotein Carries Markers That Track with Recovery from Stroke. *Circ. Res.* **2020**, *127*, 1274–1287. [CrossRef]

128. Spronk, E.; Sykes, G.; Falcione, S.; Munsterman, D.; Joy, T.; Kamtchum-Tatuene, J.; Jickling, G.C. Hemorrhagic Transformation in Ischemic Stroke and the Role of Inflammation. *Front. Neurol.* **2021**, *12*, 597. [CrossRef]

129. Levard, D.; Buendia, I.; Lanquetin, A.; Glavan, M.; Vivien, D.; Rubio, M. Filling the gaps on stroke research: Focus on inflammation and immunity. *Brain Behav. Immun.* **2021**, *91*, 649–667. [CrossRef]

130. Vezzani, B.; Carinci, M.; Patergnani, S.; Pasquin, M.P.; Guarino, A.; Aziz, N.; Pinton, P.; Simonato, M.; Giorgi, C. The Dichotomous Role of Inflammation in the CNS: A Mitochondrial Point of View. *Biomolecules* **2020**, *10*, 1437. [CrossRef] [PubMed]

131. Han, B.; Jiang, W.; Cui, P.; Zheng, K.; Dang, C.; Wang, J.; Li, H.; Chen, L.; Zhang, R.; Wang, Q.M.; et al. Microglial PGC-1α protects against ischemic brain injury by suppressing neuroinflammation. *Genome Med.* **2021**, *13*, 47. [CrossRef] [PubMed]

132. Chamorro, A.; Urra, X.; Planas, A.M. Infection after acute ischemic stroke: A manifestation of brain-induced immunodepression. *Stroke* **2007**, *38*, 1097–1103. [CrossRef] [PubMed]

133. Urra, X.; Mirã, F.; Chamorro, A.; Planas, A.M.; Miró, F. Antigen-specific immune reactions to ischemic stroke. *Front. Cell. Neurosci.* **2014**, *8*, 278. [CrossRef]

134. Dave, K.R.; Pileggi, A.; Raval, A.P. Recurrent hypoglycemia increases oxygen glucose deprivation-induced damage in hippocampal organotypic slices. *Neurosci. Lett.* **2011**, *496*, 25–29. [CrossRef]

135. Nitzsche, A.; Poittevin, M.; Benarab, A.; Bonnin, P.; Faraco, G.; Uchida, H.; Favre, J.; Garcia-Bonilla, L.; Garcia, M.C.; Leger, P.-L.P.; et al. Endothelial S1P 1 Signaling Counteracts Infarct Expansion in Ischemic Stroke. *Circ. Res.* **2021**, *128*, 363–382. [CrossRef]

136. Andjelkovic, A.V.; Xiang, J.; Stamatovic, S.M.; Hua, Y.; Xi, G.; Wang, M.M.; Keep, R.F. Endothelial Targets in Stroke: Translating Animal Models to Human. *Arterioscler. Thromb. Vasc. Biol.* **2019**, *39*, 2240–2247. [CrossRef]

137. Di Lazzaro, V.; Profice, P.; Dileone, M.; Della Marca, G.; Colosimo, C.; Pravatà, E.; Pavone, A.; Pennisi, M.; Maviglia, R.; Pilato, F. Delayed hypothermia in malignant ischaemic stroke. *Neurol. Sci.* **2012**, *33*, 661–664. [CrossRef]

138. Li, X.; Li, Y.; Zhang, Z.; Bian, Q.; Gao, Z.; Zhang, S. Mild hypothermia facilitates mitochondrial transfer from astrocytes to injured neurons during oxygen-glucose deprivation/reoxygenation. *Neurosci. Lett.* **2021**, *756*, 135940. [CrossRef]

139. Russo, E.; Napoli, E.; Borlongan, C. V Healthy mitochondria for stroke cells. *Brain Circ.* **2018**, *4*, 95–98.

140. Wang, X.; Li, H.; Ding, S. The Effects of NAD+ on Apoptotic Neuronal Death and Mitochondrial Biogenesis and Function after Glutamate Excitotoxicity. *Int. J. Mol. Sci.* **2014**, *15*, 20449–20468. [CrossRef] [PubMed]

141. Wang, S.-D.; Fu, Y.-Y.; Han, X.-Y.; Yong, Z.-J.; Li, Q.; Hu, Z.; Liu, Z.-G. Hyperbaric Oxygen Preconditioning Protects Against Cerebral Ischemia/Reperfusion Injury by Inhibiting Mitochondrial Apoptosis and Energy Metabolism Disturbance. *Neurochem. Res.* **2021**, *46*, 866–877. [CrossRef] [PubMed]

142. Smith, G.; Lawson, D.; Renfrew, S.; Ledingham, I.; Sharp, G. Preservation of cerebral cortical activity by breathing oxygen at two atmospheres of pressure during cerebral ischemia. *Surg. Gynecol. Obstet.* **1961**, *113*, 13.

143. Hentia, C.; Rizzato, A.; Camporesi, E.; Yang, Z.; Muntean, D.M.; Săndesc, D.; Bosco, G. An overview of protective strategies against ischemia/reperfusion injury: The role of hyperbaric oxygen preconditioning. *Brain Behav.* **2018**, *8*, e00959. [CrossRef]

144. Engin, A.; Engin, A.B. N-Methyl-D-Aspartate Receptor Signaling-Protein Kinases Crosstalk in Cerebral Ischemia. *Adv. Exp. Med. Biol.* **2021**, *1275*, 259–283.

145. Huo, Y.; Feng, X.; Niu, M.; Wang, L.; Xie, Y.; Wang, L.; Ha, J.; Cheng, X.; Gao, Z.; Sun, Y. Therapeutic time windows of compounds against NMDA receptors signaling pathways for ischemic stroke. *J. Neurosci. Res.* **2021**, *99*, 3204–3221. [CrossRef]

146. Teng, J.-F.; Ma, Y.-H.; Deng, W.-J.; Luo, Z.-Y.; Jing, J.; Pan, P.-W.; Yao, Y.-B.; Fang, Y.-B. Inhibition of microRNA-29b suppresses oxidative stress and reduces apoptosis in ischemic stroke. *Neural Regen. Res.* **2022**, *17*, 433–439. [CrossRef]

147. Harraz, M.M.; Eacker, S.M.; Wang, X.; Dawson, T.M.; Dawson, V.L. MicroRNA-223 is neuroprotective by targeting glutamate receptors. *Proc. Natl. Acad. Sci. USA* **2012**, *109*, 18962–18967. [CrossRef]

148. Verma, P.; Augustine, G.J.; Ammar, M.-R.; Tashiro, A.; Cohen, S.M. A neuroprotective role for microRNA miR-1000 mediated by limiting glutamate excitotoxicity. *Nat. Neurosci.* **2015**, *18*, 379–385. [CrossRef]

149. Majdi, A.; Mahmoudi, J.; Sadigh-Eteghad, S.; Farhoudi, M.; Shotorbani, S.S. The interplay of microRNAs and post-ischemic glutamate excitotoxicity: An emergent research field in stroke medicine. *Neurol. Sci.* **2016**, *37*, 1765–1771. [CrossRef]

150. Zeyen, T.; Noristani, R.; Habib, S.; Heinisch, O.; Slowik, A.; Huber, M.; Schulz, J.B.; Reich, A.; Habib, P. Microglial-specific depletion of TAK1 is neuroprotective in the acute phase after ischemic stroke. *J. Mol. Med.* **2020**, *98*, 833–847. [CrossRef]

151. Wu, X.; Lin, L.; Qin, J.; Wang, L.; Wang, H.; Zou, Y.; Zhu, X.; Hong, Y.; Zhang, Y.; Liu, Y.; et al. CARD3 Promotes Cerebral Ischemia-Reperfusion Injury Via Activation of TAK1. *J. Am. Heart Assoc.* **2020**, *9*, e014920. [CrossRef]

152. Chrostek, M.; Fellows, E.G.; Crane, A.T.; Grande, A.; Low, W.C. Efficacy of stem cell-based therapies for stroke. *Brain Res.* **2019**, *1722*, 146362. [CrossRef]

153. Yu, F.; Huang, T.; Ran, Y.; Li, D.; Ye, L.; Tian, G.; Xi, J.; Liu, Z. New Insights into the Roles of Microglial Regulation in Brain Plasticity-Dependent Stroke Recovery. *Front. Cell. Neurosci.* **2021**, *15*, 299. [CrossRef]

154. Feng, L.; Dou, C.; Xia, Y.; Li, B.; Zhao, M.; Yu, P.; Zheng, Y.; El-Toni, A.M.; Atta, N.F.; Galal, A.; et al. Neutrophil-like Cell-Membrane-Coated Nanozyme Therapy for Ischemic Brain Damage and Long-Term Neurological Functional Recovery. *ACS Nano* **2021**, *15*, 2263–2280. [CrossRef]

155. Hatakeyama, M.; Ninomiya, I.; Otsu, Y.; Omae, K.; Kimura, Y.; Onodera, O.; Fukushima, M.; Shimohata, T.; Kanazawa, M. Cell Therapies under Clinical Trials and Polarized Cell Therapies in Pre-Clinical Studies to Treat Ischemic Stroke and Neurological Diseases: A Literature Review. *Int. J. Mol. Sci.* **2020**, *21*, 6194. [CrossRef]

156. ClinicalTrials.gov Identifier (NCT Number): NCT01955707. Available online: https://clinicaltrials.gov/ct2/show/NCT01955707 (accessed on 7 October 2013).

157. Chavda, V.; Madhwani, K.; Chaurasia, B. Stroke and immunotherapy: Potential mechanisms and its implications as immune-therapeutics. *Eur. J. Neurosci.* **2021**, *54*, 4338–4357. [CrossRef]

158. Yu, C.Y.; Ng, G.; Liao, P. Therapeutic Antibodies in Stroke. *Transl. Stroke Res.* **2013**, *4*, 477–483. [CrossRef]

159. Hou, K.; Li, G.; Yu, J.; Xu, K.; Wu, W. Receptors, Channel Proteins, and Enzymes Involved in Microglia-mediated Neuroinflammation and Treatments by Targeting Microglia in Ischemic Stroke. *Neuroscience* **2021**, *460*, 167–180. [CrossRef]

160. Cao, Z.; Harvey, S.S.; Chiang, T.; Foltz, A.G.; Lee, A.G.; Cheng, M.Y.; Steinberg, G.K. Unique Subtype of Microglia in Degenerative Thalamus After Cortical Stroke. *Stroke* **2021**, *52*, 687–698. [CrossRef]

161. Xu, H.; Jia, B.; Huo, X.; Mo, D.; Ma, N.; Gao, F.; Yang, M.; Miao, Z. Predictors of Futile Recanalization After Endovascular Treatment in Patients with Acute Ischemic Stroke in a Multicenter Registry Study. *J. Stroke Cerebrovasc. Dis.* **2020**, *29*, 105067. [CrossRef] [PubMed]

162. Lyden, P.D.; Jackson-Friedman, C.; Shin, C.; Hassid, S. Synergistic Combinatorial Stroke Therapy: A Quantal Bioassay of a GABA Agonist and a Glutamate Antagonist. *Exp. Neurol.* **2000**, *163*, 477–489. [CrossRef] [PubMed]

163. Bryll, A.; Krzyściak, W.; Karcz, P.; Pilecki, M.; Śmierciak, N.; Szwajca, M.; Skalniak, A.; Popiela, T. Determinants of Schizophrenia Endophenotypes Based on Neuroimaging and Biochemical Parameters. *Biomedicines* **2021**, *9*, 372. [CrossRef]
164. Cai, W.; Liu, H.; Zhao, J.; Chen, L.; Chen, J.; Lu, Z.; Hu, X. Pericytes in Brain Injury and Repair After Ischemic Stroke. *Transl. Stroke Res.* **2017**, *8*, 107–121. [CrossRef] [PubMed]
165. Cai, W.; Zhang, K.; Li, P.; Zhu, L.; Xu, J.; Yang, B.; Hu, X.; Lu, Z.; Chen, J. Dysfunction of the neurovascular unit in ischemic stroke and neurodegenerative diseases: An aging effect. *Ageing Res. Rev.* **2017**, *34*, 77–87. [CrossRef]
166. Tymianski, M. Can molecular and cellular neuroprotection be translated into therapies for patients?: Yes, but not the way we tried it before. *Stroke* **2010**, *41*, S87–S90. [CrossRef]

Journal of
Clinical Medicine

Review

Neuroimaging Techniques as Potential Tools for Assessment of Angiogenesis and Neuroplasticity Processes after Stroke and Their Clinical Implications for Rehabilitation and Stroke Recovery Prognosis

Lidia Włodarczyk [1,*], Natalia Cichon [2], Joanna Saluk-Bijak [3], Michal Bijak [2], Agata Majos [4] and Elzbieta Miller [1,*]

[1] Department of Neurological Rehabilitation, Medical University of Lodz, Poland Milionowa 14, 93-113 Lodz, Poland
[2] Biohazard Prevention Centre, Faculty of Biology and Environmental Protection, University of Lodz, Pomorska, 141/143, 90-236 Lodz, Poland; natalia.cichon@biol.uni.lodz.pl (N.C.); michal.bijak@biol.uni.lodz.pl (M.B.)
[3] Department of General Biochemistry, Faculty of Biology and Environmental Protection, University of Lodz, Pomorska, 141/143, 90-236 Lodz, Poland; joanna.saluk@biol.uni.lodz.pl
[4] Department of Radiological and Isotopic Diagnosis and Therapy, Medical University of Lodz, 92-213 Lodz, Poland; agata.majos@umed.lodz.pl
* Correspondence: lidia.wlodarczyk@umed.lodz.pl (L.W.); elzbieta.dorota.miller@umed.lodz.pl (E.M.); Tel.: +48-(0)4-2666-77461 (E.M.); Fax: +48-(0)4-2676-1785 (E.M.)

Citation: Włodarczyk, L.; Cichon, N.; Saluk-Bijak, J.; Bijak, M.; Majos, A.; Miller, E. Neuroimaging Techniques as Potential Tools for Assessment of Angiogenesis and Neuroplasticity Processes after Stroke and Their Clinical Implications for Rehabilitation and Stroke Recovery Prognosis. *J. Clin. Med.* **2022**, *11*, 2473. https://doi.org/10.3390/jcm11092473

Academic Editors: Gabriel Broocks and Lukas Meyer

Received: 20 March 2022
Accepted: 26 April 2022
Published: 28 April 2022

Abstract: Stroke as the most frequent cause of disability is a challenge for the healthcare system as well as an important socio-economic issue. Therefore, there are currently a lot of studies dedicated to stroke recovery. Stroke recovery processes include angiogenesis and neuroplasticity and advances in neuroimaging techniques may provide indirect description of this action and become quantifiable indicators of these processes as well as responses to the therapeutical interventions. This means that neuroimaging and neurophysiological methods can be used as biomarkers—to make a prognosis of the course of stroke recovery and define patients with great potential of improvement after treatment. This approach is most likely to lead to novel rehabilitation strategies based on categorizing individuals for personalized treatment. In this review article, we introduce neuroimaging techniques dedicated to stroke recovery analysis with reference to angiogenesis and neuroplasticity processes. The most beneficial for personalized rehabilitation are multimodal panels of stroke recovery biomarkers, including neuroimaging and neurophysiological, genetic-molecular and clinical scales.

Keywords: stroke; recovery; neurorehabilitation; neuroimaging; neuroplasticity; angiogenesis

1. Introduction

Stroke is an acute cerebral, spinal, or retinal vascular accident with neurological dysfunction that persists longer than 24 h, or one of any duration when infarction or hemorrhage corresponding to symptoms is demonstrated by imaging (computed tomography/magnetic resonance scans) or autopsy [1]. The clinical symptoms are very heterogenous and conditional on the topography of damage [2]. There is effective, specific treatment for acute ischemic stroke available, such as thrombolysis (intravenous administration of tissue plasminogen activator) or endovascular thrombectomy [3]. Furthermore, ischemic stroke patients should be administrated with antiplatelet or anticoagulant drugs (depending on the etiology of ischemia) as secondary stroke prevention [4]. The procedure for acute hemorrhage stroke is based on intensive blood pressure and intracranial pressure reduction. In addition, early open-surgery evacuation or drainage of hematoma might be beneficial and should be considered in specific cases [5]. Following the onset of stroke, natural, spontaneous processes of recovery occur, but are generally incomplete and difficult to predict amongst individuals. By fostering angiogenic and neuroplastic changes, restitution of function in damaged, nearby, or distant (but functionally or structurally connected) regions to the lesion may become more efficient [6]. Therefore, novel therapeutic approaches of stroke rehabilitation should be a focus of attention among researchers. Neurorehabilitation is intended to aid stroke patients to become as independent as possible and should be introduced soon after the patient's condition stabilizes. Novel rehabilitation strategies are specific to the individual's therapeutic goal, so directed at affected functional areas including motor impairments (most frequent), gait disturbance, speech disorders, cognitive failure, vision disturbances, etc. Therefore, it is appropriate to provide post-stroke rehabilitation by an interdisciplinary medical team together with physicians, psychiatrists or psychologists, neurologists, physical and occupational therapists, speech-language pathologists, nutritionists, and others [7]. Increasingly, novel technology applications are employed in post-stroke procedures, such as rehabilitation robots—as therapy devices (to train lost motor function) as well as assistive devices (to compensate lost skills) [8]. When considering the strategies of modern post-stroke rehabilitation, the importance of non-invasive neuromodulation techniques including repetitive transcranial magnetic stimulation and transcranial direct current stimulation should also be emphasized [9–12]. Furthermore, virtual reality programs on mobile devices may constitute a time-efficient, clinically effective, easy-to-implement and goal-oriented tool for upper extremity stroke rehabilitation [13]. A relevant issue in the context of this article is that the implementation of neuroimaging measurements in clinical studies will enable a multimodal approach to brain investigation and prediction of functional recovery after stroke. Indeed, neuroimaging and brain mapping provide a wide perspective on brain repair by showing modulations in cortical and subcortical activity or changes in functional brain connectivity [14]. The aim of this article is to review the literature and analyze neuroimaging measurements and their role in predicting recovery after stroke. Simple brain imaging measures have proved value in total infarct volume, correlating with predicting overall neurological status. Saver et al. indicate that subacute CT infarct volume correlates moderately with the 3-month clinical outcome, as determined by widely used neurological and functional assessment scales [15]. Nowadays, there are more complex imaging measures used as biomarkers of recovery. We use them to analyze both structure (anatomical assessment of key pathways injury, particularly corticospinal fiber number by diffusion tensor imaging) as well as function (for example, imaging of region pertinent to specific neurotransmitters or evaluation of cortical excitation) of brain tissue [16]. In this review, we have decided to briefly introduce the field of neuroimaging possibilities and analyze selected neuroimaging biomarkers in the context of recovery processes such as angiogenesis and neuroplasticity.

For outcome prognosis, recovery processes such as angiogenesis and neuroplasticity should be analyzed. Angiogenesis is a renewed growth of blood vessels to restore blood supply to the damaged brain tissue and it occurs after stroke [17,18]. This restorative action is undoubtedly beneficial after stroke; it was proved that there is a stroke-related period of heightened vascular plasticity that is correlated with restoration of blood flow and then predictive of motor function recovery [19]. However, it is also worth remembering that angiogenic factors, such as VEGF and MMPs, contribute to blood brain barrier injury and may lead to oedema formation or haemorrhagic transformation [20]. Neuroplasticity definition contains adaptive structural and functional processes, including synaptogenesis, neurogenesis and neuroprotection. The critical period for post-stroke recovery and neuroplasticity is considered the acute (0–7 days) and the early subacute (7 days-3 months) phase [8]. In the acute phase, secondary neuronal networks are exploited to preserve function, whereas in the subacute stage, new synaptic connections are formed, and in the chronic phase, remodeling by axonal sprouting and then reorganization occurs [21]. There is also enhanced expression of growth associated genes and proteins observed [22], changes in GABA (gamma-aminobutyric acid), NMDA (N-methyl-D-aspartate) receptor subtypes and upregulation of NMDA receptors to increase brain excitability [23]. Worth noticing for this review is also the theory of diaschisis, the assumed function decrease in spatially discrete brain areas, functionally related to the site of injury [24]. Diaschisis can be demonstrated by neuroimaging techniques that evaluate changes in cerebral blood flow, neurotransmitters, and metabolism action in regions distant from the lesion.

2. Neuroimaging Techniques Dedicated to Stroke

Neuroimaging is usually associated with differentiating ischemic from hemorrhage stroke in the acute phase; it also has a crucial role in ischemic stroke patients' selection for novel treatment options, including late window thrombectomy. In this review, we emphasized the role and potential of neuroimaging techniques for post-stroke recovery prediction. Correlating recovery prediction with measurements entitles researchers to use determination of biomarkers. The FDA-NIH biomarker working group presented the following definition of biomarker: "A defined characteristic that is measured as an indicator of normal biological processes, pathogenic processes or responses to an exposure or intervention" [25], considering the harmonization of terms used in science as a requisite priority. As stroke studies develop, the role of term biomarkers has changed from diagnosis to therapeutic mechanisms and currently includes a wide group of factors with various origin: genetic, molecular, clinical scales, neuroimaging and neurophysiological. Figure 1 shows the emerging roles of stroke biomarkers [26]. Figure 2 presents a classification of neuroimaging techniques with potential for use in stroke recovery prognosis. Neuroimaging stroke recovery biomarkers include measures of structure and function [16]. To evaluate structure, we commonly use characteristics including infarct volume, extent of cortical or white matter injury, white matter integrity, and percentage of corticospinal tract injury. Functional assessment should be focused on features including activation within ipsilesional and contralesional sites, interhemispheric balance, resting state functional connectivity, task-based synchronization, and desynchronization, as well as cortical excitability, facilitation, and inhibition [27]. In the next section, we briefly discuss techniques dedicated to stroke recovery analysis with reference to angiogenesis and neuroplasticity processes. Table 1 summarizes the techniques discussed in the text.

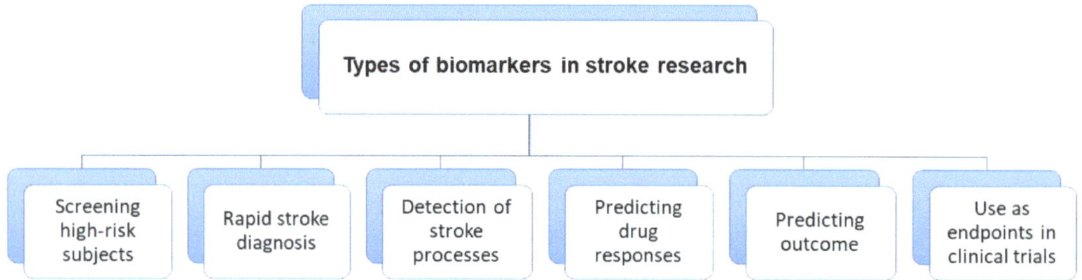

Figure 1. Roles of stroke biomarkers.

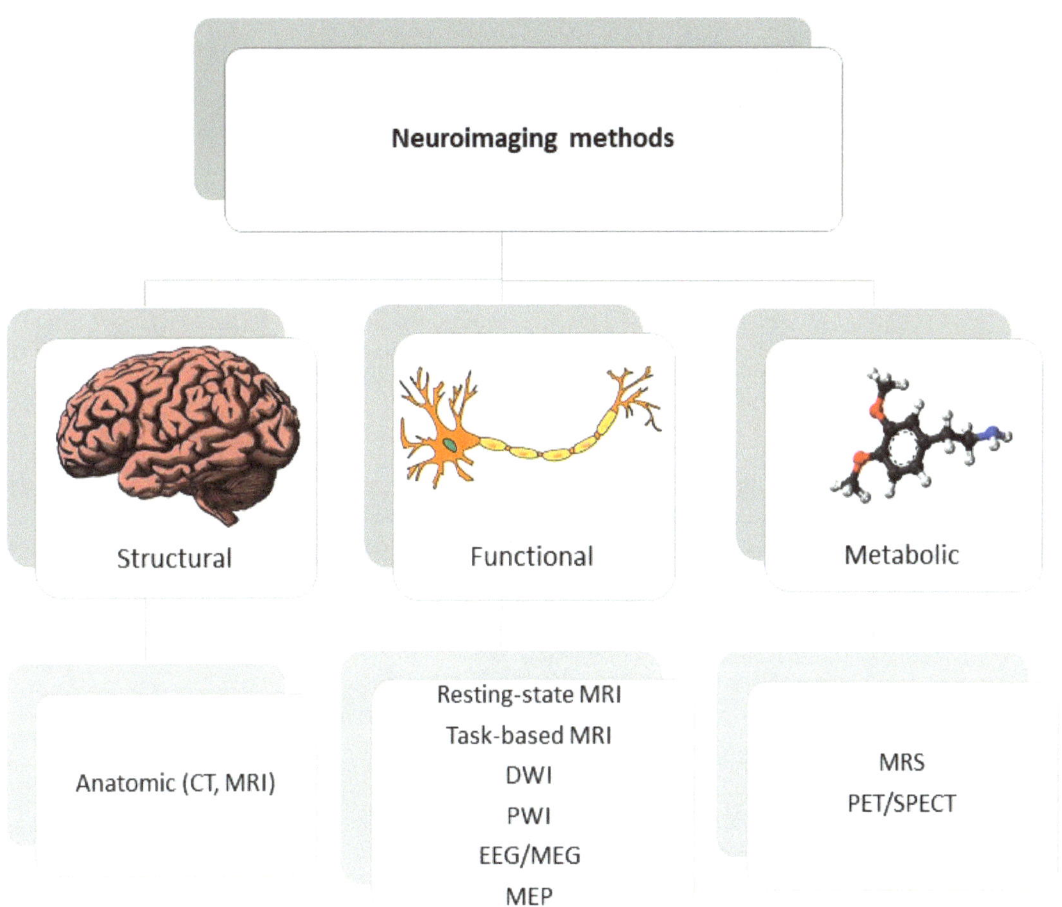

Figure 2. Classification of neuroimaging methods; CT—computer tomography, EEG—electroencephalography, MEP—motor evoked potential, MRI—magnetic resonance imaging, MRS—magnetic spectroscopy, PET/SPECT—positron emission tomography/ single photon emission computed tomography.

3. Stroke Recovery Prognosis Based on Selected Neuroimaging Measurements—Review of Literature

3.1. Potential Angiogenic Biomarkers of Stroke Recovery

Angiogenesis after stroke is a complex and multistep process, involving (after gene transcription and releasing proangiogenic factors) endothelial cell proliferation, vascular sprouting, and finally microvessel formation [28]. In the current state of knowledge, imaging techniques allow evaluation of a wide spectrum of structural and functional tissue features. Recent developments in magnetic resonance imaging (MRI) techniques presented the possibility of assessing tissue perfusion and estimating the quantification of various features in the vascular network, including microvascular cerebral blood volume (CBV) or microvascular density [29]. In an experimental study of Yanev et al., steady-state contrast-enhanced (ssCE-) MRI with a long-circulating blood pool was applied to characterize the model of vascular reorganization within the ischemic lesion and in secondarily affected areas, from subacute to chronic stages, after focal cerebral stroke [30]. Results showed that dynamic revascularization in the perilesional area and progressive neovascularization in non-ischemic connected areas were found. These phenomena may contribute to non-neuronal tissue remodeling and be relevant in post-stroke recovery. In turn, the initial stage of angiogenesis might be observed by MRI techniques as blood-brain barrier leakage, because BBB permeability is associated with endothelial cell proliferation and vascular sprouting [31]. To quantify BBB integrity, the dynamic contrast-enhanced MRI (DCE-MRI) with gadolinium chelates could be included in patients with changes in MRI signals, which occur as a consequence of leakage of an intravenously injected contrast into the interstitial area [32]. According to Pradillo et al., DCE-MRI was shown to be a useful and non-invasive technique for evaluation of vascular function and angiogenesis processes [33]. The most prevalent investigation into direct assessment of vessels is angiography MR. However, the direct depiction of cerebrovascular remodeling or angiogenesis is very rare in clinical usage, because of the relatively low spatial resolution of MRI. Endothelial sprouting and microvessel formation proceed at length scales that are at minimum one order of magnitude lower than the current technology provides. Nevertheless, in the recent experimental study of Kang et al., high-resolution T1-contrast based on ultra-short echo time MR angiography (UTE-MRA) was carried out to visualize macro- and microvasculature and their association with ischemic edema status in transient middle cerebral artery occlusion rat models [34]. Although the presented results are very promising, the limitations associated with MRI as a tool for demonstrating post-stroke angiogenesis should be taken into consideration. Elevation of CBV is not only specific for neovascularization but may also contribute to the collateral flow and vasodilatation of existing vessels. Likewise, contrast agent leakage may arise in relation to angiogenesis as well as a result of BBB disruption in response to vascular pathology and hemorrhage [31]. To ensure new insights into the role of vascular remodeling in functional recovery after stroke, more studies need to be performed. The use of modern technologies may be very valuable as a monitoring tool for possible future therapies designed to support neovascularization in post-stroke patients, thus providing the possibility of introducing personalized therapies.

3.2. Neuroplastic/Neurogenic Markers of Stroke Recovery

In many current studies, neuroimaging techniques are employed to assess and understand post-stroke impairment. Understanding neural mechanisms of brain tissue damage as well as regeneration processes could be essential for predicting recovery and monitoring therapy. Such neural mechanisms of recovery involve in particular the perilesional tissue in the injured hemisphere, but also the contralateral hemisphere, subcortical and spinal regions [35]. All those processes that support recovery, termed neuroplasticity, are possible to identify as structural and functional brain changes in various neuroimaging techniques, such as magnetic resonance imaging (MRI), functional MRI, and functional magnetic spectroscopy (MRS) [36]. What is more, by using neurophysiological agents such as electroencephalography (EEG), we can map the brain activity and by using transcranial magnetic stimulation (TMS), it is also possible to test the influence of specific brain regions on motor learning and post-stroke recovery [37]. Described below are actionable techniques to predict post-stroke recover, that might be used for a personalized strategy of stroke treatment.

MRI is an essential clinical tool for diagnosing stroke severity, implementing treatment and predicting outcome [38]. Multimodal MRI reveals various parameters that help determine stroke mechanisms which affect recovery, such as differentiation of ischemic core from ischemic penumbra. The ischemic core is an area of infarction, which develops rapidly after artery occlusion. The differentiating parameter is the cBV, which is kept in the penumbra zone and decreased in the ischemic core. Using MRI technology, ischemic lesions can be identified with high precision, using diffusion-weighted image (DWI). Perfusion-weighted image (PWI) in turn can identify ischemic penumbral tissue [39]. To assign areas with PWI-DWI mismatch (the area difference when the perfusion lesion is larger than the diffusion lesion) is to identify representative salvageable tissue that may be responsible in recovery [40]. There is agreement as to the usefulness of characterizing the ischemic penumbra at the acute stage in relation to predicting motor outcomes. However, there are also data which suggest that the location of ischemic penumbra, instead of volume, could predict outcome and affect motor recovery [41].

MRI also delivers a method for assessing indices of white matter integrity and remodeling following ischemic stroke through diffusion-based methods. Measures of corticospinal tract (CST) white matter integrity is possible by diffusion tensor imaging (DTI) that uses anisotropic diffusion to estimate the axonal (white matter) organization of the brain. Ratio and asymmetry index of fractional anisotropy (FA) between ipsi- and contralesional corticospinal tracts (CSTs) is a very popular predictor in DTI studies. In general, a lower FA value of the ipsilesional CST may indicate greater damage tp the CST that can lead to more Wallerian degeneration of CST axons [42]. It has been proved that corticospinal tract injury is a valuable predictor of motor recovery in acute and post-acute stages [43–45]. In turn, Doughty et al. discovered that FA reduction of the CST (detected in the acute phase of stroke) present fractional predictive value to motor outcomes at 3 months [46]. Essentially, FA value can also be influenced by other factors (not only damage of CST), such as white matter architecture, so it should be carefully considered as a biomarker of brain impairment and poor recovery. Furthermore, according to Cassidy et al., CST-related atrophy and CST integrity are not useful in predicting treatment-related behavioral gains [47]. Only percentage of CST injury significantly predicted motor gains in response to therapy in the setting of subacute-chronic stroke, and might be used to stratify variables in restorative stroke trials. Likewise, Lim et al. proved that CTS injury provides low levels of prediction of upper extremity motor recovery and only in patients with severe initial motor deficits [48]. In a recently published study from Mattos [49], white matter plasticity effected by intensive task-specific upper limb training in chronic stroke was investigated and there were no changes in white matter integrity in patients with motor improvements. There was significant diversity in the response to test-specific training allowing for the specification of responder and non-responder groups. In responders, larger motor recovery was correlated with integrity in contralesional fibers and non-responders had severe damage of transcallosal fibers (more than responders). These results suggest the engagement of interhemispheric processes in responder groups. In many different stroke studies, it has also been proved that brain tissue reorganization in motor tracts and subsystems extends beyond the corticospinal tract [50–52].

After stroke damage, we can observe a dynamic process of changing brain activation patterns. Measurement of brain function presents complexities that do not increase with the measurement of anatomy. The way to follow this complexity is functional MRI (fMRI), which measures brain activity by detecting changes associated with blood flow. It is considered that neuronal activation and cerebral blood flows are coupled [53]. The most common form of fMRI is based on the blood-oxygen-level dependent (BOLD) contrast, which can indirectly measure neural activity based on changes in blood flow and deoxyhemoglobin concentration [54]. To activate the brain with fMRI, a specific behavioral paradigm must be executed by a patient; it should be performed on command, correctly and on time. Therefore, the behavioral paradigm should also be carefully selected to investigate the brain's functional field of interest. However, post-stroke motor impairments can make even simple motor performance difficult; thus, resting-state imaging is an attractive method for studying stroke network activity. With this technique, the functional connectivity represents the synchrony of intrinsic blood oxygen level-dependent (BOLD) signal fluctuations among different brain regions [55]. Functional connectivity that reflects the integrity of various motor and non-motor networks is associated with stroke outcome [56,57]. In a study by Puig, it has also been proved that patients with good outcome had greater functional connectivity than patients with bad outcome [58]. It has been shown by preserved bilateral interhemispheric connectivity between the anterior inferior temporal gyrus and superior frontal gyrus and decreased connectivity between the caudate and anterior inferior temporal gyrus in the left hemisphere. A larger potential for post-stroke recovery is also linked to the dynamic connectivity (time-varying functional network connectivity) between the bilateral intraparietal lobule and left angular gyrus [59]. Dynamic connectivity analysis is relevant in showing transiently increased information exchange between motor domains in moderate motor stroke and more isolated information processing in severe motor stroke [60]. To generalize, it is suggested that a better recovery is linked to keeping or recreating normal brain activation patterns and a bad recovery is linked to persistent contralesional fMRI response [61].

The next technique that allows assessment of the function of brain tissue is MRS (magnetic resonance spectroscopy). The presence and concentration of various metabolites is analyzed based on the principle that the distribution of electrons within an atom cause nuclei in different molecules to experience a slightly different magnetic field. In a study by Blicher et al., the higher GABA level in ipsilesional M1 was related to better motor function improvements after constraint-induced therapy [62]. In a recent report [63], it has been demonstrated that progressive falls in NAA and late increases in choline, myoinositol and lactate may indicate progressive non-ischemic neuronal damage, which has a harmful effect on motor recovery. Undoubtedly more studies linking MRS to functional outcomes would be beneficial in the context of personalized treatment choices.

Electroencephalography (EEG) is the most common, non-invasive method to record spontaneous or evoked electrical oscillation at various frequencies of the brain and, importantly, is one of the few mobile techniques available, unlike CT and MRI. Assessment of neuronal oscillations with electro or magneto-encephalography may supply an easy, available method to evaluate the balance between excitatory and inhibitory cortical actions [64]. EEG signals can identify sensitive changes in brain activity that cannot be detected by clinical measures. Furthermore, quantification of the EEG signal before and after treatment (rehabilitation) may evaluate neuroplasticity near the lesion and within whole-brain networks. The general findings, suggesting bad recovery in post-stroke patients investigated with EEG or MEG at the acute or subacute stages, indicate predominant inhibitory processes in the perilesional areas of cortex, shown by increased low-frequency oscillations [65,66]. Simultaneously, good recovery after stroke is associated with increased sensorimotor excitability in acute stages, presented by lower beta-rebound as a reaction to tactile finger stimulation [67]. Regarding the chronic post-stroke stages, in Mane's study, changes in the cortical activity after different types of motor rehabilitation, and its possible outcome, were analyzed using quantitative electroencephalography (QEEG) [68]. They obtained the relative theta power and interactions between the theta, alpha, and beta power as monitory biomarkers of motor recovery. In a currently ongoing study, EEG has been employed as a biomarker able to predict rehabilitation outcomes, providing a novel individual strategy of rehabilitation protocol (based on action observation treatment) for chronic stroke outpatients; no results have been publicized yet, but the idea appears to be promising [69]. Interesting conclusions have arisen from a study using MEG to assess cortical gamma synchronization (>30 Hz), which proved to be a predictor of recovery in patients undergoing intensive rehabilitation after stroke [70]. The next method useful as a post-stroke recovery biomarker is the assessment/presence of motor evoked potential (MEP) recorded by surface electromyography from muscles in response to TMS (transcranial magnetic stimulation). TMS is a non-invasive brain stimulation technique that allows to investigate as well as to modulate cortical excitability. There is clearly an understandable agreement that the presence of MEP (due to TMS) predicts good motor recovery as well as shorter MEP latencies, which means shorter central motor conduction times are associated with improved outcome [71]. In a retrospective study from Korea, 113 participants underwent TMS-induced MEP to estimate corticospinal excitability within 3 weeks after stroke onset [72]. They confirmed the MEP responsiveness value as one of the strong instruments of predicting motor function at 3 months after stroke. The next valuable method is to evoke cortical activity by TMS and afterwards measure cortical reorganization by EEG. Pellicciari et al. proved that TMS-evoked alpha oscillatory activity recorded just after stroke was associated with better functional recovery at 40- and 60-day follow-up evaluations [73]. Therefore, the power of the alpha rhythm is not only to predict recovery, but could also be helpful in determining the temporal window for enhancing neuroplasticity.

Table 1. Selected neuroimaging biomarkers of stroke recovery.

Biomarker	Type of Imaging	Usefulness Depending on the Stroke Phase	References
MRI-DTI (diffusion tensor imaging)	assess white matter integrity	acute, subacute, chronic	[42,44,45,48]
Ultra-short echo time MRI angiography	visualize macro- and microvasculature	acute, subacute	[34]
Steady-state contrast-enhanced MRI	assess vascular reorganization	subacute, chronic	[30]
Dynamic contrast-enhanced MRI	assess blood-brain barrier integrity	acute, subacute	[31]
Resting-state functional MRI	functional connectivity	subacute	[57,58]
Magnetic Resonance Spectroscopy	assess metabolic changes	subacute, chronic	[62,63]
EEG (electroencephalography)	assess balance between excitatory and inhibitory cortical actions	acute, subacute, chronic	[65–67]
TMS (transcranial magnetic stimulation) with MEP (motor evoked potential)	assess motor corticospinal excitability	subacute, chronic	[71,72]
TMS with EEG	assess cortical reorganization	subacute	[73]

4. Future Directions—Multimodal Panels of Neuroimaging Biomarkers and Application of Machine Learning Models

The aim of this article is not only to review neuroimaging stroke recovery biomarkers but also to analyze their application in treatment strategies. Therefore, it seems to be most beneficial to categorize individuals for personalized treatment based on multimodal panels including various biomarkers of stroke recovery. Recently introduced by Stinear et al., the Predict Recovery Potential (PREP) algorithm suggests combining clinical scales, transcranial magnetic stimulation, and diffusion-weighted magnetic resonance imaging to provide accurate predictions of upper limb function [71]. This information was used to modify therapy and enhance rehabilitation efficiency. Firstly, the Medical Research Council grades for shoulder abduction and finger extension strength within 3 days of stroke symptom onset were assessed. For patients with a summed score below 8 (out of 10), the functional integrity of the corticospinal tract was evaluated by identification of the presence or absence of paretic upper limb MEPs using TMS. The presence of MEPs indicated good prognosis. For individuals without MEPs, the characteristics of corticomotor pathways were assessed with diffusion tensor imaging. Low asymmetry in FA of the corticospinal tract (greater integrity of the ipsilesional corticospinal tract) was correlated with a prognosis of limited functional improvement; and high asymmetry in FA of the corticospinal tract (less integrity of the ipsilesional corticospinal tract) was correlated with the poorest prognosis for functional improvement. In this way, the algorithm predicts one of four possible outcomes for each patient: excellent, good, limited, or none. In accordance with this stratification, a suitable rehabilitation strategy was performed. Based only on clinical scales (without biomarker information about the corticospinal tract), the potential for a good recovery of function might go unrecognized by clinicians. Referring to multimodal panels of stroke recovery biomarkers, it is necessary to mention a very promising ongoing study of Picelli et al., where the study protocol for a randomized controlled trial is being developed [74]. The study will allow definition of a set of biomarkers (including neuroimaging: diffusion tensor imaging (DTI) and functional magnetic resonance imaging (fMRI); and neurophysiological: motor evoked potentials (MEP) and intracortical excitability measured by single and paired transcranial magnetic stimulation (TMS); somatosensory evoked potentials (SEP); and brain connectivity measured by electroencephalographic (EEG) phase synchrony) related to stroke recovery and rehabilitation result in order to discover patients with stronger potential for improvement and define personalized rehabilitation programs.

Future directions towards predicting post-stroke recovery and treatment outcomes could be machine-learning algorithms, which classify and predict the participants' responsiveness to therapy [75]. In this pilot study, 64 post-stroke aphasia patients were analyzed, the predictive framework was created based on collected data—demographic, brain structure (MRI) and behavioral (clinical scales of aphasia severity)—and then Random Forest models were used to evaluate the importance of these features. Preliminary results of this study suggest the potential of their framework to predict individualized rehabilitation. Furthermore, the utility of machine learning approaches has been newly explored to predict post-stroke recovery relying on multi-channel electroencephalographic recordings and clinical scales [76]. In this study, nonlinear support vector regressor (SVR) was exploited to predict recovery in the acute phase of stroke, which might allow early and personalized therapies.

5. Conclusions

The use of neuroimaging techniques allows prediction of recovery potential more objectively, based not solely on functional clinical scales which represent a measure of impairment. However, it is relevant for stroke recovery to also consider other aspects, such as age, gender, type of stroke, type of treatment, as well as environmental factors. The importance of these factors should be assessed individually, but very often it is impossible to isolate one critical variable in clinical studies and therefore heterogenous groups might cancel individual benefits. To overcome this challenge, multimodal panels of biomarkers defining patients with greater recovery potential appear to be an appealing tool. Likewise, understanding the processes of angiogenesis and neuroplasticity that occur during stroke recovery, as well as monitoring these processes (by neuroimaging and neurophysiological techniques), is extremely important for planning effective treatment. Mentioned above, research using various types of biomarkers (clinical, imaging, neurophysiological, and genetic-molecular) is very promising for defining stroke rehabilitation protocol in future studies [74]. To further improve poststroke recovery, neuroimaging techniques may also be implemented to enhance robot-assisted rehabilitation [77]. This allows for monitoring of the rehabilitation process, providing precise feedback from therapy and adapting to the specific needs of patients. Nevertheless, there are still important challenges to overcome in future studies. Considerations should include issues such as suitable sample sizes, standardized methods and patient stratification. Neuroimaging and neurophysiological techniques are already well-proven as diagnostic tools for stroke but still require establishment of their prognostic value for stroke recovery.

Author Contributions: Conceptualization, L.W., A.M. and E.M.; methodology, L.W. and N.C.; software, L.W. and J.S.-B.; validation, E.M. and A.M. and J.S.-B.; formal analysis, E.M.; investigation, L.W.; resources, L.W. and N.C.; data curation, L.W. and M.B.; writing—original draft preparation, L.W. and N.C.; writing—review and editing, M.B. and L.W.; visualization, N.C.; supervision, E.M. and J.S.-B.; project administration, A.M.; funding acquisition, E.M. All authors have read and agreed to the published version of the manuscript.

Funding: This study was supported by grant from the Medical University of Lodz no 503/6-127-05/503-51-001-19.

Institutional Review Board Statement: Not applicable.

Informed Consent Statement: Not applicable.

Conflicts of Interest: The authors declare no conflict of interest.

References

1. Sacco, R.L.; Kasner, S.E.; Broderick, J.P.; Caplan, L.R.; Connors, J.J.; Culebras, A.; Elkind, M.S.; George, M.G.; Hamdan, A.D.; Higashida, R.T.; et al. An updated definition of stroke for the 21st century: A statement for healthcare professionals from the American Heart Association/American Stroke Association. *Stroke* **2013**, *44*, 2064–2089. [CrossRef] [PubMed]
2. Hankey, G.J. Stroke. *Lancet* **2017**, *389*, 641–654. [CrossRef]
3. Powers, W.J.; Rabinstein, A.A.; Ackerson, T.; Adeoye, O.M.; Bambakidis, N.C.; Becker, K.; Biller, J.; Brown, M.; Demaerschalk, B.M.; Hoh, B.; et al. Guidelines for the Early Management of Patients With Acute Ischemic Stroke: 2019 Update to the 2018 Guidelines for the Early Management of Acute Ischemic Stroke: A Guideline for Healthcare Professionals From the American Heart Association/American Stroke Association. *Stroke* **2019**, *50*, e344–e418. [CrossRef]

4. Kleindorfer, D.O.; Towfighi, A.; Chaturvedi, S.; Cockroft, K.M.; Gutierrez, J.; Lombardi-Hill, D.; Kamel, H.; Kernan, W.N.; Kittner, S.J.; Leira, E.C.; et al. 2021 Guideline for the Prevention of Stroke in Patients With Stroke and Transient Ischemic Attack: A Guideline From the American Heart Association/American Stroke Association. *Stroke* **2021**, *52*, e364–e467. [CrossRef] [PubMed]
5. De Oliveira Manoel, A.L. Surgery for spontaneous intracerebral hemorrhage. *Crit. Care* **2020**, *24*, 45. [CrossRef] [PubMed]
6. Dancause, N.; Barbay, S.; Frost, S.B.; Plautz, E.J.; Chen, D.; Zoubina, E.V.; Stowe, A.M.; Nudo, R.J. Extensive cortical rewiring after brain injury. *J. Neurosci.* **2005**, *25*, 10167–10179. [CrossRef] [PubMed]
7. Winstein, C.J.; Stein, J.; Arena, R.; Bates, B.; Cherney, L.R.; Cramer, S.C.; Deruyter, F.; Eng, J.J.; Fisher, B.; Harvey, R.L.; et al. Guidelines for Adult Stroke Rehabilitation and Recovery: A Guideline for Healthcare Professionals from the American Heart Association/American Stroke Association. *Stroke* **2016**, *47*, e98–e169. [CrossRef]
8. Klamroth-Marganska, V. Stroke Rehabilitation: Therapy Robots and Assistive Devices. *Adv. Exp. Med. Biol.* **2018**, *1065*, 579–587. [CrossRef]
9. Chieffo, R.; Comi, G.; Leocani, L. Noninvasive Neuromodulation in Poststroke Gait Disorders: Rationale, Feasibility, and State of the Art. *Neurorehabil. Neural Repair* **2016**, *30*, 71–82. [CrossRef]
10. Breining, B.L.; Sebastian, R. Neuromodulation in Post-stroke Aphasia Treatment. *Curr. Phys. Med. Rehabil. Rep.* **2020**, *8*, 44–56. [CrossRef]
11. Beuter, A.; Balossier, A.; Vassal, F.; Hemm, S.; Volpert, V. Cortical stimulation in aphasia following ischemic stroke: Toward model-guided electrical neuromodulation. *Biol. Cybern.* **2020**, *114*, 5–21. [CrossRef] [PubMed]
12. Powell, K.; White, T.G.; Nash, C.; Rebeiz, T.; Woo, H.H.; Narayan, R.K.; Li, C. The Potential Role of Neuromodulation in Subarachnoid Hemorrhage. *Neuromodulation* **2022**. [CrossRef] [PubMed]
13. Choi, Y.H.; Paik, N.J. Mobile Game-based Virtual Reality Program for Upper Extremity Stroke Rehabilitation. *J. Vis. Exp.* **2018**, *133*, 56241. [CrossRef]
14. Grefkes, C.; Fink, G.R. Reorganization of cerebral networks after stroke: New insights from neuroimaging with connectivity approaches. *Brain* **2011**, *134*, 1264–1276. [CrossRef] [PubMed]
15. Saver, J.L.; Johnston, K.C.; Homer, D.; Wityk, R.; Koroshetz, W.; Truskowski, L.L.; Haley, E.C. Infarct volume as a surrogate or auxiliary outcome measure in ischemic stroke clinical trials. The RANTTAS Investigators. *Stroke* **1999**, *30*, 293–298. [CrossRef] [PubMed]
16. Boyd, L.A.; Hayward, K.S.; Ward, N.S.; Stinear, C.M.; Rosso, C.; Fisher, R.J.; Carter, A.R.; Leff, A.P.; Copland, D.A.; Carey, L.M.; et al. Biomarkers of stroke recovery: Consensus-based core recommendations from the Stroke Recovery and Rehabilitation Roundtable. *Int. J. Stroke* **2017**, *12*, 480–493. [CrossRef] [PubMed]
17. Arai, K.; Jin, G.; Navaratna, D.; Lo, E.H. Brain angiogenesis in developmental and pathological processes: Neurovascular injury and angiogenic recovery after stroke. *FEBS J.* **2009**, *276*, 4644–4652. [CrossRef]
18. Yanev, P.; Dijkhuizen, R.M. In vivo imaging of neurovascular remodeling after stroke. *Stroke* **2012**, *43*, 3436–3441. [CrossRef]
19. Williamson, M.R.; Franzen, R.L.; Fuertes, C.J.A.; Dunn, A.K.; Drew, M.R.; Jones, T.A. A Window of Vascular Plasticity Coupled to Behavioral Recovery after Stroke. *J. Neurosci.* **2020**, *40*, 7651–7667. [CrossRef]
20. Adamczak, J.; Hoehn, M. Poststroke angiogenesis, con: Dark side of angiogenesis. *Stroke* **2015**, *46*, e103–e104. [CrossRef]
21. Grefkes, C.; Fink, G.R. Recovery from stroke: Current concepts and future perspectives. *Neurol. Res. Pract.* **2020**, *2*, 17. [CrossRef] [PubMed]
22. Li, S.; Carmichael, S.T. Growth-associated gene and protein expression in the region of axonal sprouting in the aged brain after stroke. *Neurobiol. Dis.* **2006**, *23*, 362–373. [CrossRef]
23. Redecker, C.; Wang, W.; Fritschy, J.M.; Witte, O.W. Widespread and long-lasting alterations in GABA(A)-receptor subtypes after focal cortical infarcts in rats: Mediation by NMDA-dependent processes. *J. Cereb. Blood Flow Metab.* **2002**, *22*, 1463–1475. [CrossRef] [PubMed]
24. Carmichael, S.T.; Tatsukawa, K.; Katsman, D.; Tsuyuguchi, N.; Kornblum, H.I. Evolution of diaschisis in a focal stroke model. *Stroke* **2004**, *35*, 758–763. [CrossRef]
25. FDA-NIH Biomarker Working Group. *BEST (Biomarkers, EndpointS, and Other Tools) Resource*; Food and Drug Administration (US): Silver Spring, MD, USA, 2016.
26. Bang, O.Y. Advances in biomarker for stroke patients: From marker to regulator. *Precis. Future Med.* **2017**, *1*, 32–42. [CrossRef]
27. Jones, T.A.; Kleim, J.A.; Greenough, W.T. Synaptogenesis and dendritic growth in the cortex opposite unilateral sensorimotor cortex damage in adult rats: A quantitative electron microscopic examination. *Brain Res.* **1996**, *733*, 142–148. [CrossRef]
28. Zhu, H.; Zhang, Y.; Zhong, Y.; Ye, Y.; Hu, X.; Gu, L.; Xiong, X. Inflammation-Mediated Angiogenesis in Ischemic Stroke. *Front. Cell. Neurosci.* **2021**, *15*, 652647. [CrossRef]
29. Callewaert, B.; Jones, E.A.V.; Himmelreich, U.; Gsell, W. Non-Invasive Evaluation of Cerebral Microvasculature Using Pre-Clinical MRI: Principles, Advantages and Limitations. *Diagnostics* **2021**, *11*, 926. [CrossRef]
30. Yanev, P.; Seevinck, P.R.; Rudrapatna, U.S.; Bouts, M.J.; van der Toorn, A.; Gertz, K.; Kronenberg, G.; Endres, M.; van Tilborg, G.A.; Dijkhuizen, R.M. Magnetic resonance imaging of local and remote vascular remodelling after experimental stroke. *J. Cereb. Blood Flow Metab.* **2017**, *37*, 2768–2779. [CrossRef]
31. Yang, Y.; Torbey, M.T. Angiogenesis and Blood-Brain Barrier Permeability in Vascular Remodeling after Stroke. *Curr. Neuropharmacol.* **2020**, *18*, 1250–1265. [CrossRef]

32. Merali, Z.; Huang, K.; Mikulis, D.; Silver, F.; Kassner, A. Evolution of blood-brain-barrier permeability after acute ischemic stroke. *PLoS ONE* **2017**, *12*, e0171558. [CrossRef] [PubMed]

33. Pradillo, J.M.; Hernández-Jiménez, M.; Fernández-Valle, M.E.; Medina, V.; Ortuño, J.E.; Allan, S.M.; Proctor, S.D.; Garcia-Segura, J.M.; Ledesma-Carbayo, M.J.; Santos, A.; et al. Influence of metabolic syndrome on post-stroke outcome, angiogenesis and vascular function in old rats determined by dynamic contrast enhanced MRI. *J. Cereb. Blood Flow Metab.* **2020**, *41*, 1692–1706. [CrossRef] [PubMed]

34. Kang, M.; Jin, S.; Lee, D.; Cho, H. MRI Visualization of Whole Brain Macro- and Microvascular Remodeling in a Rat Model of Ischemic Stroke: A Pilot Study. *Sci. Rep.* **2020**, *10*, 4989. [CrossRef]

35. Alia, C.; Spalletti, C.; Lai, S.; Panarese, A.; Lamola, G.; Bertolucci, F.; Vallone, F.; Di Garbo, A.; Chisari, C.; Micera, S.; et al. Neuroplastic Changes Following Brain Ischemia and their Contribution to Stroke Recovery: Novel Approaches in Neurorehabilitation. *Front. Cell. Neurosci.* **2017**, *11*, 76. [CrossRef] [PubMed]

36. Auriat, A.M.; Neva, J.L.; Peters, S.; Ferris, J.K.; Boyd, L.A. A Review of Transcranial Magnetic Stimulation and Multimodal Neuroimaging to Characterize Post-Stroke Neuroplasticity. *Front. Neurol.* **2015**, *6*, 226. [CrossRef]

37. Rolle, C.E.; Baumer, F.M.; Jordan, J.T.; Berry, K.; Garcia, M.; Monusko, K.; Trivedi, H.; Wu, W.; Toll, R.; Buckwalter, M.S.; et al. Mapping causal circuit dynamics in stroke using simultaneous electroencephalography and transcranial magnetic stimulation. *BMC Neurol.* **2021**, *21*, 280. [CrossRef]

38. Pinto, A.; Mckinley, R.; Alves, V.; Wiest, R.; Silva, C.A.; Reyes, M. Stroke Lesion Outcome Prediction Based on MRI Imaging Combined With Clinical Information. *Front. Neurol.* **2018**, *9*, 1060. [CrossRef]

39. González, R.G. Clinical MRI of acute ischemic stroke. *J. Magn. Reson. Imaging* **2012**, *36*, 259–271. [CrossRef]

40. Bracard, S. Predicting Tissue Viability in Ischemic Stroke with Diffusion and Perfusion MRI. *Open Access J. Neurol. Neurosurg.* **2019**, *10*, 555799. [CrossRef]

41. Rosso, C.; Samson, Y. The ischemic penumbra: The location rather than the volume of recovery determines outcome. *Curr. Opin. Neurol.* **2014**, *27*, 35–41. [CrossRef]

42. Puig, J.; Pedraza, S.; Blasco, G.; Daunis-I-Estadella, J.; Prats, A.; Prados, F.; Boada, I.; Castellanos, M.; Sánchez-González, J.; Remollo, S.; et al. Wallerian degeneration in the corticospinal tract evaluated by diffusion tensor imaging correlates with motor deficit 30 days after middle cerebral artery ischemic stroke. *Am. J. Neuroradiol.* **2010**, *31*, 1324–1330. [CrossRef] [PubMed]

43. Bigourdan, A.; Munsch, F.; Coupé, P.; Guttmann, C.R.; Sagnier, S.; Renou, P.; Debruxelles, S.; Poli, M.; Dousset, V.; Sibon, I.; et al. Early Fiber Number Ratio Is a Surrogate of Corticospinal Tract Integrity and Predicts Motor Recovery After Stroke. *Stroke* **2016**, *47*, 1053–1059. [CrossRef] [PubMed]

44. Lin, D.J.; Cloutier, A.M.; Erler, K.S.; Cassidy, J.M.; Snider, S.B.; Ranford, J.; Parlman, K.; Giatsidis, F.; Burke, J.F.; Schwamm, L.H.; et al. Corticospinal Tract Injury Estimated From Acute Stroke Imaging Predicts Upper Extremity Motor Recovery After Stroke. *Stroke* **2019**, *50*, 3569–3577. [CrossRef] [PubMed]

45. Wen, H.; Alshikho, M.J.; Wang, Y.; Luo, X.; Zafonte, R.; Herbert, M.R.; Wang, Q.M. Correlation of Fractional Anisotropy With Motor Recovery in Patients With Stroke After Postacute Rehabilitation. *Arch. Phys. Med. Rehabil.* **2016**, *97*, 1487–1495. [CrossRef]

46. Doughty, C.; Wang, J.; Feng, W.; Hackney, D.; Pani, E.; Schlaug, G. Detection and Predictive Value of Fractional Anisotropy Changes of the Corticospinal Tract in the Acute Phase of a Stroke. *Stroke* **2016**, *47*, 1520–1526. [CrossRef]

47. Cassidy, J.M.; Tran, G.; Quinlan, E.B.; Cramer, S.C. Neuroimaging Identifies Patients Most Likely to Respond to a Restorative Stroke Therapy. *Stroke* **2018**, *49*, 433–438. [CrossRef]

48. Lim, J.Y.; Oh, M.K.; Park, J.; Paik, N.J. Does Measurement of Corticospinal Tract Involvement Add Value to Clinical Behavioral Biomarkers in Predicting Motor Recovery after Stroke? *Neural Plast.* **2020**, *2020*, 8883839. [CrossRef]

49. Mattos, D.J.S.; Rutlin, J.; Hong, X.; Zinn, K.; Shimony, J.S.; Carter, A.R. White matter integrity of contralesional and transcallosal tracts may predict response to upper limb task-specific training in chronic stroke. *NeuroImage Clin.* **2021**, *31*, 102710. [CrossRef]

50. Li, Y.; Wu, P.; Liang, F.; Huang, W. The microstructural status of the corpus callosum is associated with the degree of motor function and neurological deficit in stroke patients. *PLoS ONE* **2015**, *10*, e0122615. [CrossRef]

51. Wadden, K.P.; Peters, S.; Borich, M.R.; Neva, J.L.; Hayward, K.S.; Mang, C.S.; Snow, N.J.; Brown, K.E.; Woodward, T.S.; Meehan, S.K.; et al. White Matter Biomarkers Associated with Motor Change in Individuals with Stroke: A Continuous Theta Burst Stimulation Study. *Neural Plast.* **2019**, *2019*, 7092496. [CrossRef]

52. Lindenberg, R.; Zhu, L.L.; Rüber, T.; Schlaug, G. Predicting functional motor potential in chronic stroke patients using diffusion tensor imaging. *Hum. Brain Mapp.* **2012**, *33*, 1040–1051. [CrossRef] [PubMed]

53. Zilles, K.; Amunts, K. Anatomical Basis for Functional Specialization. In *fMRI: From Nuclear Spins to Brain Functions*; Uludag, K., Ugurbil, K., Berliner, L., Eds.; Springer: Boston, MA, USA, 2015; pp. 27–66.

54. Crofts, A.; Kelly, M.E.; Gibson, C.L. Imaging Functional Recovery Following Ischemic Stroke: Clinical and Preclinical fMRI Studies. *J. Neuroimaging* **2020**, *30*, 5–14. [CrossRef] [PubMed]

55. Glover, G.H. Overview of functional magnetic resonance imaging. *Neurosurg. Clin. N. Am.* **2011**, *22*, 133–139. [CrossRef] [PubMed]

56. Grefkes, C.; Fink, G.R. Connectivity-based approaches in stroke and recovery of function. *Lancet Neurol.* **2014**, *13*, 206–216. [CrossRef]

57. Almeida, S.R.; Vicentini, J.; Bonilha, L.; De Campos, B.M.; Casseb, R.F.; Min, L.L. Brain Connectivity and Functional Recovery in Patients With Ischemic Stroke. *J. Neuroimaging* **2017**, *27*, 65–70. [CrossRef]

58. Puig, J.; Blasco, G.; Alberich-Bayarri, A.; Schlaug, G.; Deco, G.; Biarnes, C.; Navas-Martí, M.; Rivero, M.; Gich, J.; Figueras, J.; et al. Resting-State Functional Connectivity Magnetic Resonance Imaging and Outcome After Acute Stroke. *Stroke* **2018**, *49*, 2353–2360. [CrossRef]

59. Bonkhoff, A.K.; Schirmer, M.D.; Bretzner, M.; Etherton, M.; Donahue, K.; Tuozzo, C.; Nardin, M.; Giese, A.-K.; Wu, O.; Calhoun, V.D.; et al. Abnormal dynamic functional connectivity is linked to recovery after acute ischemic stroke. *Hum. Brain Mapp.* **2021**, *42*, 2278–2291. [CrossRef]

60. Bonkhoff, A.K.; Rehme, A.K.; Hensel, L.; Tscherpel, C.; Volz, L.J.; Espinoza, F.A.; Gazula, H.; Vergara, V.M.; Fink, G.R.; Calhoun, V.D.; et al. Dynamic connectivity predicts acute motor impairment and recovery post-stroke. *Brain Commun.* **2021**, *3*, fcab227. [CrossRef]

61. Buma, F.E.; Lindeman, E.; Ramsey, N.F.; Kwakkel, G. Functional neuroimaging studies of early upper limb recovery after stroke: A systematic review of the literature. *Neurorehabil. Neural Repair* **2010**, *24*, 589–608. [CrossRef]

62. Blicher, J.U.; Near, J.; Næss-Schmidt, E.; Stagg, C.J.; Johansen-Berg, H.; Nielsen, J.F.; Østergaard, L.; Ho, Y.C. GABA levels are decreased after stroke and GABA changes during rehabilitation correlate with motor improvement. *Neurorehabil. Neural Repair* **2015**, *29*, 278–286. [CrossRef]

63. Mazibuko, N.; Tuura, R.O.; Sztriha, L.; O'Daly, O.; Barker, G.J.; Williams, S.C.R.; O'Sullivan, M.; Kalra, L. Subacute Changes in N-Acetylaspartate (NAA) Following Ischemic Stroke: A Serial MR Spectroscopy Pilot Study. *Diagnostics* **2020**, *10*, 482. [CrossRef] [PubMed]

64. Rabiller, G.; He, J.-W.; Nishijima, Y.; Wong, A.; Liu, J. Perturbation of Brain Oscillations after Ischemic Stroke: A Potential Biomarker for Post-Stroke Function and Therapy. *Int. J. Mol. Sci.* **2015**, *16*, 5605. [CrossRef] [PubMed]

65. Zappasodi, F.; Pasqualetti, P.; Rossini, P.M.; Tecchio, F. Acute Phase Neuronal Activity for the Prognosis of Stroke Recovery. *Neural Plast.* **2019**, *2019*, 1971875. [CrossRef]

66. Laaksonen, K.; Helle, L.; Parkkonen, L.; Kirveskari, E.; Mäkelä, J.P.; Mustanoja, S.; Tatlisumak, T.; Kaste, M.; Forss, N. Alterations in spontaneous brain oscillations during stroke recovery. *PLoS ONE* **2013**, *8*, e61146. [CrossRef] [PubMed]

67. Laaksonen, K.; Kirveskari, E.; Mäkelä, J.P.; Kaste, M.; Mustanoja, S.; Nummenmaa, L.; Tatlisumak, T.; Forss, N. Effect of afferent input on motor cortex excitability during stroke recovery. *Clin. Neurophysiol.* **2012**, *123*, 2429–2436. [CrossRef]

68. Mane, R.; Chew, E.; Phua, K.S.; Ang, K.K.; Vinod, A.P.; Guan, C. Quantitative EEG as Biomarkers for the Monitoring of Post-Stroke Motor Recovery in BCI and tDCS Rehabilitation. In Proceedings of the 2018 40th Annual International Conference of the IEEE Engineering in Medicine and Biology Society (EMBC), Honolulu, HI, USA, 18–21 July 2018; Volume 2018, pp. 3610–3613. [CrossRef]

69. Neuroimaging Biomarkers Toward a Personalized Upper Limb Action Observation Treatment in Chronic Stroke Patients (BE-TOP). Available online: https://clinicaltrials.gov/ct2/show/study/NCT04047134 (accessed on 22 April 2022).

70. Pellegrino, G.; Arcara, G.; Cortese, A.M.; Weis, L.; Di Tomasso, S.; Marioni, G.; Masiero, S.; Piccione, F. Cortical gamma-synchrony measured with magnetoencephalography is a marker of clinical status and predicts clinical outcome in stroke survivors. *Neuroimage Clin.* **2019**, *24*, 102092. [CrossRef] [PubMed]

71. Stinear, C.M.; Byblow, W.D.; Ackerley, S.J.; Barber, P.A.; Smith, M.C. Predicting Recovery Potential for Individual Stroke Patients Increases Rehabilitation Efficiency. *Stroke* **2017**, *48*, 1011–1019. [CrossRef]

72. Jo, J.Y.; Lee, A.; Kim, M.S.; Park, E.; Chang, W.H.; Shin, Y.I.; Kim, Y.H. Prediction of Motor Recovery Using Quantitative Parameters of Motor Evoked Potential in Patients With Stroke. *Ann. Rehabil. Med.* **2016**, *40*, 806–815. [CrossRef]

73. Pellicciari, M.C.; Bonnì, S.; Ponzo, V.; Cinnera, A.M.; Mancini, M.; Casula, E.P.; Sallustio, F.; Paolucci, S.; Caltagirone, C.; Koch, G. Dynamic reorganization of TMS-evoked activity in subcortical stroke patients. *Neuroimage* **2018**, *175*, 365–378. [CrossRef]

74. Picelli, A.; Filippetti, M.; Del Piccolo, L.; Schena, F.; Chelazzi, L.; Della Libera, C.; Donadelli, M.; Donisi, V.; Fabene, P.F.; Fochi, S.; et al. Rehabilitation and Biomarkers of Stroke Recovery: Study Protocol for a Randomized Controlled Trial. *Front. Neurol.* **2021**, *11*, 1800. [CrossRef]

75. Gu, Y.; Bahrani, M.; Billot, A.; Lai, S.; Braun, E.J.; Varkanitsa, M.; Bighetto, J.; Rapp, B.; Parrish, T.B.; Caplan, D.; et al. A machine learning approach for predicting post-stroke aphasia recovery: A pilot study. In Proceedings of the 13th ACM International Conference on PErvasive Technologies Related to Assistive Environments, Corfu, Greece, 30 June–3 July 2020; p. 22.

76. Chiarelli, A.M.; Croce, P.; Assenza, G.; Merla, A.; Granata, G.; Giannantoni, N.M.; Pizzella, V.; Tecchio, F.; Zappasodi, F. Electroencephalography-Derived Prognosis of Functional Recovery in Acute Stroke Through Machine Learning Approaches. *Int. J. Neural. Syst.* **2020**, *30*, 2050067. [CrossRef] [PubMed]

77. Iandolo, R.; Marini, F.; Semprini, M.; Laffranchi, M.; Mugnosso, M.; Cherif, A.; De Michieli, L.; Chiappalone, M.; Zenzeri, J. Perspectives and Challenges in Robotic Neurorehabilitation. *Appl. Sci.* **2019**, *9*, 3183. [CrossRef]

Journal of
Clinical Medicine

Review

Admission Severity of Atrial-Fibrillation-Related Acute Ischemic Stroke in Patients under Anticoagulation Treatment: A Systematic Review and Meta-Analysis

Catarina Garcia [1], Marcelo Silva [1], Mariana Araújo [1], Mariana Henriques [1], Marta Margarido [1], Patrícia Vicente [1], Hipólito Nzwalo [1,2,3,*] and Ana Macedo [1,2]

[1] Faculty of Medicine and Biomedical Sciences, University of Algarve, 8005-139 Faro, Portugal; a63572@ualg.pt (C.G.); a63504@ualg.pt (M.S.); a63584@ualg.pt (M.A.); a63507@ualg.pt (M.H.); a63508@ualg.pt (M.M.); a63510@ualg.pt (P.V.); amamacedo@ualg.pt (A.M.)
[2] Algarve Biomedical Center, 8005-139 Algarve, Portugal
[3] Stroke Unit, Algarve University Hospital Center, 8000-386 Algarve, Portugal
* Correspondence: nzwalo@gmail.com

Abstract: Background: In non-valvular-associated atrial fibrillation (AF), direct oral anticoagulants (DOAC) are as effective as vitamin K antagonists (VKA) for the prevention of acute ischemic stroke (AIS). DOAC are associated with decreased risk and severity of intracranial hemorrhage. It is unknown if different pre-admission anticoagulants impact the prognosis of AF related AIS (AF-AIS). We sought to analyze the literature to assess the association between pre-admission anticoagulation (VKA or DOAC) and admission severity of AF-AIS. Methods: A Systematic literature search (PubMed and ScienceDirect) between January 2011 to April 2021 was undertaken to identify studies describing the outcome of AF-AIS. Results: A total of 128 articles were identified. Of 9493 patients, 1767 were on DOAC, 919 were on therapeutical VKA, 792 were on non-therapeutical VKA and 6015 were not anticoagulated. In comparison to patients without anticoagulation, patients with therapeutical VKA and under DOAC presented with less severe stroke (MD -1.69; 95% CI [-2.71, -0.66], $p = 0.001$ and MD -2.96; 95% Cl [-3.75, -2.18], $p < 0.00001$, respectively). Patients with non-therapeutical VKA presented with more severe stroke (MD 1.28; 95% Cl [0.45, 2.12], $p = 0.003$). Conclusions: In AF-AIS, patients under therapeutical VKA or DOAC have reduced stroke severity on admission in comparison to patients without any anticoagulation, with higher magnitude of protection for DOAC.

Keywords: ischemic stroke; atrial fibrillation; oral anticoagulation; stroke severity

Citation: Garcia, C.; Silva, M.; Araújo, M.; Henriques, M.; Margarido, M.; Vicente, P.; Nzwalo, H.; Macedo, A. Admission Severity of Atrial-Fibrillation-Related Acute Ischemic Stroke in Patients under Anticoagulation Treatment: A Systematic Review and Meta-Analysis. *J. Clin. Med.* **2022**, *11*, 3563. https://doi.org/10.3390/jcm11123563

Academic Editor: Peter Sporns

Received: 19 May 2022
Accepted: 15 June 2022
Published: 20 June 2022

Publisher's Note: MDPI stays neutral with regard to jurisdictional claims in published maps and institutional affiliations.

1. Introduction

Acute ischemic stroke (AIS) is a major cause of disability and death worldwide [1]. Atrial fibrillation (AF) is the most common cardioembolic etiology, associated with 15 to 30% of all cases of AIS [2,3]. Without treatment, AF is associated with a five-fold increased risk of AIS [4–6]. The burden of AF related AIS (AF-AIS) is well characterized. AF-AIS is associated with higher mortality, greater disability, prolonged hospitalization, and a lower chance of being discharged home [7–15]. Oral anticoagulants (OAC), namely vitamin K antagonists (VKA) and direct oral anticoagulants (DOAC), reduce the risk of AIS [4,7,16–18]. In non-valvular AF, the efficacy of DOAC parallels that of other OAC. Moreover, in comparison to patients taking VKA, patients under treatment with DOAC have reduced frequency and severity of intracerebral hemorrhage, the most lethal complication of anticoagulation therapy [19].

Reflecting the presence of residual thromboembolic risk, AIS can occur even in patients under adequate OAC. Nevertheless, the annual incidence of AF-AIS in patients without OAC (43.4 per 1000 person-years) is much higher than the 9.4 and 10.4 per 1000 person-years from patients under DOAC and warfarin, respectively [20].

It is unknown if different pre-admission OAC can have different impacts on AF-AIS prognosis. We sought to review the literature to assess the association between pre-admission anticoagulation (VKA or DOAC) therapy and initial severity in patients with AF-AIS.

2. Materials and Methods

This protocol follows the Preferred Reporting Items for Systematic Reviews and Meta-Analysis Protocols (PRISMA-P) guidelines [21,22]. The review was registered within the International Prospective Register of Systematic Reviews on 17 April 2021 (PROSPERO ID: CRD42021249531).

2.1. Literature Research

A systematic literature review was performed using PubMed and ScienceDirect databases. The search strategy in this study was as follows: ("atrial fibrillation") AND ("prior anticoagulation" OR "prestroke anticoagulation" OR "anticoagulation before stroke" OR "preceding anticoagulation") AND ("stroke severity" OR "stroke outcome").

2.2. Data Extraction

Two investigators independently screened articles (title, abstract, full articles) to assess eligibility. Titles, names of authors, year of publication, study design, sample and control group size, study characteristics, outcomes and results were extracted and recorded in a Microsoft Excel spreadsheet. Therapeutical and non-therapeutical dosages of the drugs involved were defined based on the criteria of each author in the respective articles. All duplicated articles were removed. Any disagreement was resolved by a third reviewer.

2.3. Study Selection and Quality Assessment

To assess the impact of preadmission anticoagulation on stroke severity, analysis was limited to studies specifically designed or aimed to compare AF-AIS outcomes between patients taking or not taking any specific OAC. To reduce the risk of information or selection bias as well as socioeconomic status as a confounding bias, studies not specifically designed to compare the outcome with a control group were excluded.

We also excluded articles written in a language other than English. The methodological quality of each study was assessed in a qualitative way. Risk of bias was assessed according to the Cochrane Handbook for Systematic Reviews of Interventions guidelines [23].

2.4. Statistical Analysis

The Review Manager (RevMan) software (version 5.4, Cochrane Collaboration) was used for data analysis. For the meta-analysis, the included outcomes were measured using the same metrics in a minimum of three studies. Stroke severity score on admission, measured by National Institutes of Health Stroke Scale (NIHSS), was analyzed in subgroups (VKA therapeutical, VKA non-therapeutical and DOAC). For continuous variables, we used Hozo et al.'s [24,25] formula to estimate mean and variance values from the median, interquartile range and size of the samples. Results were reported with effect sizes and 95% confidence intervals (CI). For continuous variables, this was determined using the inverse variance random-effects model. For dichotomous variables, CI and odds ratio (OR) were measured using the Mantel–Haenszel random-effects model. The OR and mean differences were represented on a forest plot. Heterogeneity across studies was quantified based on Higgins statistics, considering $I^2 \geq 50\%$ evidence of a substantial heterogeneity in our meta-analysis. Studies not reporting AIS stroke severity on admission with NIHSS as a mean or a median were excluded from the meta-analysis. In addition, a random-effects approach was selected to extrapolate results as a small number of studies—not functionally identical (performed by researchers operating independently)—were included. This model considers the existence of differences between the studies that go beyond simple sample variability [26]. We intended to assess publication bias through funnel plot asymmetry [23].

3. Results

3.1. Study Selection

PRISMA flowchart diagram (Figure 1) outlines the selection and inclusion process. The initial search yielded 128 studies. The number of studies after deduplication was 111, and 31 fulfilled criteria for full-text reading. From those, 25 were excluded. Reasons for exclusion were the absence of a control group ($n = 9$), prior anticoagulation due to conditions other than AF ($n = 4$), post-stroke intervention ($n = 2$), absence of data regarding AIS outcome ($n = 1$) and lack of quantitative data reporting stroke severity (NIHSS) as a mean or a median ($n = 9$).

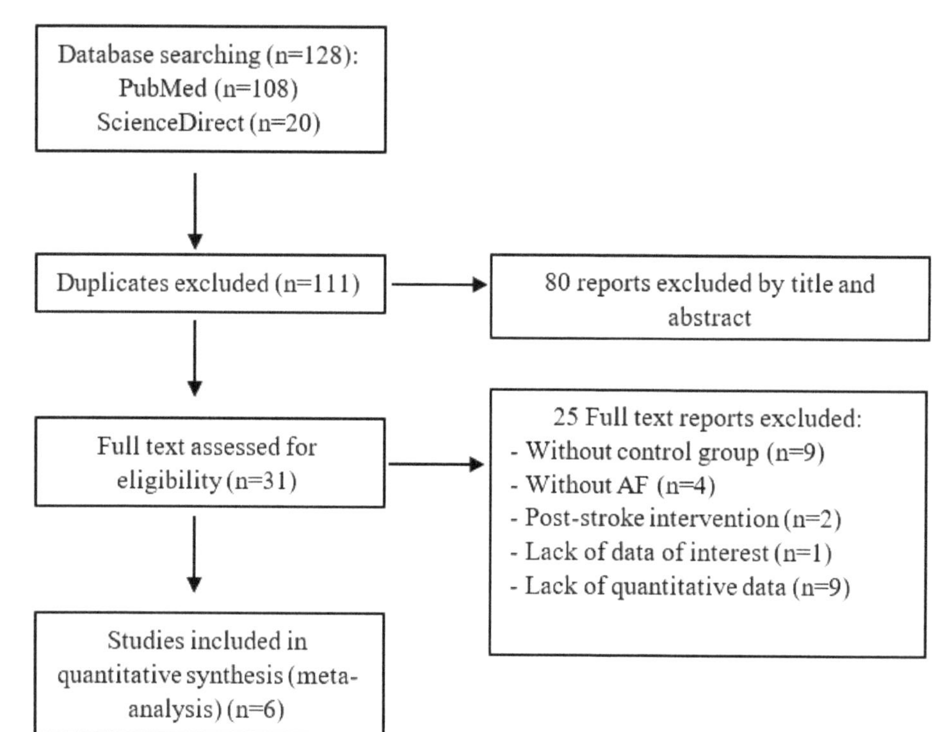

Figure 1. PRISMA-P flowchart of the study inclusion process. PRISMA indicates Preferred Reporting Items for Systematic Reviews and Meta-Analyses.

3.2. Study Characteristics

Six eligible studies reporting on stroke severity, enrolling a total of 9493 participants, 3478 anticoagulated patients and 6015 controls, were included. Regarding anticoagulated patients, 1767 were on DOAC and 1711 on VKA. From the latter, only 919 were therapeutically anticoagulated, considering data provided by the authors. Baseline characteristics of the participants involved are summarized in Table 1.

Table 1. Description of the studies included in the meta-analysis.

Study	Year	Type of Study	Inclusion Criteria	Exposure				Outcomes	Risk of Bias
				Without OAC	DOAC	Therapeutical VKA	Non-Therapeutical VKA		
Meinel et al. [13]	2020	Observational retrospective	AIS patients with AF (AF diagnosed either before or after stroke onset) aged ≥18 years	n = 4964	n = 1603	INR > 1.7 n = 854	INR < 1.7 n = 604	NIHSS at admission	Low
Sakamoto et al. [27]	2017	Observational retrospective	Patients with AIS or TIA with known AF	n = 352	n = 43	n = 35	n = 54	NIHSS at admission	Moderate
Yamashiro et al. [28]	2018	Observational retrospective	Patients with AF who developed AIS or TIA	n = 248	n = 22	INR ≥ 2 n = 16	INR < 2 n = 95	NIHSS at admission	Low
Hoyer et al. [12]	2018	Observational retrospective	Patients with newly detected AF or known AF admitted for AIS	n = 277	n = 99		-	NIHSS at admission	Low
Hannon et al. [29]	2011	Observational prospective	Patients with new stroke events and AF (known or new)	n = 62	-	INR 2–3 n = 14	INR < 2 or > 3 n = 13	NIHSS < 72 h	Low
Matsumoto et al. [30]	2017	Observational retrospective	AF patients who suffered AIS	n = 112	-	-	n = 26	NIHSS at admission	Low

VKA: Vitamin K Antagonists; AIS: Acute Ischemic Stroke; AF: Atrial Fibrillation; DOAC: Direct Oral Anticoagulants; NIHSS: National Institute of Health Stroke Scale; OAC: Oral Anticoagulants; TIA: Transient Ischemic Stroke; IS: Ischemic Stroke.

Table 2. Population and vascular risk factors characteristics in studies included in the meta-analysis.

	DOAC					VKA Therapeutical				VKA Non-Therapeutical					Control Group				
	Meinel et al. [13]	Sakamoto et al. [27]	Yamashiro et al. [28]	Hoyer et al. [12]	Meinel et al. [13]	Sakamoto et al. [27]	Yamashiro et al. [28]	Hannon et al. [29]	Meinel et al. [13]	Sakamoto et al. [27]	Yamashiro et al. [28]	Matsumoto et al. [30]	Hannon et al. [29]	Meinel et al. [13]	Sakamoto et al. [27]	Yamashiro et al. [28]	Matsumoto et al. [30]	Hoyer et al. [12]	Hannon et al. [29]
Mean age, y	79.8	-	-	79.2	80.7	-	-	-	82.1	-	-	76	-	79.4	-	-	76	79.4	-
Median age, y	-	78 (71–82)	71 (63–81)	-	-	80 (74–88)	80 (73–84)	77 (70–80)	-	79 (75–86)	80 (72–85)	-	76 (69–80)	-	78 (70–85)	74 (67–82)	-	-	77 (66–83)
Sex, (% female)	46%	49%	40.6%	52.5%	44%	54%	31.3%	64.3%	54%	41%	39%	45%	46%	49%	42%	36.7%	49%	54.5%	50%
Hypertension (%)	87%	74%	77.3%	86.9%	87%	69%	87.5%	71.4%	90%	67%	67.4%	86%	38.5%	82%	61%	64.9%	73%	85.6%	53%
Diabetes mellitus (%)	25%	26%	45.5%	28.3%	28%	17%	37.5%	21.4%	27%	20%	35.8%	24%	23.1%	21%	16%	20.2%	27%	31.8%	3.3%
Hyperlipidemia (%)	66%	33%	45.5%	30.3%	67%	54%	43.8%	71.4%	64%	29%	39%	41%	46.2%	60%	29%	36.7%	35%	27.1%	29.5%
Smokers (%)	12%	5%	10.5%	1%	10%	6%	28.6%	42.9%	8%	15%	25.6%	22%	61.5%	13%	17%	24.3%	30%	51%	57.9%
CHADS2 score	-	3 (2–4)	2 (1–3.3)	-	-	3 (2–4)	3 (2.3–4)		-	3 (2–4)	2 (2–3)	3.1 ± 1.5	-	-	2 (1–3)	2 (1–2)	2.8 ± 1.4	-	-

CG: control group; DOAC: Direct Oral Anticoagulants; tVKA: therapeutical Vitamin K Antagonists; ntVKA: non-therapeutical Vitamin K Antagonists.

3.3. Quality Assessment

The quality assessment of each study is described in Table A1 and Appendix A. We considered that five studies had a low risk of bias and one had a moderate risk. None of the included articles had a high risk of bias. According to the Cochrane Handbook for Systematic Reviews of Interventions, the funnel plot asymmetry can only be used when there is a minimum of 10 studies included in the meta-analysis, otherwise the test has a low power to distinguish chance from real asymmetry [23]. As we have only included 6 studies in our meta-analysis, publication bias was not assessed.

3.4. General Sociodemographic and Risk Factors

As seen in Table 2, the median age (range 71–80) as well the distribution of gender, vascular risk factors and the thrombotic risk (CHADS2 score 2–3) was in general comparable between the studies and between treatment arms (DOAC versus therapeutic VKA versus non-therapeutic VKA versus controls).

3.5. Stroke Severity

Six studies were included in the quantitative analysis (Figure 2). Of these, five studies reported results on VKA therapy [13,27–30] and four studies reported results on DOAC therapy [12,13,27,28]. In five studies, there was a difference of AIS severity between therapeutical VKA and non-therapeutical VKA [13,27,28,30]. None of the studies presented a comparison of admission severity between patients taking DOAC and VKA.

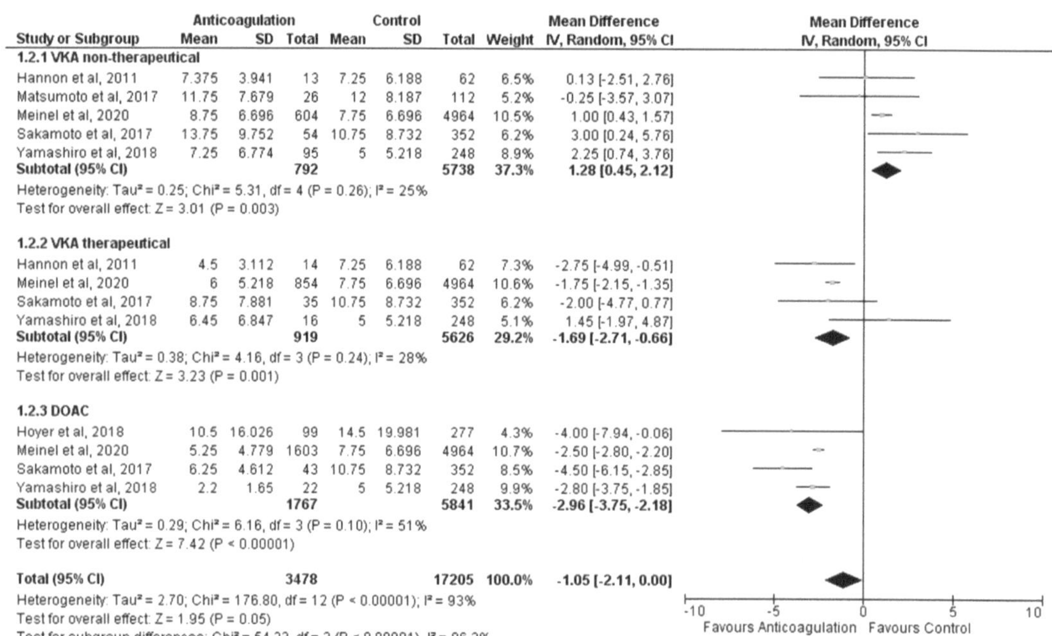

Study or Subgroup	Anticoagulation Mean	SD	Total	Control Mean	SD	Total	Weight	Mean Difference IV, Random, 95% CI
1.2.1 VKA non-therapeutical								
Hannon et al, 2011	7.375	3.941	13	7.25	6.188	62	6.5%	0.13 [-2.51, 2.76]
Matsumoto et al, 2017	11.75	7.679	26	12	8.187	112	5.2%	-0.25 [-3.57, 3.07]
Meinel et al, 2020	8.75	6.696	604	7.75	6.696	4964	10.5%	1.00 [0.43, 1.57]
Sakamoto et al, 2017	13.75	9.752	54	10.75	8.732	352	6.2%	3.00 [0.24, 5.76]
Yamashiro et al, 2018	7.25	6.774	95	5	5.218	248	8.9%	2.25 [0.74, 3.76]
Subtotal (95% CI)			792			5738	37.3%	**1.28 [0.45, 2.12]**
Heterogeneity: Tau² = 0.25; Chi² = 5.31, df = 4 (P = 0.26); I² = 25%								
Test for overall effect: Z = 3.01 (P = 0.003)								
1.2.2 VKA therapeutical								
Hannon et al, 2011	4.5	3.112	14	7.25	6.188	62	7.3%	-2.75 [-4.99, -0.51]
Meinel et al, 2020	6	5.218	854	7.75	6.696	4964	10.6%	-1.75 [-2.15, -1.35]
Sakamoto et al, 2017	8.75	7.881	35	10.75	8.732	352	6.2%	-2.00 [-4.77, 0.77]
Yamashiro et al, 2018	6.45	6.847	16	5	5.218	248	5.1%	1.45 [-1.97, 4.87]
Subtotal (95% CI)			919			5626	29.2%	**-1.69 [-2.71, -0.66]**
Heterogeneity: Tau² = 0.38; Chi² = 4.16, df = 3 (P = 0.24); I² = 28%								
Test for overall effect: Z = 3.23 (P = 0.001)								
1.2.3 DOAC								
Hoyer et al, 2018	10.5	16.026	99	14.5	19.981	277	4.3%	-4.00 [-7.94, -0.06]
Meinel et al, 2020	5.25	4.779	1603	7.75	6.696	4964	10.7%	-2.50 [-2.80, -2.20]
Sakamoto et al, 2017	6.25	4.612	43	10.75	8.732	352	8.5%	-4.50 [-6.15, -2.85]
Yamashiro et al, 2018	2.2	1.65	22	5	5.218	248	9.9%	-2.80 [-3.75, -1.85]
Subtotal (95% CI)			1767			5841	33.5%	**-2.96 [-3.75, -2.18]**
Heterogeneity: Tau² = 0.29; Chi² = 6.16, df = 3 (P = 0.10); I² = 51%								
Test for overall effect: Z = 7.42 (P < 0.00001)								
Total (95% CI)			3478			17205	100.0%	**-1.05 [-2.11, 0.00]**
Heterogeneity: Tau² = 2.70; Chi² = 176.80, df = 12 (P < 0.00001); I² = 93%								
Test for overall effect: Z = 1.95 (P = 0.05)								
Test for subgroup differences: Chi² = 54.23, df = 2 (P < 0.00001), I² = 96.3%								

Figure 2. Stroke severity. Forest plot showing the effects of anticoagulation versus control (no anticoagulation) on stroke severity, measured by NIHSS.

3.6. VKA Non-Therapeutical

Five studies reported on stroke severity in patients with non-therapeutical VKA. The studies by Meinel et al. [13] (MD 1.00; 95% CI [0.43, 1.57]), Sakamoto et al. [27] (MD 3.00; 95% CI [0.24, 5.76]) and Yamashiro et al. [28] (MD 2.25; 95% CI [0.74, 3.76]) showed that the severity was lower in the group of patients without any OAC compared to patients with non-therapeutical VKA. Hannon et al. [29] and Matsumoto et al. [30] did not find any significant statistical differences between the two groups, with MD 0.13; 95% CI [−2.51, 2.76] and MD −0.25; 95% CI [−3.57, 3.07], respectively.

Overall, compared with control groups (87.87%, n = 5738), patients with non-therapeutical VKA on admission (12.13%, n = 792) had significantly higher NIHSS scores (MD 1.28; 95% CI [0.45, 2.12], p = 0.003). I^2 was 25%, showing a low heterogeneity in this dataset.

3.7. VKA Therapeutical

Therapeutical INR levels were provided in four studies. Hannon et al. [29] and Meinel et al. [13] revealed differences between patients with therapeutical VKA and those without OAC, MD −2.75; 95% CI [−4.99, −0.51] and MD −1.75; 95% CI [−2.15, −1.35], respectively. The other two studies, by Sakamoto et al. [27] and Yamashiro et al. [28], did not show a significant difference between the two groups, MD −2.00; 95% CI [−4.77, 0.77] and MD 1.45; 95% CI [−1.97, 4.87], respectively.

Overall, in AF-AIS, patients with therapeutical VKA (14.04%, n = 919) had significantly lower NIHSS scores (MD −1.69; 95% CI [−2.71; −0.66], p = 0.001) compared to control groups (89.96%, n = 5626), suggesting a protective effect associated with therapeutical VKA. I^2 was 28%, showing a low heterogeneity in this dataset.

3.8. DOAC

Four studies reported results about anticoagulation therapy with DOAC. In all studies, DOAC therapy (23.23%, n = 1767) was associated with lower NIHSS scores (MD −2.96; 95% CI [−3.75, −2.18], p < 0.00001) compared to control groups (76.77%, n = 5841) [12,13,27,28]. I^2 was 51%, showing a substantial heterogeneity in this dataset.

The meta-analysis taking into account all subgroups of OAC (VKA non-therapeutical, VKA therapeutical and DOAC) did not show any significant statistical difference between the anticoagulated and control groups (MD −1.05; 95% CI [−2.11, 0.00], p = 0.05). However, I^2 was 93%, showing a substantial heterogeneity in this dataset.

4. Discussion

In this study we assessed associations between pre-admission use of OAC and severity of AF-AIS. The main finding of our analysis was that, despite differences in the mechanism of action, prior use of OAC is associated with lower severity of AF-AIS on admission. Clot formation inhibition depends on the international normalized ratio (INR) for VKA and on plasma concentration for DOAC [31]. It is known that patients on VKA [32] or DOAC [33] therapy present with smaller infarct volumes on diffusion-weighted magnetic resonance imaging. There is some evidence showing an increased susceptibility to fibrinolysis in the presence of VKA or DOAC [31,32]. Surprisingly, patients with non-therapeutic INR on admission present with higher AIS severity. One might speculate that suboptimal INR control may reflect a higher burden of medical comorbidities, poor medication compliance (other than anticoagulation), worse health literacy or even later hospital presentation. All these factors may adversely influence stroke outcome [34]. Another factor that might help to explain these results is VKA's mechanism of action: these drugs inhibit the production of several factors dependent on vitamin K, including protein C and protein S [35]. In subtherapeutic INR, there is an insufficient suppression of coagulation such that it cannot overcome the reduction in protein C anticoagulant activity [27]. As a result, when compared with no treatment, the use of subtherapeutic VKA may result in larger thrombi formation [27]. Although no direct comparison between DOAC use and VKA use was presented, the magnitude of the protection was higher for patients taking DOAC in comparison to patients taking VKA (MD −2.96; 95% CI [−3.75, −2.18], p < 0.00001 versus MD −1.69; 95% CI [−2.71; −0.66], p = 0.001). Real life meta-analysis to compare VKA to DOAC impact on AF-AIS outcomes would be interesting as cost-effectiveness is an important factor for decision.

There are some limitations that are worthwhile to discuss. Population heterogeneity is a reality as we analyzed observational studies using different patient selection. For instance, the samples were not controlled for pre-hospital delay which can determine the extent of severity at the time of admission to the hospital. There is a discrepancy between the number of controls and anticoagulated patients, and the number of therapeutically anticoagulated patients is considerably small. Standardized blood plasma or serum concentration therapeutic ranges for DOAC have not been established and quantitative test results have not been correlated with clinical outcomes [36]. Assumption of anticoagulation status for the DOAC group was based on interview or medical prescription as blood drug concentration has not been measured. Nevertheless, adequate adherence to DOAC therapy has been shown to have a beneficial effect on stroke severity on admission [28] and concurrent pro-thrombotic risk factors appear to be relevant when AIS occurs despite adequate OAC [37].

Only six studies were suitable for meta-analysis; publication bias was not assessed (funnel plot asymmetry has low power if $n < 10$) [23]. This means that we cannot exclude that such bias might have affected our results.

5. Conclusions

In conclusion, our systematic review demonstrated that pre-stroke anticoagulation is associated with less severe presentation of AF-AIS with apparent higher protection in patients under DOAC.

Author Contributions: Conceptualization, C.G., M.S., M.A., M.H., M.M. and P.V.; methodology, M.A. and A.M.; validation, M.S. and M.H.; formal analysis, P.V. and A.M.; investigation, C.G., M.S., M.A., M.H., M.M. and P.V.; writing—original draft preparation, C.G., M.S., M.A., M.H., M.M. and P.V.; writing—review and editing, H.N. and A.M.; supervision, A.M. and H.N. All authors have read and agreed to the published version of the manuscript.

Funding: This research was funded by the Algarve Biomedical Center with funds from the Municipio de Loulé, in Algarve, Faro, Portugal.

Institutional Review Board Statement: Not applicable.

Informed Consent Statement: Not applicable.

Data Availability Statement: Not applicable.

Conflicts of Interest: The authors declare no conflict of interest.

Appendix A

Table A1. Risk of bias Assessment.

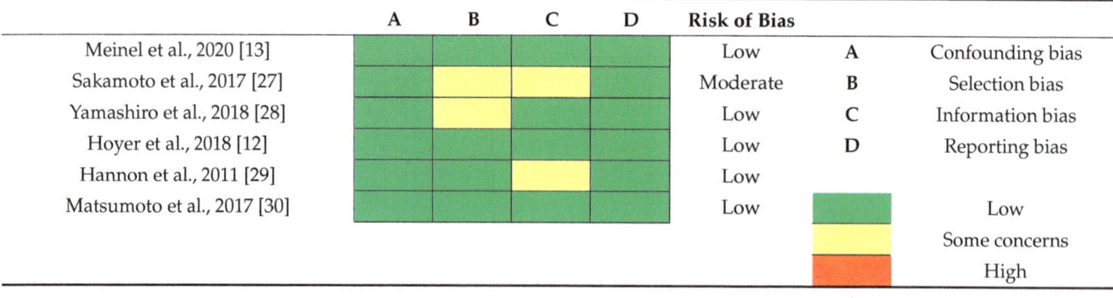

	A	B	C	D	Risk of Bias		
Meinel et al., 2020 [13]					Low	A	Confounding bias
Sakamoto et al., 2017 [27]					Moderate	B	Selection bias
Yamashiro et al., 2018 [28]					Low	C	Information bias
Hoyer et al., 2018 [12]					Low	D	Reporting bias
Hannon et al., 2011 [29]					Low		
Matsumoto et al., 2017 [30]					Low		Low
							Some concerns
							High

References

1. Yi, X.; Lin, J.; Han, Z.; Luo, H.; Shao, M.; Fan, D.; Qiang, Z. Preceding Antithrombotic Treatment is Associated With Acute Ischemic Stroke Severity and Functional Outcome at 90 Days Among Patients With Atrial Fibrillation. *J. Stroke Cerebrovasc. Dis.* **2019**, *28*, 2003–2010. [CrossRef] [PubMed]
2. Jung, Y.H.; Choi, H.Y.; Lee, K.Y.; Cheon, K.; Han, S.W.; Park, J.H.; Cho, H.J.; Park, H.J.; Nam, H.S.; Heo, J.H.; et al. Stroke Severity in Patients on Non-Vitamin K Antagonist Oral Anticoagulants with a Standard or Insufficient Dose. *Thromb. Haemost.* **2018**, *118*, 2145–2151. [CrossRef]
3. Hellwig, S.; Grittner, U.; Audebert, H.; Endres, M.; Haeusler, K.G. Non-Vitamin K-dependent oral anticoagulants have a positive impact on ischaemic stroke severity in patients with atrial fibrillation. *EP Europace* **2018**, *20*, 569–574. [CrossRef] [PubMed]
4. January, C.T.; Wann, L.S.; Calkins, H.; Chen, L.Y.; Cigarroa, J.E.; Cleveland, J.C.; Ellinor, P.T.; Ezekowitz, M.D.; Field, M.E.; Furie, K.L.; et al. 2019 AHA/ACC/HRS Focused Update of the 2014 AHA/ACC/HRS Guideline for the Management of Patients With Atrial Fibrillation: A Report of the American College of Cardiology/American Heart Association Task Force on Clinical Practice Guidelines and the Heart Rhythm Society. *Circulation* **2019**, *140*, e125–e151. [CrossRef] [PubMed]
5. Boursier-Bossy, V.; Zuber, M.; Emmerich, J. Ischemic stroke and non-valvular atrial fibrillation: When to introduce anticoagulant therapy? *J. Med. Vasc.* **2020**, *45*, 72–80. [CrossRef] [PubMed]
6. Wolf, P.A.; Abbott, R.D.; Kannel, W.B. Atrial fibrillation as an independent risk factor for stroke: The framingham study. *Stroke* **1991**, *22*, 983–988. [CrossRef]
7. Lip, G.Y.; Lane, D.A. Stroke prevention in atrial fibrillation: A systematic review. *JAMA J. Am. Med. Assoc.* **2015**, *313*, 1950–1962. [CrossRef]
8. Wodchis, W.P.; Bhatia, R.S.; Leblanc, K.; Meshkat, N.; Morra, D. A review of the cost of atrial fibrillation. *Value Health* **2012**, *15*, 240–248. [CrossRef] [PubMed]
9. Tziomalos, K.; Giampatzis, V.; Bouziana, S.D.; Spanou, M.; Kostaki, S.; Papadopoulou, M.; Dourliou, V.; Sofogianni, A.; Savopoulos, C.; Hatzitolios, A.I. Adequacy of preadmission oral anticoagulation with vitamin K antagonists and ischemic stroke severity and outcome in patients with atrial fibrillation. *J. Thromb. Thrombolysis* **2015**, *41*, 336–342. [CrossRef] [PubMed]
10. Tavares, S.F.; Ferreira, I.; Chaves, V.; Flores, L.; Correia, C.; Almeida, J.; Fonseca, L.; Chaves, P.C. Acute Ischemic Stroke Outcome and Preceding Anticoagulation: Direct Oral Anticoagulants Versus Vitamin K Antagonists. *J. Stroke Cerebrovasc. Dis.* **2020**, *29*, 104691. [CrossRef] [PubMed]
11. Fernandes, L.; Sargento-Freitas, J.; Milner, J.; Silva, A.; Novo, A.; Gonçalves, T.; Marinho, A.V.; Pego, G.M.; Cunha, L.; António, N. Ischemic stroke in patients previously anticoagulated for non-valvular atrial fibrillation: Why does it happen? *Rev. Port. Cardiol.* **2019**, *38*, 117–124. [CrossRef] [PubMed]
12. Hoyer, C.; Filipov, A.; Neumaier-Probst, E.; Szabo, K.; Ebert, A.; Alonso, A. Impact of pre-admission treatment with non-vitamin K oral anticoagulants on stroke severity in patients with acute ischemic stroke. *J. Thromb. Thrombolysis* **2018**, *45*, 529–535. [CrossRef] [PubMed]
13. Meinel, T.R.; Branca, M.; Marchis, G.M.; Nedeltchev, K.; Kahles, T.; Bonati, L.; Arnold, M.; Heldner, M.R.; Jung, S.; Carrera, E.; et al. Prior Anticoagulation in Patients with Ischemic Stroke and Atrial Fibrillation. *Ann. Neurol.* **2021**, *89*, 42–53. [CrossRef] [PubMed]
14. Seiffge, D.J.; Marchis, G.D.; Koga, M.; Paciaroni, M.; Wilson, D.; Capellari, M.; Md, K.M.; Tsivgoulis, G.; Ambler, G.; Arihiro, S.; et al. Ischemic stroke despite oral anticoagulant therapy in patients with atrial fibrillation. *Ann. Neurol.* **2020**, *87*, 677–687. [CrossRef]
15. Xian, Y.; O'Brien, E.C.; Liang, L.; Xu, H.; Schwamm, L.; Fonarow, G.; Bhatt, D.L.; Smith, E.E.; Olson, D.M.; Maisch, L.; et al. Association of preceding antithrombotic treatment with acute ischemic stroke severity and in-hospital outcomes among patients with atrial fibrillation. *JAMA J. Am. Med. Assoc.* **2017**, *317*, 1057–1067. [CrossRef] [PubMed]
16. Jame, S.; Barnes, G. Stroke and thromboembolism prevention in atrial fibrillation. *Heart* **2020**, *106*, 10–17. [CrossRef]
17. Aguilar, M.I.; Hart, R.; Pearce, L.A. Oral anticoagulants versus antiplatelet therapy for preventing stroke in patients with non-valvular atrial fibrillation and no history of stroke or transient ischemic attacks (Review). *Cochrane Database Syst. Rev.* **2007**, *3*, CD006186. [CrossRef]
18. Hindricks, G.; Potpara, T.; Dagres, N.; Arbelo, E.; Bax, J.; Blomström-Lundqvist, C.; Boriani, G.; Castella, M.; Dan, G.A.; Dilaveris, P.E.; et al. 2020 ESC Guidelines for the diagnosis and management of atrial fibrillation developed in collaboration with the European Association for Cardio-Thoracic Surgery (EACTS). *Eur. Heart J.* **2021**, *42*, 373–498. [CrossRef]
19. Adachi, T.; Hoshino, H.; Takagi, M.; Fujioka, S. Volume and Characteristics of Intracerebral Hemorrhage with Direct Oral Anticoagulants in Comparison with Warfarin. *Cerebrovasc. Dis. Extra.* **2017**, *7*, 62–71. [CrossRef]
20. Bakhai, A.; Petri, H.; Vahidnia, F.; Wolf, C.; Ding, Y.; Foskett, N.; Sculpher, M. Real-world data on the incidence, mortality, and cost of ischaemic stroke and major bleeding events among non-valvular atrial fibrillation patients in England. *J. Eval. Clin. Pract.* **2021**, *27*, 119–133. [CrossRef] [PubMed]
21. Shamseer, L.; Moher, D.; Clarke, M.; Ghersi, D.; Liberati, A.; Petticrew, M.; Shekelle, P.; Stewart, L.A. Preferred reporting items for systematic review and meta-analysis protocols (prisma-p) 2015: Elaboration and explanation. *Br. Med. J.* **2015**, *349*, g7647. [CrossRef] [PubMed]
22. Donato, H.; Donato, M. Stages for undertaking a systematic review. *Acta Médica Portuguesa.* **2019**, *32*, 227–235. [CrossRef] [PubMed]

23. Higgins, J.P.T.; Thomas, J.; Chandler, J.; Cumpston, M.; Li, T.; Page, M.J.; Welch, V. Cochrane Handbook for Systematic Reviews of Interventions Version 6.2. Available online: www.training.cochrane.org/handbook (accessed on 6 May 2021).
24. Hozo, S.P.; Djulbegovic, B.; Hozo, I. Estimating the mean and variance from the median, range, and the size of a sample. *BMC Med. Res. Methodol.* **2005**, *5*, 13. [CrossRef]
25. Higgins, J.P.T.; Thompson, S.G. Quantifying heterogeneity in a meta-analysis. *Stat. Med.* **2002**, *21*, 1539–1558. [CrossRef] [PubMed]
26. Sousa-Pinto, B.; Azevedo, L. Avaliação Crítica de uma Revisão Sistemática e Meta-Análise: Da Execução da Meta-análise à Exploração da Heterogeneidade. *Rev. Soc. Port. Anestesiol.* **2019**, *28*, 187–191. [CrossRef]
27. Sakamoto, Y.; Okubo, S.; Nito, C.; Suda, S.; Matsumoto, N.; Abe, A.; Aoki, J.; Shimoyama, T.; Takayama, Y.; Suzuki, K.; et al. The relationship between stroke severity and prior direct oral anticoagulant therapy in patients with acute ischaemic stroke and non-valvular atrial fibrillation. *Eur. J. Neurol.* **2017**, *24*, 1399–1406. [CrossRef] [PubMed]
28. Yamashiro, K.; Kurita, N.; Tanaka, R.; Ueno, Y.; Miyamoto, N.; Hira, K.; Nakajima, S.; Urabe, T.; Hattori, N. Adequate Adherence to Direct Oral Anticoagulant is Associated with Reduced Ischemic Stroke Severity in Patients with Atrial Fibrillation. *J. Stroke Cerebrovasc. Dis.* **2018**, *28*, 1773–1780. [CrossRef] [PubMed]
29. Hannon, N.; Callaly, E.; Moore, A.; Chróinín, D.N.; Sheehan, O.; Marnane, M.; Merwick, A.; Kyne, L.; Duggan, J.; McCormack, P.M.E.; et al. Improved late survival and disability after stroke with therapeutic anticoagulation for atrial fibrillation: A population study. *Stroke* **2011**, *42*, 2503–2508. [CrossRef]
30. Matsumoto, M.; Sakaguchi, M.; Okazaki, S.; Hashikawa, K.; Takahashi, T.; Matsumoto, M.; Ohtsuki, T.; Shimazu, T.; Yoshimine, T.; Mochizuki, H.; et al. Relationship between infarct volume and prothrombin time-international normalized ratio in ischemic stroke patients with nonvalvular atrial fibrillation. *Circ. J.* **2017**, *81*, 391–396. [CrossRef] [PubMed]
31. Königsbrügge, O.; Weigel, G.; Quehenberger, P.; Pabinger, I.; Ay, C. Plasma clot formation and clot lysis to compare effects of different anticoagulation treatments on hemostasis in patients with atrial fibrillation. *Clin. Exp. Med.* **2018**, *18*, 325–336. [CrossRef]
32. Ay, H.; Arsava, E.M.; Gungor, L.; Greer, D.; Singhal, A.B.; Furie, K.L.; Koroshetz, W.J.; Sorensen, A.G. Admission international normalized ratio and acute infarct volume in ischemic stroke. *Ann. Neurol.* **2008**, *64*, 499–506. [CrossRef] [PubMed]
33. Yavasoglu, N.G.; Eren, Y.; Tatar, I.G.; Yalcinkaya, I. Infarct volumes of patients with acute ischemic stroke receiving direct oral anticoagulants due to non-valvular atrial fibrillation. *Ann. Indian Acad. Neurol.* **2021**, *24*, 27–31. [CrossRef] [PubMed]
34. Hannon, N.; Arsava, E.M.; Audebert, H.J.; Ay, H.; Crowe, M.; Chróinín, D.N.; Furie, K.; McGorrian, C.; Molshatzki, N.; Murphy, S.; et al. Antithrombotic treatment at onset of stroke with atrial fibrillation, functional outcome, and fatality: A systematic review and meta-analysis. *Int. J. Stroke* **2015**, *10*, 808–814. [CrossRef] [PubMed]
35. Takano, K.; Iino, K.; Ibayashi, S.; Tagawa, K.; Sadoshima, S.; Fujishima, M. Hypercoagulable state under low-intensity warfarin anticoagulation assessed with hemostatic markers in cardiac disorders. *Am. J. Cardiol.* **1994**, *74*, 935–939. [CrossRef]
36. Chen, A.; Stecker, E.A.; Warden, B. Direct Oral Anticoagulant Use: A Practical Guide to Common Clinical Challenges. *J. Am. Heart Assoc.* **2020**, *9*, e017559. [CrossRef] [PubMed]
37. Wańkowicz, P.; Nowacki, P.; Gołąb-Janowska, M. Risk factors for ischemic stroke in patients with non-valvular atrial fibrillation and therapeutic international normalized ratio range. *Arch Med. Sci.* **2019**, *15*, 1217–1222. [CrossRef] [PubMed]

Journal of
Clinical Medicine

Article

Iron Deficiency and Reduced Muscle Strength in Patients with Acute and Chronic Ischemic Stroke

Nadja Scherbakov [1,2,3,4,*], Anja Sandek [5,6], Miroslava Valentova [5,6], Antje Mayer [1], Stephan von Haehling [5,6], Ewa Jankowska [7,8], Stefan D. Anker [1,4,9] and Wolfram Doehner [1,2,3,4]

[1]   Berlin Institute of Health, Center for Regenerative Therapies (BCRT), Charité—Universitätsmedizin Berlin, 10117 Berlin, Germany; antje.mayer@charite.de (A.M.); stefan.anker@charite.de (S.D.A.); wolfram.doehner@charite.de (W.D.)

[2]   Center for Stroke Research Berlin (CSB), Charité—Universitätsmedizin Berlin, 10117 Berlin, Germany

[3]   Department of Internal Medicine and Cardiology, Campus Virchow-Klinikum, Charité—Universitätsmedizin Berlin, 10117 Berlin, Germany

[4]   German Centre for Cardiovascular Research (DZHK), Partner Site Berlin, 10785 Berlin, Germany

[5]   Department of Cardiology and Pneumology, University of Göttingen, 37073 Göttingen, Germany; anja.sandek@med.uni-goettingen.de (A.S.); Miroslava.Valentova@med.uni-goettingen.de (M.V.); stephan.von.haehling@med.uni-goettingen.de (S.v.H.)

[6]   German Centre for Cardiovascular Research (DZHK), Partner Site Göttingen, 37075 Göttingen, Germany

[7]   Institute of Heart Disease, Wroclaw Medical University, 50-367 Wroclaw, Poland; ewa.jankowska@umed.wroc.pl

[8]   Institute of Heart Disease, University Hospital, 50-367 Wroclaw, Poland

[9]   Division of Cardiology and Metabolism-Heart Failure, Cachexia & Sarcopenia, Department of Cardiology (CVK), Charité-Universitätsmedizin Berlin, 10117 Berlin, Germany

*   Correspondence: nadja.scherbakov@charite.de; Tel.: +49-(30)-450560363; Fax: +49-(30)-450553951

Citation: Scherbakov, N.; Sandek, A.; Valentova, M.; Mayer, A.; von Haehling, S.; Jankowska, E.; Anker, S.D.; Doehner, W. Iron Deficiency and Reduced Muscle Strength in Patients with Acute and Chronic Ischemic Stroke. *J. Clin. Med.* **2022**, *11*, 595. https://doi.org/10.3390/jcm11030595

Academic Editors: Anna Bersano and Timo Siepmann

Received: 2 December 2021
Accepted: 20 January 2022
Published: 25 January 2022

Abstract: (1) Introduction: Iron deficiency (ID) contributes to impaired functional performance and reduced quality of life in patients with chronic illnesses. The role of ID in stroke is unclear. The aim of this prospective study was to evaluate the prevalence of ID and to evaluate its association with long-term functional outcome in patients with ischemic stroke. (2) Patients and Methods: 140 patients (age 69 ± 13 years, BMI 27.7 ± 4.6 kg/m^2, mean $\pm$ SD) admitted to a university hospital stroke Unit, with acute ischemic stroke of the middle cerebral artery were consecutively recruited to this observational study. Study examinations were completed after admission (3 ± 2 days after acute stroke) and at one-year follow up ($N = 64$, 382 ± 27 days after stroke). Neurological status was evaluated according to the National Institute of Health Stroke Scale (NIHSS) and the modified Rankin scale (mRS). Muscle isometric strength of the non-affected limb was assessed by the maximum handgrip test and knee extension leg test. ID was diagnosed with serum ferritin levels ≤ 100 µg/L (ID Type I) or 100–300 µg/L if transferrin saturation (TSAT) < 20% (ID Type II). (3) Results: The prevalence of ID in acute stroke patients was 48% ($N = 67$), with about two-thirds of patients ($N = 45$) displaying ID Type I and one-third ($N = 22$) Type II. Handgrip strength (HGS) and quadriceps muscle strength were reduced in patients with ID compared to patients without ID at baseline (HGS: 26.5 ± 10.4 vs. 33.8 ± 13.2 kg, $p < 0.001$ and quadriceps: 332 ± 130 vs. 391 ± 143 N, $p = 0.06$). One year after stroke, prevalence of ID increased to 77% ($p = 0.001$). While an improvement of HGS was observed in patients with normal iron status, patients with ID had no improvement in HGS difference (4.6 ± 8.3 vs. -0.7 ± 6.5 kg, $p < 0.05$). Patients with ID remained with lower HGS compared to patients with normal iron status (28.2 ± 12.5 vs. 44.0 ± 8.6 kg, $p < 0.0001$). (4) Conclusions: Prevalence of ID was high in patients after acute stroke and further increased one year after stroke. ID was associated with lower muscle strength in acute stroke patients. In patients with ID, skeletal muscle strength did not improve one year after stroke.

Keywords: iron deficiency; prevalence; acute ischemic stroke; chronic stroke; muscle strength; functional outcome

1. Introduction

Stroke is one of the leading causes of disability in adult life with a global annual incidence rate over 12 million cases [1]. Clinical outcome after stroke depends among other factors on the presence of comorbidities, such as hypertension, diabetes mellitus, heart failure (HF), or chronic kidney disease (CKD) [2–4]. Growing evidence suggests a considerable impact of iron deficiency (ID) on clinical course, prognosis, and quality of life in geriatric patients, as well as in patients with chronic diseases, such as chronic HF, cancer, and CKD [5–8].

A well-balanced iron metabolism plays a central role in the maintenance of numerous biological processes, including erythropoiesis, oxidative metabolism, immune response, and neurotransmission [9]. Physiologically, the body absorbs 7–10% of dietary iron per day, which suggests that malnutrition and/or malabsorption may have a major impact on iron balance [10]. In chronic inflammatory conditions, sufficient absorption of iron by the gastrointestinal tract and cellular iron export from the body iron stores is inhibited, leading to development of functional ID [11,12]. Inadequate nutritional iron uptake or absorption, as well as excessive blood loss, are the main causes leading to the development of absolute ID [13]. Biochemically, ID is manifested when the extracellular iron of the bone marrow, ferritin plasma levels, and transferrin saturation are low [13].

Up to date, a presence of ID, ID type, and its impact on clinical outcome remained not sufficiently studied in stroke. We hypothesized that ID is associated with low functional performance in patients with stroke. In this observational study, we aimed to evaluate the prevalence of ID in patients with acute stroke and at one year after stroke. Additionally, functional outcome in relation to ID after acute stroke and at one year after the event was investigated.

2. Methods

2.1. Study Design and Population

We prospectively studied 140 patients with acute ischemic stroke (AIS) in the territory of the middle cerebral artery (MCA), participating in the longitudinal prospective observational Body Size in Stroke Study [14] (BoSSS, German registry for clinical trials number DRKS00000514). The patients with mild to moderate neurological deficit (defined by the National Institute of Health Stroke Scale (NIHSS) as ≤12 points) were consecutively enrolled within 48 h after stroke onset, being admitted to the Stroke Unit at a tertiary university center (Charité University Hospital Berlin, Campus Virchow Clinic, Berlin, Germany).

Stroke was classified according to Trial of ORG 10172 in Acute Stroke Treatment (TOAST) classification, etiology of cardio embolism, large-artery atherosclerosis, small-vessel occlusion, and stroke of undetermined etiology [15]. One year after stroke, patients were invited to a follow up examination (Follow up cohort, FU cohort).

The study was approved by the Ethics Committee of Charité University Hospital Berlin (EA2/008/09), and all patients gave written informed consent.

2.2. Assessment of Functional Status

Study examinations were performed at baseline (3 ± 2 days after acute stroke) and at one-year follow up (FU, mean 382 ± 27 days after stroke). Functional status was assessed by the modified Rankin Scale (mRS), which measures physical independency by assessment of the body function, activity, and participation in daily tasks on the scale ranging from "0" (no symptoms) to "6" (death) [16].

Short physical performance battery (SPPB) was performed to assess functional capacity. SPPB includes examinations of standing ability with both feet together in the side-by-side, semi-tandem, and tandem positions; 4-m walk; and time taken to rise five times from the chair and return to the seated position. The score ranges from 0 (not attended) to 12 points (completed) [17].

Isometric muscle strength of the hand was assessed by the handgrip strength (HGS) test using a handgrip dynamometer (Saehan Corporation, Changwon, Korea). The highest of three handgrip measurements of the non-paretic or strongest arm was used for analysis. HGS below mean was considered as low muscle strength (low HGS).

Maximal isometric muscle strength of the quadriceps muscle (Newton, *N*) was measured as described previously [18]. Briefly, the freely hanging legs of the sitting patients were connected at the ankle with a pressure transducer (Multitrace 2, Lectromed, Jersey, Channel Islands), and maximal isometric strength was assessed from the best of three contractions on each leg, with a resting period of at least 60 s in between.

2.3. Body Composition and Nutritional Status

Appetite (the subjective desire to eat) was assessed according to the visual analogue scale (VAS) ranging from "0" (no appetite at all) to "10" (always a very good appetite) [19]. None of the patients included in our study had dysphagia on a clinically relevant level (preventing oral feeding) and none were fed enterally or parenterally. The nutritional status was assessed by Mini Nutritional Assessment (MNA) at 12 months as follows: normal nutritional status was considered if the patients achieved $\geq$24 points, and all other patients were considered to have low nutritional status [20].

The body mass index (BMI) was calculated as a ratio of body weight to height squared (kg/m^2).

2.4. Blood Sampling and Iron Deficiency

Venous blood samples we obtained in all patients after 12 h of overnight fasting. Standard biochemical parameters were assessed by routine laboratory measurement. Normal iron status was defined by a serum ferritin > 100 µg/mL and transferrin saturation (TSAT) $\geq$ 20% [21]. Iron deficiency type I (ID I) was considered when plasma ferritin levels were $\leq$100 µg/mL, and iron deficiency type II (ID II) was considered for plasma ferritin levels of 100–300 µg/L and TSAT < 20% [22]. Anemia was defined by hemoglobin plasma levels of fewer than 12 g/L in females and 13 g/L in males [13]. Systemic inflammation was present if the C-reactive protein (CrP) plasma level was over 6.1 mg/dL, as defined previously [23].

2.5. Statistical Analysis

All data are presented as means $\pm$ standard deviation (SD), median [interquartile range (IQR)], or percentage as appropriate. All variables were tested for normal distribution using the Kolmogorov-Smirnov test. Serum levels of CrP were non-normally distributed, and statistical comparisons between subgroups were made using analysis of variance (ANOVA), followed by Fisher's post hoc test, Mann–Whitney or Kruskal–Wallis test, or analysis of covariance (ANCOVA). The Chi square test was used to assess categorical distribution between the groups. Multivariable models for associations of risk factors with low HGS were applied (logistic regression analysis), including all factors showing a *p*-value $\leq$ 0.1 in univariable analysis. Furthermore, age and BMI were added. Odds ratios with 95% confidence intervals (OR [95% CI]) were reported. A value of $p < 0.05$ was considered statistically significant. A total of 76 patients (54%) without data at follow up were excluded from follow-up analyses. Statistical analyses were performed with the StatView 5.0 software package (SAS Institute Inc., Cary, NC, USA).

3. Results

3.1. Baseline

Baseline clinical characteristics of all patients with acute ischemic stroke ($N = 140$) are presented in Table 1. Study examinations were performed at mean 3 ± 2 days after stroke.

Table 1. Baseline characteristics of study population.

Clinical Parameters	All Patients $N = 140$	Normal Iron $N = 73$	ID $N = 67$	ID Type I $N = 45$	ID Type II $N = 22$	p Value ID vs. Normal Iron	p Value ID I vs. ID II vs. Normal Iron
Age, y, mean $\pm$ SD	69 ± 13	67 ± 12	70 ± 14	66 ± 14	77 ± 12 *	n.s.	<0.01
Body mass index, kg/m^2, mean $\pm$ SD	27.7 ± 4.6	28.3 ± 4.8	27.1 ± 4.2	26.0 ± 3.8	27.5 ± 4.3	n.s.	n.s.
Systolic RR, mmHg, mean $\pm$ SD	136 ± 28	137 ± 33	135 ± 21	146 ± 18	136 ± 22	n.s.	n.s.
Diastolic RR, mmHg, mean $\pm$ SD	79 ± 14	81 ± 13	77 ± 14	84 ± 10	75 ± 16	n.s.	n.s.
Female sex N, %	55 (39)	17 (23)	38 (57)	26 (58)	12 (55)	<0.001	<0.001
Self-reported appetite	6.5 ± 2.2	6.7 ± 2.1	6.3 ± 2.3	6.8 ± 2.1	5.4 ± 2.4 *	n.s.	0.05
Stroke severity National Institute of Health Stroke Scale (NIHSS)							
Mean score $\pm$ SD	4.7 ± 3.4	4.8 ± 3.6	4.6 ± 3.1	4.1 ± 2.7	5.6 ± 3.6	n.s	n.s.
0–4, N (%)	83 (59)	42 (57)	41 (62)	32 (71)	9 (41)	n.s.	n.s.
Trial of ORG 10172 in Acute Stroke Treatment							
Cardioembolic, N (%)	44 (31)	20 (27)	24 (36)	12 (27)	12 (55) *	n.s.	<0.05
Large-artery atherosclerosis, N (%)	49 (35)	26 (36)	23 (35)	16 (36)	7 (32)	n.s.	n.s.
Small-vessel occlusion, N (%)	25 (18)	14 (19)	11 (16)	9 (20)	2 (9)	n.s.	n.s.
Stroke of undetermined etiology	22 (16)	13 (18)	9 (13)	8 (17)	1 (4)	n.s.	n.s.
Physical status Modified Rankin Scale (mRS)							
Mean score $\pm$ SD	2.4 ± 1.5	2.4 ± 1.6	2.4 ± 1.6	2.1 ± 1.3	2.9 ± 1.6	n.s	n.s.
0–1, N (%)	58 (41)	32 (44)	26 (39)	19 (42)	7 (32)	n.s.	n.s.
Low Handgrip strength, N (%)	61 (44)	28 (38)	33 (49)	19 (42)	14 (64)	n.s.	n.s.
Comorbidities							
Diabetes mellitus, N (%)	40 (29)	19 (26)	21 (38)	13 (36)	7 (33)	n.s.	n.s.
Arterial hypertension, N (%)	96 (69)	49 (67)	47 (70)	31 (69)	16 (76)	n.s.	n.s.
Dyslipidemia, N (%)	45 (32)	22 (30)	23 (34)	17 (38)	6 (27)	n.s.	n.s.
Anemia, N (%)	25 (19)	7 (10)	18 (27)	9 (20)	9 (41)	<0.01	<0.01
Cardiovascular disease, N (%)	56 (40)	29 (40)	29 (43)	15 (33)	5 (23)	n.s.	n.s.
Biochemistry							
Hemoglobin, mg/dL, mean $\pm$ SD	14.0 ± 1.9	14.6 ± 1.7	13.3 ± 1.8	13.4 ± 1.8	13.1 ± 1.9	n.s.	0.0001
White blood cells count	8.3 ± 2.5	8.1 ± 1.9	8.7 ± 3.1	8.1 ± 2.3	10.1 ± 4.2 **	n.s.	<0.01
Creatinine, mg/dL, mean $\pm$ SD	1.0 ± 0.4	1.0 ± 0.4	1.0 ± 0.4	1.0 ± 0.4	0.9 ± 1.9	n.s.	n.s.
Cholesterol, mg/dL, mean $\pm$ SD	186 ± 43	189 ± 37	182 ± 49	190 ± 50	166 ± 43 *	n.s.	n.s.
High density lipoprotein, mg/dL, mean $\pm$ SD	49 ± 15	46 ± 12	51 ± 16	54 ± 17 *	46 ± 12	<0.05	<0.05
Low density lipoprotein, mg/dL, mean $\pm$ SD	110 ± 38	115 ± 34	104 ± 41	109 ± 39	93 ± 45	n.s.	n.s.
Hemoglobin A1c, %, median [IQR]	5.9 [5.4–6.5]	5.8 [5.4–6.5]	5.9 [5.5–6.5]	5.8 [5.5–6.7]	5.6 [5.6–6.6]	n.s	n.s.
C-reactive protein, mg/L, median [IQR]	4.8 [1.7–11.8]	4.1 [1.7–12]	6.6 [1.7–10.35]	4.1 [1.7–7]	16.7 [7.2–26.2] *	n.s	<0.01
Systemic inflammation, N (%)	63 (45)	30 (41)	33 (49)	16 (35)	17 (77) *	n.s.	n.s.
Medication							
Antiplatelet drugs, N (%)	120 (86)	61 (84)	59 (88)	41 (91)	18 (82)	n.s.	n.s.
Anticoagulants, N (%)	29 (21)	16 (22)	13 (19)	6 (13)	7 (32)	n.s.	n.s.
Proton pump inhibitors, N (%)	34 (24)	16 (22)	18 (27)	13 (29)	5 (23)	n.s.	n.s.
β-blocker, N (%)	57 (41)	28 (38)	29 (43)	17 (38)	12 (55)	n.s.	n.s.
ACE-inhibitors, N (%)	65 (46)	39 (53)	26 (39)	15 (45)	11 (50)	n.s.	n.s.
Ca^{2+}-channel antagonists, N (%)	14 (10)	7 (10)	7 (19)	5 (11)	2 (9)	n.s.	n.s.
Angiotensin II receptor blockers, N (%)	4 (3)	2 (3)	2 (3)	1 (2)	1 (5)	n.s.	n.s.
Diuretics, N (%)	29 (21)	12 (16)	17 (25)	10 (22)	7 (32)	n.s.	n.s.
Statins, N (%)	100 (71)	53 (73)	47 (70)	37 (82)	10 (45)	n.s.	n.s.

ACE, angiotensin converting enzyme; IQR, interquartile range; LDL, high-density lipoprotein; SD, standard deviation. * $p < 0.05$ vs. Normal Iron; ** $p < 0.01$ vs. Normal Iron. n.s., non-significant.

3.1.1. Iron Status at Baseline

Iron deficiency was present in 67 patients (48%) (Table 1, Figure 1A). ID was more frequently observed in women (57%) compared to men (43%, $p < 0.001$). There were no significant differences regarding clinical characteristics, medication, and frequency of comorbidities, except anemia (Table 1). Patients with and without ID showed a similar severity of neurologic deficit after stroke, as indicated by the NIHSS scale, and functional dependency, as indicated by the mRS.

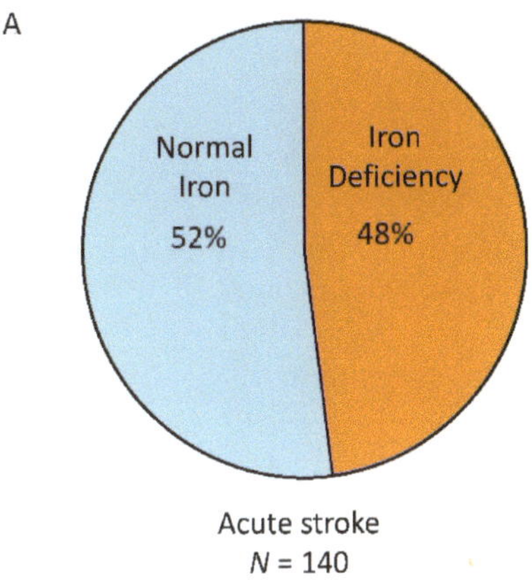

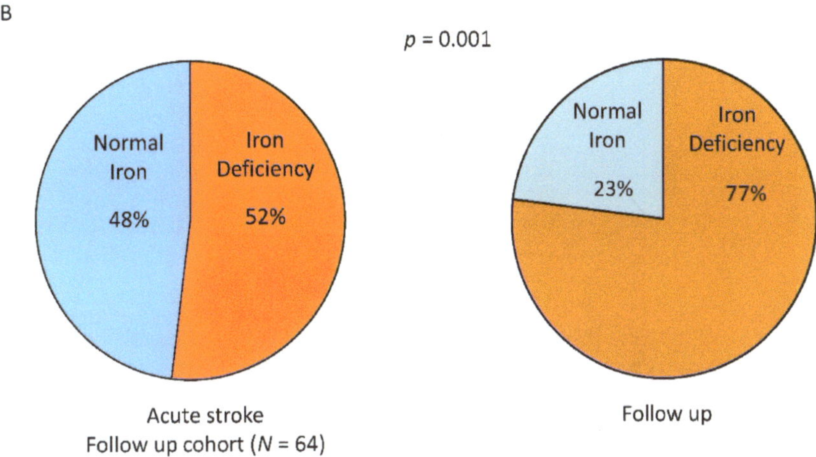

Figure 1. The prevalence of iron deficiency in acute ischemic stroke assessed at baseline (**A**) and in the follow up cohort at baseline and at one-year follow up (**B**).

ID type I was found in 32% of patients, and ID type II was found in 16% of patients. Patients with ID II were older, more frequently had a cardioembolic type of stroke, and more frequently had systemic inflammation and elevated white blood cell counts compared to patients with ID type I and patients with normal iron status (Table 1).

3.1.2. Physical Status at Baseline

At baseline, patients with ID showed lower maximal HGS compared to patients with normal iron status (mean 26.5 ± 10.4 vs. 33.8 ± 13.3 kg, $p < 0.001$). In subgroup analysis, the lower HGS was observed in patients with ID type II, followed by patients with ID type I vs. patients with Normal Iron (mean 23.0 ± 8.8 vs. 28.0 ± 10.7 vs. 33.8 ± 13.3 kg, $p < 0.01$, Figure 2).

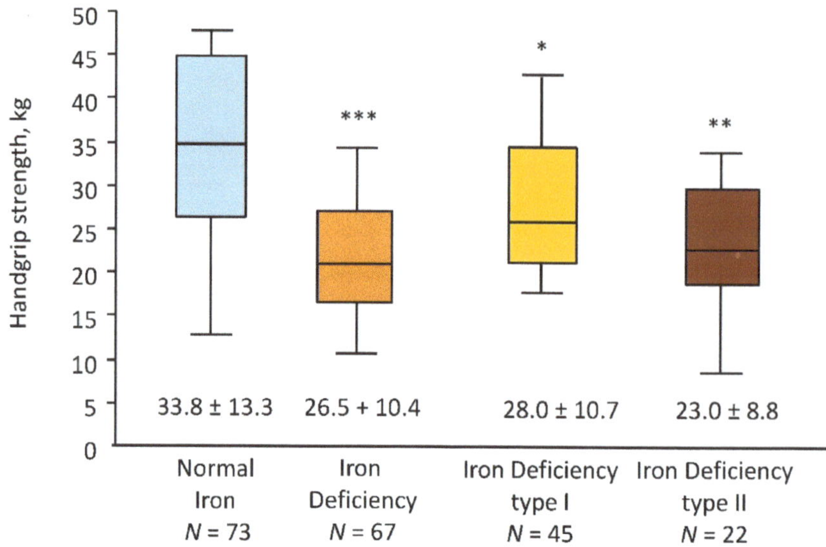

Figure 2. Handgrip strength at baseline in the study cohort divided into groups according to the presence of iron deficiency: normal iron status (normal iron), iron deficiency (ID), and categories of iron deficiency (ID type I and ID type II). * $p < 0.05$ vs. Normal Iron; ** $p < 0.01$ vs. Normal Iron; *** $p < 0.001$ vs. Normal Iron.

Of all patients, 78 patients (56%) were able to perform the quadriceps strength test. Patients with ID showed a trend towards lower quadriceps muscle strength compared to the patients with normal iron status (mean 332 ± 130 vs. $391 \pm 143 \, N$, $p = 0.06$) independently of their type of ID (I or II) (not shown).

SPPB was performed in 50% of patients with ID and with normal iron status. There was no significant difference either in frequency of attendance to the SPPB or in score values between both subgroups (Supplementary Materials Table S3).

3.1.3. Nutritional Status at Baseline

At baseline, there was no difference in self-reported appetite according to the visual analogue scale between the patients with ID and normal iron status. However, patients with ID type II reported the lowest appetite compared to patients with ID type I and Normal Iron (mean 5.4 ± 2.4 vs. 6.8 ± 2.1 vs. 6.7 ± 2.1, $p = 0.05$, respectively, Table 1).

3.1.4. Regression Analyses of Handgrip Strength at Baseline

Mean HGS in female patients was 21 ± 9 kg and in male patients was 36 ± 11 kg. Notably, those patients with ID and TSAT < 20% showed more often a lower HGS in comparison to patients with ID and TSAT $\geq$ 20% (64% vs. 36%, $p = 0.04$).

In univariable logistic regression, low HGS was associated with the presence of ID, ID type II, TSAT < 20%, and age in the whole patients' cohort (Table 2). Low HGS remained associated with ID type II after adjustment for BMI (Model 1). When analyzing low iron status only by marker TSAT < 20% low HGS was independently associated with low iron status after multivariable adjustment for BMI, age, and inflammation (Model 2).

Table 2. Logistic regression analyses applying presence of handgrip strength below mean as a dependent variable at baseline.

Parameter	OR	95% CI	p	OR	95% CI	p	OR	95% CI	p
		Univariate			Model 1			Model 2	
Transferrin saturation < 20%	3.81	1.74–8.33	<0.001				3.0	1.24–7.18	<0.05
Presence of ID	2.04	1.00–4.15	<0.05						
Presence of ID I	0.96	0.44–2.08	0.9						
Presence of ID II	4.42	1.25–15.65	0.02	4.35	1.23–15.45	0.03			
BMI (per kg/m² increase)	1.07	0.98–1.16	0.1	0.96	0.88–1.04	0.3	0.97	0.88–1.06	0.5
Age (per year increase)	1.07	1.04–1.11	<0.001				1.06	1.03–1.10	<0.001
NIHSS (per point increase)	1.06	0.95–1.19	0.3						
Hemoglobin, per mg/dL	0.88	0.72–1.07	0.2						
Presence of Inflammation	1.89	0.92–3.86	0.08				1.16	0.51–2.64	0.7

BMI, body mass index; ID, iron deficiency; NIHSS, National Institute of Health Stroke Scale. CI, confidence interval; OR, odds ratio.

3.2. Follow Up

One year after stroke (mean 382 ± 27 days), only 64 patients (46%) participated in the FU examination (FU cohort). Thus, 16 patients (11%) had problems traveling to the study center, 14 patients (10%) declined to continue the study, 6 patients (4%) had died, and 40 patients (29%) were lost to FU.

Comparing the entire study cohort with the FU cohort showed no difference in demographic, clinical, and biochemical parameters; stroke type and severity; comorbidities; and medication (Supplementary Materials Table S1).

3.2.1. Iron Status

At baseline, ID was found in 31 patients (52%) of the FU cohort (Figure 1B). This proportion increased up to 77% at FU examination ($p = 0.001$, Figure 1B). According to the ID subtypes, the proportion of patients with ID type I in this cohort increased from 36% to 49% and patients with ID type II from 13% to 28% ($p < 0.05$) from baseline to FU, respectively. All of the female patients that participated in the FU examination (38%) were found to have ID.

3.2.2. Muscle Strength

Patients with ID remained with lower HGS at FU (mean 28.8 ± 12.0 vs. 44.5 ± 8.4 kg, $p < 0.001$, Figure 3A) and in a sensitivity analysis performed in male patients only (mean 37.1 ± 9.9 vs. 44.5 ± 8.4 kg, $p < 0.05$).

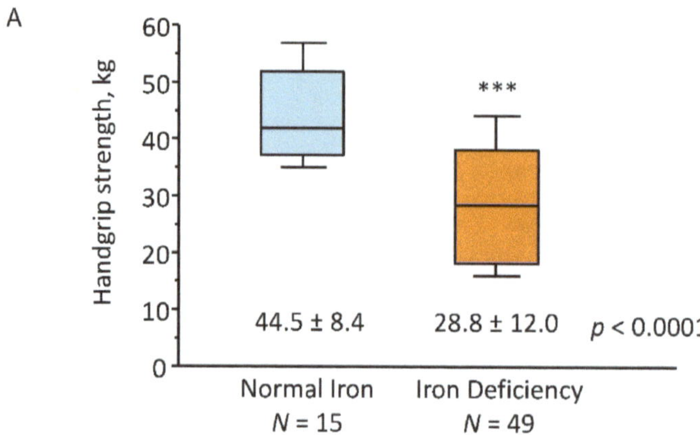

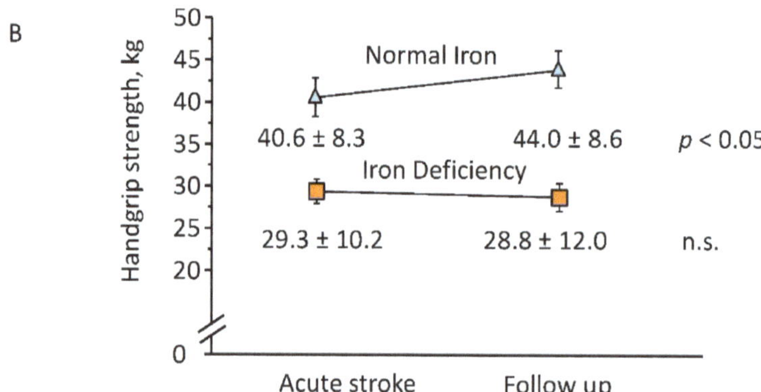

Figure 3. Handgrip strength at one-year follow up in patients with normal iron status (Normal Iron) and with iron deficiency (**A**). Changes in handgrip strength in patients with normal iron status (Normal Iron) and with iron deficiency (ID) within one year of follow up (**B**). n.s., non-significant. *** $p < 0.001$ vs. Normal Iron.

While patients with normal iron status showed an improvement of HGS at FU compared to baseline (mean 40.6 ± 8.3 vs. 44.0 ± 8.6 kg, $p < 0.05$), no improvement was observed in patients with ID (mean 29.3 ± 10.2 vs. 28.8 ± 12.0 kg, n.s., Figure 3B). The improvement of HGS was associated with normal iron status at FU (OR 3.7 [95% CI 1.01–13.5], $p < 0.05$).

3.2.3. Nutritional Status

Patients who developed ID at FU were found to have reduced appetite at baseline (mean 6.4 ± 2.1 vs. 7.6 ± 1.6, $p = 0.05$). At FU, patients with ID had more frequently low nutritional status compared to patients with normal iron status (49% vs. 20%, $p < 0.05$).

3.2.4. Logistic Regression Analyses of Handgrip Strength at Follow Up

In this restricted number of patients in the FU cohort, we found a trend in association between low HGS and reduced iron status. In invariable logistic regression analyses, ID, ID II, and TSAT < 20% all showed trends towards an association with low HGS (Table 3).

Table 3. Univariable logistic regression analyses applying low handgrip strength as a dependent variable at FU.

Parameter	OR	95% CI	p
		Univariate	
Transferrin saturation < 20%	2.86	0.97–8.42	0.06
Presence of ID	3.0	0.91–9.91	0.07
Presence of ID I	2.59	0.73–9.25	0.1
Presence of ID II	3.9	0.91–16.8	0.07
BMI (per kg/m^2 increase)	0.92	0.82–1.04	0.2
Age (per year increase)	1.08	1.02–1.14	<0.001
Hemoglobin, per mg/dL	0.98	0.75–1.29	0.9
Presence of Inflammation	4.69	0.94–23.3	0.06

BMI, body mass index; ID, iron deficiency.

4. Discussion

Up to date, ID in patients with stroke has not been recognized as relevant clinical complication that might affect the clinical outcome. This study shows a high prevalence of ID, and an association with low muscle strength in patients with acute ischemic stroke. The main findings of this study are as follows: (1) the prevalence of ID was high in patients with acute ischemic stroke and even increased within one year after the acute event; (2) patients with ID had lower muscle strength after stroke; (3) patients with ID remained with lower muscle strength one year after stroke; and (4) while patients with normal iron status showed an improvement in muscle strength at FU, such an improvement of muscle strength was not observed in patients with ID.

The present study investigated the longitudinal changes in ID in patients after stroke. We observed no spontaneous improvement of iron status within one year after stroke but rather an increase in ID prevalence. To date, we lack longitudinal studies on development of ID in patients after stroke. The majority of clinical studies investigating ID in patients with acute and chronic diseases report ID prevalence obtained as a cross-sectional value only. Previously, a longitudinal multicenter clinical trial investigating an association between ID and unspecific inflammation in otherwise healthy adults showed a 12% increase to 39% in prevalence of ID within three years in community-dwelling older individuals [24]. Previous observations in patients with acute conditions, including acute exacerbation of chronic obstructive pulmonary disease, acute coronary syndrome, or acute HF, reported a prevalence of ID ranging between 20% and 80% [25–27]. In patients with chronic conditions, such as chronic heart failure, pulmonary arterial hypertension, or cancer, the prevalence of ID ranged between 30% and 50% [13,28–30]. Therefore, our results fit into the range of ID prevalence described for other chronic conditions. However, further studies are warranted to validate our observations.

4.1. ID Categories in Stroke

We investigated the distribution of ID categories in patients with acute and chronic stroke since a difference in the prevalence of both ID subtypes has been found in cardio-oncology patients and patients with acute and chronic heart failure [21,27,29,31]. In the present study, ID type I, defined by ferritin levels < 100 suggesting depleted iron stores, was most prevalent. The proportion of patients with this type of ID increased by 12% within one year of stroke. The main causes of iron store depletion are excessive blood loss and inadequate dietary iron uptake or absorption [11,12]. Acute blood loss in the context of acute ischemic stroke is not considered relevant for exploration of the ID in these patients. Since ID might develop as a consequence of malnutrition [31], we investigated appetite and nutritional status of patients in acute and chronic stroke. Dysphagia and eating-related difficulties due to disability are frequently observed after stroke [32,33]. None of the patients included in our study had eating difficulties secondary to stroke event. However, patients who developed ID during FU were found to have the lowest appetite in acute stroke and showed the lowest nutritional status upon FU examination. In addition, prescribed medications, including aspirin and proton pump inhibitors, might contribute to the depletion of iron stores [12]. Thus, in renal transplant recipients it was observed that the use of proton pump inhibitors was associated with ID and low ferritin levels independently of potential confounders [34]. In our study, almost 90% of patients in the FU cohort received aspirin for the secondary stroke prevention, and more than 20% of patients received proton pump inhibitors, which may explain the high prevalence of ID type I in the present study.

About one third of the patients with ID were found with ID type II, which is characterized by impaired iron metabolism triggered by systemic inflammation [35]. Indeed, we observed in these patients elevated C-reactive protein serum levels and white blood cell counts, indicating inflammation. We also found these patients to have reduced appetite and lower nutritional status one year after stroke.

In the present study, ID type I was found as a predominant mechanism of ID in patients with stroke. This is in line with previous reports, which observed this type of ID as a more frequent mechanism of ID in patients with cardio-oncologic diseases and patients with heart failure [27,28,36,37]. A recent prospective multicenter trial investigating about 700 patients with acute HF reported a predominance of ID type I and its persistence after treatment of acute HF [37]. The causes leading to the onset of ID type I in chronic stroke might include an inadequate iron uptake due to dietary habits and low appetite, as well as intake of medications, e.g., aspirin and proton pump inhibitors. ID type II is commonly observed in inflammatory condition in both clinical and experimental settings [11,12]. This type of ID is also expected in patients with underlying chronic comorbidities [12]. Our study is in line with these reports, as patients with inflammatory levels more often had ID type II, whereas patients without inflammation more often had ID type I. The therapeutic consequences of ID treatment depend on the underlying mechanism. While for the treatment and prevention of iron deficiency with low tissue iron stores (type I), parenteral iron supplementation, improved nutrition, and treatment of malabsorption may be considered, patients with ID type II should be treated for inflammation first [12]. Further investigations exploring the mechanisms of ID and the effects of iron supplementation on muscle recovery and outcome in acute and chronic stroke should be carried out.

4.2. Physical Performance and Muscle Strength

Lower physical performance has been reported in patients with ID and chronic diseases [6–8]. Patients with ID in the present study had a lower muscle strength assessed by handgrip and quadriceps muscle strength tests. Importantly, stroke severity according to the NIHSS and mRS was similar in patients with ID and without ID. ID patients with TSAT < 20% more often had low HGS. Notably, this cutoff has been shown to reflect ID in the bone marrow as assessed by bone marrow biopsies in patients with chronic HF [38]. TSAT < 20% is considered as a pathophysiological marker of reduced peripheral iron availability in all organs, including cardiac and skeletal muscle [39]. Our findings suggest that this could contribute to lower HGS in the present cohort.

We also found lower HGS in patients with ID at FU. Notably, we observed that patients without ID improved in HGS one year after stroke, whereas patients with ID did not show significant improvement.

The present results are consistent with the previous clinical trial demonstrating low physical capacity and worse clinical outcomes in the presence of ID during post-stroke rehabilitation [40].

4.3. Study Limitations

The present study has several limitations. This is a small-size prospective observational study including only patients with ischemic stroke with mild to moderate stroke severity. However, the study allowed the analysis of clinical and functional parameters in relation to ID, its prevalence, and different mechanisms of ID after stroke.

5. Conclusions

The present study showed that a significant proportion of patients with acute ischemic stroke present with iron deficiency. This ID is associated with lower muscle strength in acute and chronic stroke. ID might be underdiagnosed in patients with stroke. Assessment of iron status should be regularly performed in patients with stroke in order to diagnose and treat ID.

Supplementary Materials: The following supporting information can be downloaded at https://www.mdpi.com/article/10.3390/jcm11030595/s1. Table S1. Comparison in clinical characteristic of the entire cohort and follow-up (FU) cohort at baseline. Table S2. Baseline clinical characteristics of patients available to the clinical follow up examination one year after stroke according to presence of iron deficiency (ID) at follow up. Table S3. Short physical performance battery (SPPB) at baseline in the study groups.

Author Contributions: Conceptualization, S.v.H., E.J., S.D.A. and W.D.; Data curation, N.S. and A.S.; Formal analysis, N.S. and A.S.; Investigation, N.S., A.S., M.V. and A.M.; Methodology, W.D.; Project administration, A.M.; Resources, W.D.; Supervision, S.v.H., A.S. and W.D.; Visualization, N.S.; Writing—original draft, N.S. and A.S.; Writing—review & editing, N.S., A.S., A.M., M.V., S.v.H., E.J., S.D.A. and W.D. All authors have read and agreed to the published version of the manuscript.

Funding: This research was funded by German Federal Ministry of Education and Research (BMBF), grant number EO 0801.

Institutional Review Board Statement: The study was approved by the Ethic Committee of Charité University Hospital Berlin (EA2/008/09).

Informed Consent Statement: Informed consent was obtained from all subjects involved in the study.

Data Availability Statement: The data presented in this study are available on request from the corresponding author. The data are not publicity available due to privacy.

Conflicts of Interest: A.S. was supported by a grant from Oskar–Helene–Stiftung. W.D. reports speaker fees and advisory honoraria from Aimediq, Luxembourg; Bayer, Germany; Boehringer Ingelheim, Germany; Lilly, Germany; Medtronic, Germany; Pfizer, Germany; Sanofi-Aventis, France; Sphingotec, Germany; and Vifor Pharma, Switzerland outside the submitted work and research support from Vifor Pharma, Switzerland and ZS Pharma, UK outside the submitted work. E.J. reports speaker advisory honoraria from Vifor Pharma, Switzerland outside of the submitted work and received an unrestricted grant from Vifor Pharma, Switzerland for Wroclaw Medical University outside of the submitted work. S.v.H. received consultancy and speaker honoraria from Bayer, Germany; Boehringer Ingelheim, Germany; BRAHMS, Germany; Chugai, Japan; Grünenthal, Germany; Helsinn, Ireland; Hexal, Germany; Novartis, Germany; Respicardia, USA; Roche, Switzerland; Sorin, Italy and Vifor Pharma, Switzerland. S.D.A. reports personal fees from Vifor Pharma, Switzerland; Bayer, Germany; Boehringer Ingelheim, Germany; Servier, USA; V-Wave, USA; and Novartis, Germany outside the submitted work. The authors (N.S., M.V. and A.M.) declare no conflict of interest.

References

1. Krishnamurthi, R.V.; Ikeda, T.; Feigin, V.L. Global, Regional and Country-Specific Burden of Ischaemic Stroke, Intracerebral Haemorrhage and Subarachnoid Haemorrhage: A Systematic Analysis of the Global Burden of Disease Study. *Neuroepidemiology* **2020**, *54*, 171–179. [CrossRef]
2. Burkot, J.; Kopeć, G.; Pera, J.; Slowik, A.; Dziedzic, T. Decompensated Heart Failure Is a Strong Independent Predictor of Functional Outcome After Ischemic Stroke. *J. Card. Fail.* **2015**, *21*, 642–646. [CrossRef]
3. Patlolla, S.H.; Lee, H.-C.; Noseworthy, P.A.; Wysokinski, W.E.; Hodge, D.O.; Greene, E.L.; Gersh, B.J.; Melduni, R.M. Impact of Diabetes Mellitus on Stroke and Survival in Patients with Atrial Fibrillation. *Am. J. Cardiol.* **2020**, *131*, 33–39. [CrossRef]
4. Holmqvist, L.; Boström, K.B.; Kahan, T.; Schiöler, L.; Hasselström, J.; Hjerpe, P.; Wettermark, B.; Manhem, K. Cardiovascular outcome in treatment-resistant hypertension: Results from the Swedish Primary Care Cardiovascular Database (SPCCD). *J. Hypertens.* **2018**, *36*, 402–409. [CrossRef]
5. Barandiarán Aizpurua, A.; Sanders-van Wijk, S.; Brunner-La Rocca, H.P.; Henkens, M.T.H.M.; Weerts, J.; Spanjers, M.H.A.; Knackstedt, C.; van Empel, V.P.M. Iron deficiency impacts prognosis but less exercise capacity in heart failure with preserved ejection fraction. *ESC Heart Fail.* **2021**, *8*, 1304–1313. [CrossRef]

6. Guedes, M.; Muenz, D.; Zee, J.; Lopes, M.B.; Waechter, S.; Stengel, B.; Massy, Z.; Speyer, E.; Ayav, C.; Finkelstein, F.; et al. Serum biomarkers of iron stores are associated with worse physical health-related quality of life in nondialysis-dependent chronic kidney disease patients with or without anemia. *Nephrol. Dial. Transplant.* **2021**, *36*, 1694–1703. [CrossRef]
7. Chopra, V.K.; Anker, S.D. Anaemia, ID and heart failure in 2020: Facts and numbers. *ESC Heart Fail.* **2020**, *7*, 2007–2011. [CrossRef]
8. Neidlein, S.; Wirth, R.; Pourhassan, M. Iron deficiency, fatigue and muscle strength and function in older hospitalized patients. *Eur. J. Clin. Nutr.* **2021**, *75*, 456–463. [CrossRef]
9. Muñoz, I.V.M.; Villar, I.; García-Erce, J.A. An update on iron physiology. *World J. Gastroenterol.* **2009**, *15*, 4617–4626. [CrossRef]
10. Kulkarni, A.; Khade, M.; Arun, S.; Badami, P.; Kumar, G.R.K.; Dattaroy, T.; Soni, B.; Dasgupta, S. An overview on mechanism, cause, prevention and multi-nation policy level interventions of dietary ID. *Crit. Rev. Food Sci. Nutr.* **2021**, 1–15. [CrossRef]
11. Guida, C.; Altamura, S.; Klein, F.A.; Galy, B.; Boutros, M.; Ulmer, A.J.; Hentze, M.W.; Muckenthaler, M.U. A novel inflammatory pathway mediating rapid hepcidin-independent hypoferremia. *Blood* **2015**, *125*, 2265–2275. [CrossRef] [PubMed]
12. Pasricha, S.R.; Tye-Din, J.; Muckenthaler, M.U.; Swinkels, D.W. Iron deficiency. *Lancet* **2021**, *397*, 233–248. [CrossRef]
13. Von Haehling, S.; Ebner, N.; Evertz, R.; Ponikowski, P.; Anker, S.D. ID in Heart Failure: An Overview. *JACC Heart Fail.* **2019**, *7*, 36–46. [CrossRef] [PubMed]
14. Knops, M.; Werner, C.G.; Scherbakov, N.; Fiebach, J.; Dreier, J.P.; Meisel, A.; Heuschmann, P.U.; Jungehülsing, G.J.; von Haehling, S.; Dirnagl, U.; et al. Investigation of changes in body composition, metabolic profile and skeletal muscle functional capacity in ischemic stroke patients: The rationale and design of the Body Size in Stroke Study (BoSSS). *J. Cachex Sarcopenia Muscle* **2013**, *4*, 199–207. [CrossRef]
15. Adams, H.P., Jr.; Bendixen, B.H.; Kappelle, L.J.; Biller, J.; Love, B.B.; Gordon, D.L.; Marsh, E.E., 3rd. Classification of subtype of acute ischemic stroke. Definitions for use in a multicenter clinical trial. TOAST. Trial of Org 10172 in Acute Stroke Treatment. *Stroke* **1993**, *24*, 35–41. [CrossRef]
16. Kasner, S. Clinical interpretation and use of stroke scales. *Lancet Neurol.* **2006**, *5*, 603–612. [CrossRef]
17. Guralnik, J.M.; Simonsick, E.M.; Ferrucci, L.; Glynn, R.J.; Berkman, L.F.; Blazer, D.G.; Scherr, P.A.; Wallace, R.B. A Short Physical Performance Battery Assessing Lower Extremity Function: Association with Self-Reported Disability and Prediction of Mortality and Nursing Home Admission. *J. Gerontol.* **1994**, *49*, M85–M94. [CrossRef]
18. Doehner, W.; Turhan, G.; Leyva, F.; Rauchhaus, M.; Sandek, A.; Jankowska, E.; Von Haehling, S.; Anker, S.D. Skeletal muscle weakness is related to insulin resistance in patients with chronic heart failure. *ESC Heart Fail.* **2015**, *2*, 85–89. [CrossRef]
19. Koehler, F.; Doehner, W.; Hoernig, S.; Witt, C.; Anker, S.D.; John, M. Anorexia in chronic obstructive pulmonary disease—Association to cachexia and hormonal derangement. *Int. J. Cardiol.* **2007**, *119*, 83–89. [CrossRef]
20. Vellas, B.; Guigoz, Y.; Garry, P.J.; Nourhashemi, F.; Bennahum, D.; Lauque, S.; Albarede, J.L. The Mini Nutritional Assessment (MNA) and its use in grading the nutritional state of elderly patients. *Nutrition* **1999**, *15*, 116–122. [CrossRef]
21. Moliner, P.; Jankowska, E.; van Veldhuisen, D.J.; Farre, N.; Rozentryt, P.; Enjuanes, C.; Polonski, L.; Meroño, O.; Voors, A.A.; Ponikowski, P.; et al. Clinical correlates and prognostic impact of impaired iron storage versus impaired iron transport in an international cohort of 1821 patients with chronic heart failure. *Int. J. Cardiol.* **2017**, *243*, 360–366. [CrossRef] [PubMed]
22. Loncar, G.; Obradovic, D.; Thiele, H.; von Haehling, S.; Lainscak, M. Iron deficiency in heart failure. *ESC Heart Fail.* **2021**, 2368–2379. [CrossRef] [PubMed]
23. Schaap, L.A.; Pluijm, S.M.; Deeg, D.J.; Visser, M. Inflammatory Markers and Loss of Muscle Mass (Sarcopenia) and Strength. *Am. J. Med.* **2006**, *119*, 526.e9–526.e17. [CrossRef] [PubMed]
24. Wieczorek, M.; Schwarz, F.; Sadlon, A.; Abderhalden, L.A.; Molino, C.D.G.R.C.; Spahn, D.R.; Schaer, D.J.; Orav, E.J.; Egli, A.; Bischoff-Ferrari, H.A.; et al. Iron deficiency and biomarkers of inflammation: A 3-year prospective analysis of the DO-HEALTH trial. *Aging Clin. Exp. Res.* **2021**, 1–11. [CrossRef] [PubMed]
25. Silverberg, D.S.; Mor, R.; Weu, M.T.; Schwartz, D.; Schwartz, I.F.; Chernin, G. Anemia and iron deficiency in COPD patients: Prevalence and the effects of correction of the anemia with erythropoiesis stimulating agents and intra-venous iron. *BMC Pulm. Med.* **2014**, *14*, 24. [CrossRef]
26. Silva, C.; Martins, J.; Campos, I.; Arantes, C.; Braga, C.G.; Salomé, N.; Gaspar, A.; Azevedo, P.; Pereira, M.Á.; Marques, J.; et al. Prognostic impact of iron deficiency in acute coronary syndromes. *Rev. Port. Cardiol.* **2021**, *40*, 525–536. [CrossRef]
27. Jacob, J.; Miró, Ò.; Ferre, C.; Borraz-Ordás, C.; Llopis-García, G.; Comabella, R.; Fernández-Cañadas, J.M.; Mercado, A.; Roset, A.; Richard-Espiga, F.; et al. ID and safety of ferric carboxymaltose in patients with acute heart failure. AHF-ID study. *Int. J. Clin. Pract.* **2020**, *74*, e13584. [CrossRef] [PubMed]
28. Jankowska, E.A.; Rozentryt, P.; Witkowska, A.; Nowak, J.; Hartmann, O.; Ponikowska, B.; Borodulin-Nadzieja, L.; Banasiak, W.; Polonski, L.; Filippatos, G.; et al. Iron deficiency: An ominous sign in patients with systolic chronic heart failure. *Eur. Heart J.* **2010**, *31*, 1872–1880. [CrossRef]
29. Quatredeniers, M.; Mendes-Ferreira, P.; Santos-Ribeiro, D.; Nakhleh, M.K.; Ghigna, M.R.; Cohen-Kaminsky, S.; Perros, F. ID in Pulmonary Arterial Hypertension: A Deep Dive into the Mechanisms. *Cells* **2021**, *10*, 477. [CrossRef]
30. Čiburienė, E.; Čelutkienė, J.; Aidietienė, S.; Ščerbickaitė, G.; Lyon, A.R. The prevalence of ID and anemia and their im-pact on survival in patients at a cardio-oncology clinic. *Cardiooncology* **2020**, *6*, 29.
31. Camaschella, C. ID. *Blood* **2019**, *133*, 30–39. [CrossRef] [PubMed]
32. Scherbakov, N.; Dirnagl, U.; Doehner, W. Body weight after stroke: Lessons from the obesity paradox. *Stroke* **2011**, *42*, 3646–3650. [CrossRef]

33. Perry, L. Eating and dietary intake in communication-impaired stroke survivors: A cohort study from acute-stage hospital admission to 6 months post-stroke. *Clin. Nutr.* **2004**, *23*, 1333–1343. [CrossRef] [PubMed]
34. Douwes, R.M.; Gomes-Neto, A.W.; Eisenga, M.F.; Vinke, J.S.J.; De Borst, M.H.; van den Berg, E.; Berger, S.P.; Touw, D.J.; Hak, E.; Blokzijl, H.; et al. Chronic Use of Proton-Pump Inhibitors and Iron Status in Renal Transplant Recipients. *J. Clin. Med.* **2019**, *8*, 1382. [CrossRef] [PubMed]
35. Van Veldhuisen, D.J.; Anker, S.D.; Ponikowski, P.; Macdougall, I.C. Anemia and iron deficiency in heart failure: Mechanisms and therapeutic approaches. *Nat. Rev. Cardiol.* **2011**, *8*, 485–493. [CrossRef]
36. Pozzo, J.; Fournier, P.; Delmas, C.; Vervueren, P.-L.; Roncalli, J.; Elbaz, M.; Galinier, M.; Lairez, O. Absolute iron deficiency without anaemia in patients with chronic systolic heart failure is associated with poorer functional capacity. *Arch. Cardiovasc. Dis.* **2017**, *110*, 99–105. [CrossRef] [PubMed]
37. Van Dalen, D.H.; Kragten, J.A.; Emans, M.E.; van Ofwegen-Hanekamp, C.E.E.; Klaarwater, C.C.R.; Spanjers, M.H.A.; Hendrick, R.; van Deursen, C.T.B.M.; Rocca, H.B. Acute heart failure and iron deficiency: A prospective, multicentre, observational study. *ESC Hear. Fail.* **2021**. [CrossRef]
38. Grote Beverborg, N.; Klip, I.T.; Meijers, W.C.; Voors, A.A.; Vegter, E.L.; van der Wal, H.H.; Swinkels, D.W.; van Pelt, J.; Mulder, A.B.; Bulstra, S.K.; et al. Definition of Iron Defi-ciency Based on the Gold Standard of Bone Marrow Iron Staining in Heart Failure Patients. *Circ. Heart Fail.* **2018**, *11*, e004519. [CrossRef]
39. Wish, J.B. Assessing Iron Status: Beyond Serum Ferritin and Transferrin Saturation. *Clin. J. Am. Soc. Nephrol.* **2006**, *1*, S4–S8. [CrossRef]
40. Doehner, W.; Scherbakov, N.; Schellenberg, T.; Jankowska, E.A.; Scheitz, J.F.; von Haehling, S.; Joebges, M. Iron deficiency is related to low functional outcome in patients at early rehabilitation after acute stroke. 2021. in press.

Journal of
Clinical Medicine

Article

Mobile Single-Lead Electrocardiogram Technology for Atrial Fibrillation Detection in Acute Ischemic Stroke Patients

Marta Leńska-Mieciek [1,*], Aleksandra Kuls-Oszmaniec [2], Natalia Dociak [2], Marcin Kowalewski [1], Krzysztof Sarwiński [3], Andrzej Osiecki [3] and Urszula Fiszer [1]

[1] Centre of Postgraduate Medical Education, Department of Neurology and Epileptology, 00-416 Warsaw, Poland; marcin.kowalewski@cmkp.edu.pl (M.K.); ufiszer@cmkp.edu.pl (U.F.)
[2] Department of Neurology and Epileptology, Professor Orlowski's Hospital, 00-416 Warsaw, Poland; aleksandra.m.kuls@gmail.com (A.K.-O.); dociak7@tlen.pl (N.D.)
[3] Department of Cardiology, Bielański Hospital, 01-809 Warsaw, Poland; krzysztof.sarwinski@gmail.com (K.S.); mcosiek@gmail.com (A.O.)
* Correspondence: mlenska@cmkp.edu.pl; Tel.: +48-22-6294349

Abstract: (1) Background: AliveCor KardiaMobile (KM) is a portable electrocardiography recorder for detection of atrial fibrillation (AF). The aim of the study was to define the group of acute ischemic stroke (AIS) patients who can use the KM device and assess the diagnostic test accuracy. (2) Methods: the AIS patients were recruited to the study. Thirty-second single-lead electrocardiogram (ECG) usages were recorded on demand for three days using KM portable device. Each KM ECG record was verified by a cardiologist. The feasibility was evaluated using operationalization criteria. (3) Results: the recruitment rate among AIS patients was 26.3%. The withdrawal rate before the start of the intervention was 26%. The withdrawal rate after the start of the intervention was 6%. KM device detected AF in 2.8% of AIS patients and in 2.2% of ECG records. Cardiologist confirmed the AF in 0.3% AIS patients. Sensitivity and specificity of KM for AF was 100% and 98.3%, respectively. (4) Conclusions: the results of this study suggest that it is feasible to use KM device to detect AF in the selected AIS patients (younger and in better neurological condition). KM detected AF in the selected AIS patients with high specificity and sensitivity.

Keywords: mobile electrocardiography; atrial fibrillation; acute ischemic stroke

Citation: Leńska-Mieciek, M.; Kuls-Oszmaniec, A.; Dociak, N.; Kowalewski, M.; Sarwiński, K.; Osiecki, A.; Fiszer, U. Mobile Single-Lead Electrocardiogram Technology for Atrial Fibrillation Detection in Acute Ischemic Stroke Patients. *J. Clin. Med.* **2022**, *11*, 665. https://doi.org/10.3390/jcm11030665

Academic Editor: Aaron S. Dumont

Received: 15 December 2021
Accepted: 24 January 2022
Published: 27 January 2022

Publisher's Note: MDPI stays neutral with regard to jurisdictional claims in published maps and institutional affiliations.

1. Introduction

Thirty percent of ischemic strokes are of an unknown cause (cryptogenic). Several mechanisms are implicated with cryptogenic stroke. Nearly 65% of patients have cortical infarcts on brain imaging, a characteristic typically suggestive of embolism [1,2]. Atrial fibrillation (AF) is not only the most common cardiac arrhythmia in adults, but also the most common cause of cardioembolism resulting in ischemic stroke. Identifying individuals with AF in high-risk groups of patients, such as poststroke patients, could enable those patients to be treated properly [3,4].

AF diagnosis is difficult, especially in patients with paroxysmal AF and in asymptomatic (silent) AF. AF symptoms (history of palpitations, dyspnea, fatigue, chest tightness/pain, syncope/presyncope, and dizziness) could be absent during most episodes [4–7]. Up to 95% of patients with AF detected after stroke and transient ischemic attack (TIA) are asymptomatic [8–10].

Short-term electrocardiogram (ECG) recordings for at least the first 24 h after thromboembolic event, followed by continuous ECG monitoring for at least 72 h, whenever possible, are recommended in the search for AF in patients with cryptogenic stroke [4]. The ECG on admission detected new AF in 3.5–7.7% of patients, and moreover, during the hospital stay new AF was documented in another 5.1% using different routine diagnostic methods (4.4% by 6-day Holter monitoring) [11,12].

Patients in which standard diagnostic methods did not detect AF episodes might be screened using several technologies: automated blood pressure monitors, single-lead ECG devices, photopletysmography (PPG) devices, and other sensors (seismocardiography, accelerometers, and gyroscopes, etc.) used in applications for smartphones, wrist bands, and watches [13,14]. Intermittent detection of AF is possible through PPG or ECG recordings. Smartwatches and other "wearables" can passively measure pulse. To establish a definitive diagnosis of AF, a single-lead ECG tracing $\geq$ 30 s or 12-lead ECG showing episodes of AF lasting at least 30 s and analyzed by a physician with expertise in ECG rhythm interpretation is necessary [15].

AliveCor KardiaMobile (KM) ECG provides portable ECG recording for intermittent detection of AF, and it works with compatible mobile devices such as smartphones and tablets. It records on demand, stores, and transfers a single-channel, 30-s ECG. The KM device communicates with KM app, which can be downloaded on smartphones running Android OS or iOS. An automated algorithm on the KM app checks the ECG for RR wave irregularity. KM devices are Food and Drug Administration (FDA)-cleared to discriminate AF from sinus rhythm, and they demonstrated high sensitivity and specificity in screening studies [16–18]. Results from a 30-s single-lead ECG were sufficient for 42.7% of healthcare practitioners to recommend oral anticoagulation for patients with a high risk of stroke [19].

After the occurrence of ischemic stroke, the need to search for AF becomes obvious to start optimal secondary prevention. AF episodes may be missed using the standard evaluation, and therefore there is a need for improving poststroke monitoring of unknown AF detection. New technologies, such as mobile electrographic monitoring, could facilitate this in selected acute ischemic stroke (AIS) patients [20]. KM device is easy to use, noninvasive, and does not require trained healthcare staff. Its feasibility and diagnostic accuracy in AIS patients were understudied.

According to Koh et al., a 30-day smartphone ECG recording among patients with a cryptogenic stroke or TIA (index event) within the previous 12 months significantly improved the detection of AF when compared with that of the standard repeat 24-h Holter monitoring [21]. In the mentioned study, the mean duration from the index event to randomization was 87.1 days, thus patients were not in the acute phase of the disease. The data from the literature demonstrated increased detection of AF in stroke patients with earlier monitoring [22–25]. We found no literature on smartphone ECG recording to detect AF in AIS patients.

We aimed to determine the utility and feasibility of a mobile, intermittent single-channel KM ECG in hospitalized AIS patients. The aims of the study were (1) to define the group of AIS patients who can use the device and (2) to assess the diagnostic testing accuracy.

2. Materials and Methods

2.1. Ethics and Recruitment

Approval for the study by the local Ethics Committee of Centre of Postgraduate Medical Education was obtained before subject enrolment (Prot. 23/PB/2017). The study was conducted in the Department of Neurology and Epileptology, Professor Orlowski's Hospital, Warsaw, Poland. The research team discussed the study with the patients, allowed them to read the informed consent, and answered all questions. All studied patients signed the offline consent form. Ischemic stroke was diagnosed based on the neurological syndrome and results of head computed tomography examination. The inclusion criteria for the study group were as follows: (1) AIS patient (2) at least 18 years of age, who was able (3) to use the AliveCor KM ECG monitor (i.e., they were able to record at least one 30-s ECG) and (4) to provide informed consent. The exclusion criteria were: (1) AF on the initial 12-channel ECG; (2) a history of paroxysmal or persistent AF; (3) AF on 24-h Holter ECG registered during the present hospitalization period; and (4) mechanical thrombectomy treatment for AIS. Patients with AF diagnosis (AF history or AF registered on stroke unit using standard methods) were excluded to adjust the study design to the clinical practice (searching for AF only in the group of AIS patients without AF diagnosis). Patients treated with mechanical thrombectomy were excluded, as they were transferred to the other hospital for the thrombectomy procedure.

The following demographic and clinical data were obtained via a questionnaire: age, sex, length of stay in the neurological department, history of palpitation, dyspnea, fatigue, chest tightness/pain, syncope/presyncope, and dizziness.

Stroke severity and neurological deficit measures were assessed using Scandinavian Stroke Scale (SSS), which is based on various clinical observations. The assessment was done on the 1st and 7th days after admission to the stroke unit and on the last day of the hospital stay. SSS includes nine items representing consciousness, eye movements, arm, hand, and leg motor function, orientation, speech, facial palsy, and gait. Each item has 2–5 response categories, with item scores ranging from 2 to 12. The scale total score range is between 0 and 58. Higher scores indicate better neurological function [26,27].

2.2. Feasibility

The feasibility was evaluated using following operationalization criteria:

- source of recruitment: at stroke unit;
- recruitment rates: from the AIS patients treated at the stroke unit, the number of eligible AIS patients was determined using inclusion and exclusion criteria ($\geq$20%);
- withdrawal rates (before intervention): AIS patients who gave preconsent to participate but withdrew it before the start of intervention were accounted for. AIS patients for whom the device was not available were counted ($\leq$20%);
- withdrawal rates (during the course of intervention): AIS patients who received a tablet and the KM pad for three days but for whom there was no ECG registration on the device were accounted for ($\leq$10%);
- technical issues: problems that disrupted or impeded proper operation of the device [28,29].

2.3. KM ECG Monitoring

Thirty-second single-lead ECG was recorded using KM portable device (AliveCor Inc, San Francisco, CA, USA) and compatible portable tablet computer (LenovoYT3-850L and Lenovo YT3-X50L, HK Limited, China) with AliveCor application (Figure 1). The KM device is FDA- and Conformité Européenne (CE)-approved, registers ECG on-demand, and works with iOS and Android devices. The KM pocket-sized pad consists of two metal electrodes. When they are touched by the right and left hands of the user, a bipolar lead I is created. The cardiac electric signal is converted into an ultrasound FM sound signal (18–24 kHz). The iOS or Android device application demodulates the sound signal to a digital ECG tracing (300 samples/s, 16-bit resolution). The ECG tracing can be viewed in real-time, and it is also stored [30]. Patients randomized to the study were provided with a tablet and the KM pad for three days. Patients were trained by the Neurology Department staff members on how to use the KM card and how to record ECG on tablet. The device was used repeatedly for intermittent screening. Patients were instructed to record ECG if any symptoms of AF appeared (palpitation, dyspnea, fatigue, chest tightness/pain, syncope/presyncope, and dizziness) and several times during the day (if possible, every 2–3 h) [31]. Further tuition following the initial guidance were offered to patients, who had problems with operation of the KM device. The lead-I ECG was interpreted by an algorithm and verified by a cardiologist. The application classifies an ECG as a sinus rhythm, possible AF, or unclassified. If the registration time was shorter than 30 s, the analysis was not performed. In some cases (poor quality recordings; too much interference in the recording), the algorithm was not able to analyze ECG, even if the time of registration exceeded 30 s (analysis was impossible).

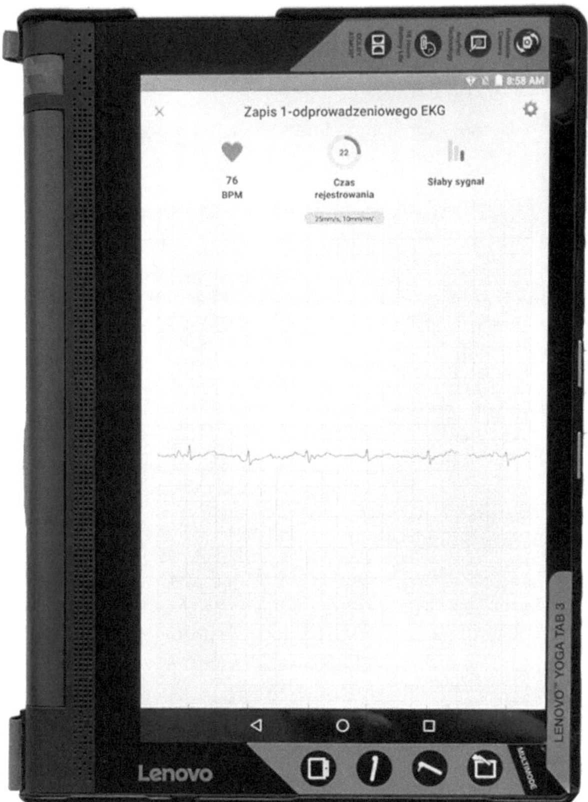

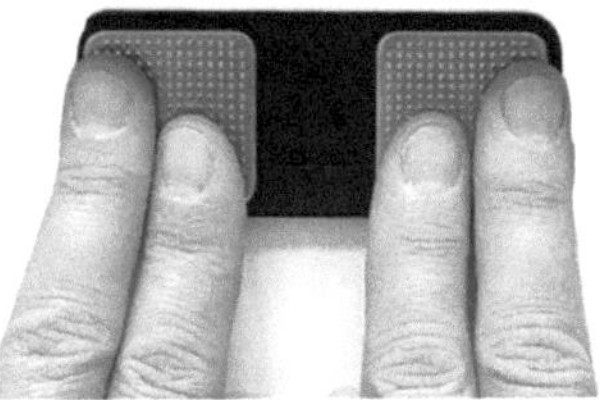

Figure 1. KardiaMobile (KM) portable device (AliveCor Inc, San Francisco, CA, USA) and compatible portable tablet computer (Lenovo YT3-X50L, HK Limited, China) with AliveCor application.

The following data from the patients' use of the KM device were analyzed: number of times device was used, number of ECGs interpreted by algorithm as AF or unclassified, and the number of ECGs for which analysis was impossible or time of registration was too short.

2.4. ECG Adjudication Analysis

The KM ECG records were printed out in 1:1 ratio and assessed in paper-based form. Each of the lead-I ECG was verified by a cardiologist, and the following data were analyzed: the number of ECGs interpreted as AF, extrasystoles, bigeminy, or bundle branch block.

2.5. Twenty-Four-Hour Holter ECG

Standard 24-h Holter ECG was performed using BTL-08 Holter H100 3-channel Holter recorder (BTL Industries Ltd., Hertfordshire, UK). The following data were analyzed: the number of the supraventricular ectopic beat (SVEB), ventricular ectopic beat (VEB), and supraventricular tachycardia (SVT).

2.6. Statistical Analysis

Demographic and clinical data are reported as mean $\pm$ SD and/or median. Pearson's correlation analyses were used to examine the associations between the number of KM ECG uses and patient's age. Pearson's chi-square test was used to determine whether there is a significant relationship between the AF clinical signs (history of palpitation, dyspnea, fatigue, chest tightness/pain, syncope/presyncope, and dizziness) and AF reporting during the KM device use. Comparisons are presented by showing the data in the respective groups (AliveCor algorithm/cardiologist interpretation) together with the p value of Pearson's chi-square test. Mann–Whitney U test was used to correlate Holter ECG data (number of SVEB/VEB/SVT) and AF reporting during KM device use. Kendall's tau b test was used to determine the correlation between the number of KM ECG uses and AF registration. Diagnostic accuracy measures (sensitivity, specificity) of KM for AF with corresponding 95% CI were calculated using 2 × 2 contingency tables. All statistical analyses were performed using PAWS Statistic 18 (SPSS). A p value < 0.05 was considered significant.

3. Results

3.1. Participants

AIS patients treated at the stroke unit were recruited over a period of 12 months. Fifty patients (26 males; 24 females; mean age 64.44, SD 10.52 years) from 285 AIS patients treated at the stroke unit Department of Neurology and Epileptology met the inclusion criteria, entered the study, and used the device properly—i.e., they were able to record at least one 30-s ECG (Figure 2, flow diagram).

Demographic and clinical features of study group and patients who were not able to use device for unclear reasons are summarized in Table 1. Patients had a history of the following symptoms: 22/50 (44%) patients—history of palpitation; 13/50 (26%)—dyspnea; 17/50 (34%)—fatigue; 16/50 (32%)—tightness/pain in the chest; 6/50 (12%)—syncope; 16/50 (32%)—presyncope; and 25/50 (50%)—dizziness.

Table 1. Summary of baseline characteristics of study cohort and group of patients not able to use device for unclear reasons.

	Study Group (*n* = 50)	Patients Not Able to Use Device for Unclear Reason (*n* = 40)	*p* Value
Age, mean (year)	64.44 (SD 10.52)	75.15 (SD 11.38)	<0.001
Gender, female (%)	24/50 (48%)	26/40 (65%)	0.11 [1]
SSS0, mean [2]	47.10 (SD 11.6)	36.33 (SD 11.88)	<0.001
SSS7, mean [3]	55.76 (SD 3.74)	45.11 (SD 11.97)	<0.001
SSSE, mean [4]	55.98 (SD 3.39)	46.83 (SD 11.65)	<0.001
Duration of hospitalization, mean (days)	9.0 (SD 1.23)	11.28 (SD 8.94)	0.22

[1] Pearson's chi-square test; [2] Scandinavian Stroke Scale (1st day after admission to the stroke unit); [3] Scandinavian Stroke Scale (7th day after admission to the stroke unit); [4] Scandinavian Stroke Scale (on the last day of the hospital stay).

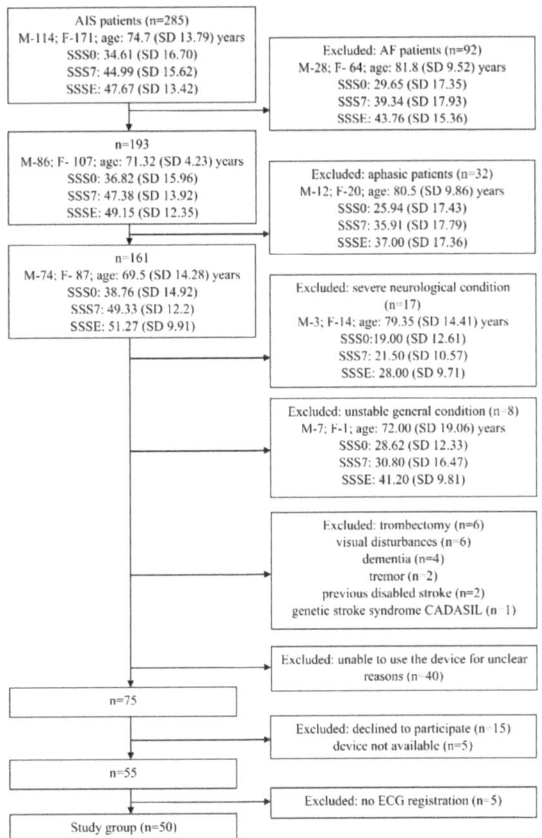

Figure 2. Flow diagram of study. AF—atrial fibrillation, AIS—acute ischemic stroke, ECG—electrocardiogram, SSS0—Scandinavian Stroke Scale (1st day after admission to the stroke unit), SSS7-Scandinavian Stroke Scale (7th day after admission to the stroke unit), SSSE—Scandinavian Stroke Scale (on the last day of the hospital stay).

3.2. Feasibility

The recruitment rate among AIS patients was 26.3% (75/285), which was higher than the target set at $\geq$20%. The withdrawal rate before the start of the intervention was 26% (20/75), which was higher than the established $\leq$20%. The withdrawal rate after the start of the intervention was 6% (5/75), which met the established criteria of $\leq$10%. Different technical issues arose during the study. The most important was the problem of connection between KM pad and the portable tablet computer. The KM pad should be localized close to the microphone of the portable tablet computer to start ECG registration; however, it was difficult for most AIS patients to comprehend this. For some patients, it was difficult to touch the metal electrodes correctly to create the bipolar lead and start ECG registration.

3.3. The KM ECG Monitoring

The KM device was used by each of the 50 patients 17.9, SD 7.98 times. There was no sex difference in the number of device use (male: 17.11, SD 7.58 times; female: 18.75, SD 8.47 times; $p = 0.47$ at t-student test). There was no correlation between patient's age and the number of the device use ($p = 0.94$ at Pearson's test).

KM device detected at least one 30-s ECG, which was interpreted by algorithm as AF in eight patients (8/50, 16% of study group; 8/285, 2.8% AIS patients; 8/193, 4.1% of non-AF AIS patients) and in 20 of 895 total ECG records (20/895, 2.2%). Fifty-one (51/895,

5.7%) ECG records were interpreted as unclassified. The analysis was impossible in nine (9/895, 1%) of the ECG records, and the time of registration was too short in 10 (10/895, 1.1%) of the ECG records.

There was no correlation between each patient's total number of ECG records and AF registration (T = -0.135, $p = 0.25$ at Kendall's tau b test). No correlation between ECG records interpreted by the AliveCor algorithm as AF and history of palpitation (χ^{2-} 0.16, $p = 0.69$ at chi-square test), dyspnea (χ^{2-} 0.90, $p = 0.34$ at chi-square test), fatigue (χ^{2-} 1.96, $p = 0.16$ at chi-square test), chest tightness/pain (χ^{2-} 1.664, $p = 0.20$ at chi-square test), presyncope (χ^{2-} 0.132, $p = 0.72$ at chi-square test), syncope (χ^{2-} 0.002, $p = 0.96$ at chi-square test), or dizziness (χ^{2-} 0.00, $p = 1.00$ at chi-square test) was noticed.

3.4. KM Automated Algorithm Versus Cardiologist Adjudication

Upon review of the single-lead ECG, it was of adequate quality for rhythm decision in 95% (19/20) of ECG records documented by AliveCor algorithm as AF. The cardiologist confirmed the AF in one patient (1/50, 2% of study group; 1/285; 0.3% AIS patients; 1/193, 0.5% non-AF AIS patients) in five KM ECG records (Figure 3). The rest of ECG records documented by AliveCor algorithm as AF were interpreted by cardiologist as supraventricular extrasystoles in three patients in a total of seven KM ECG records (Figure 4). In three patients (seven KM ECG records), the diagnosis of AF was not confirmed by a cardiologist (Figure 5). One KM ECG record interpreted by algorithm as AF was evaluated by a cardiologist as undiagnosed, although AF was a possibility (Figure 6). The data regarding cardiologist's analysis and 24 h Holter ECG are summarized in Table 2. Sensitivity and specificity of KM for AF were 100% (95% CI 47.8–100%) and 98.3% (95% CI 97.2–99.0%), respectively.

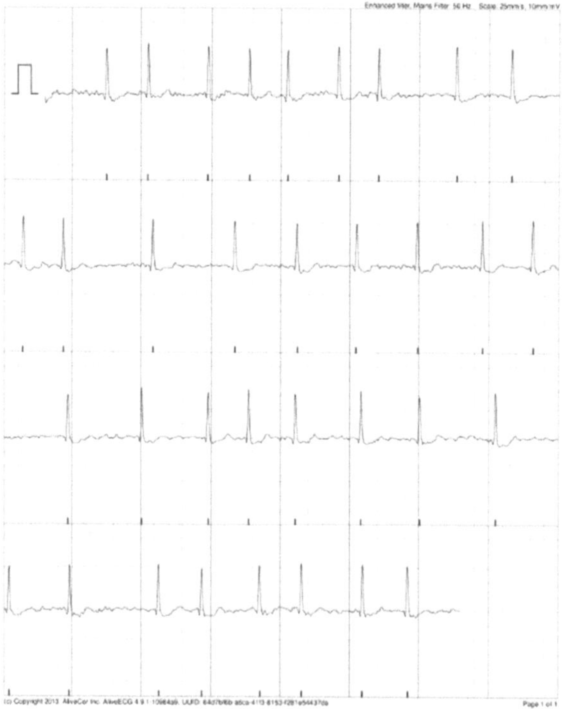

Figure 3. Possible atrial fibrillation (AF) recorded by AliveCor device, as confirmed by cardiologist.

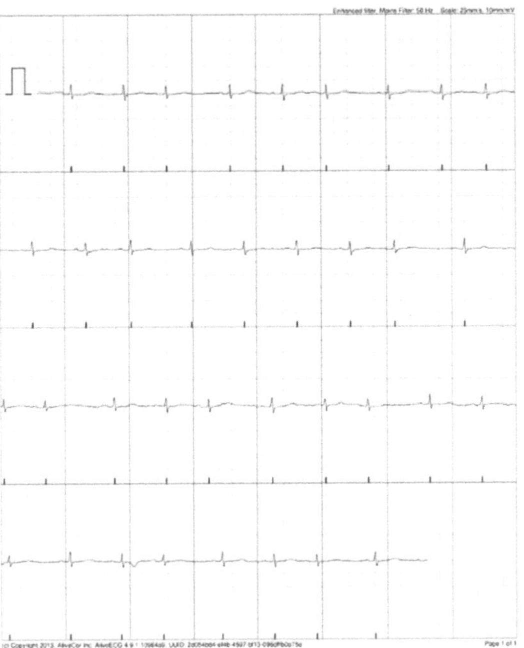

Figure 4. Possible AF recorded by AliveCor device interpreted by cardiologist as supraventricular extrasystoles.

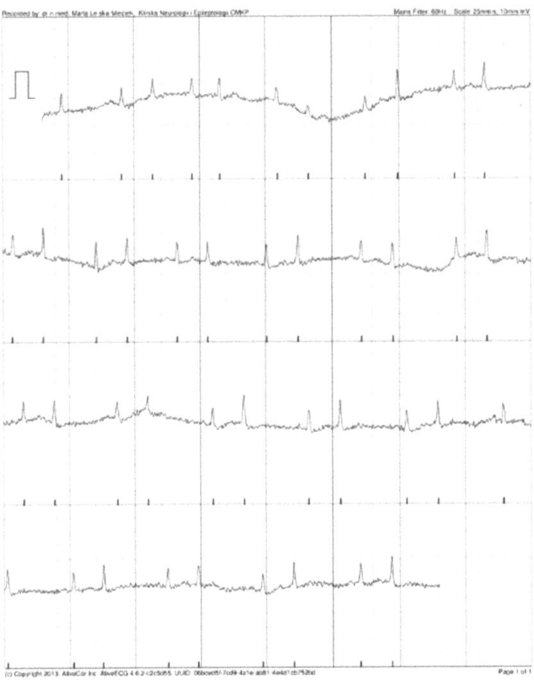

Figure 5. Possible AF recorded by AliveCor device—diagnosis was not confirmed by cardiologist.

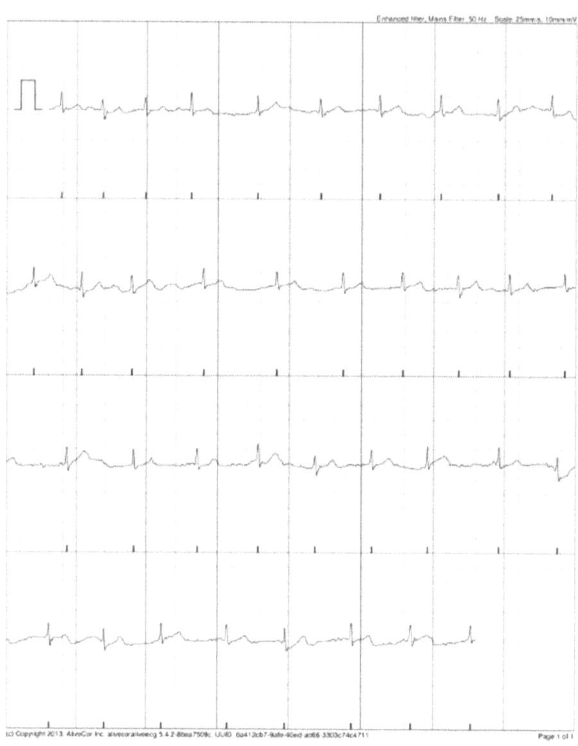

Figure 6. Possible AF recorded by AliveCor device—interpreted by cardiologist as undiagnostic with AF possibility.

Table 2. Summary of cardiologist's inspection of KM ECG records and data of standard 24-h Holter ECG registration.

Results of the Cardiologist Inspection of the KM ECG Records		
	Number of Patients (n = 50)	**Number of KM ECG Records (n = 895)**
Extrasystoles	4/50 (8%)	13/895 (1.4%)
Bigeminy	1/50 (2%)	3/895 (0.3%)
Bundle branch block	5/50 (10%)	
Data of the standard 24 h Holter ECG		
	Number of patients (n = 50)	**Number of detected arrhythmias, mean**
SVEB [1]	28/50 (56%)	564 (SD 1573.13)
VEB [2]	22/50 (44%)	124.08 (SD 341.04)
SVT [3]	19/50 (38%)	3.02 (SD 13.84)

[1] SVEB: supraventricular ectopic beat; [2] VEB: ventricular ectopic beat; [3] SVT: supraventricular tachycardia.

3.5. Twenty-Four-Hour Holter ECG

More SVEBs recorded by 24 h Holter ECG in patients with ECG records interpreted by AliveCor algorithm as AF were observed (AF: mean—2183 ± 3287.89, non-AF: mean—256.4, SD 709.64; p = 0.01 at Mann–Whitney test). There was no difference in the number of VEBs (AF: mean −200.5, SD 16.55; non-AF: mean—109.17, SD 302.46; p = 0.72 at Mann–Whitney test) and SVT (AF: mean −13.38, SD 34.23; non-AF: mean—1.05, SD 2.01; p = 0.20

at Mann–Whitney test) recorded by 24 h Holter ECG between patients, whose ECGs were interpreted by AliveCor algorithm as AF and others.

4. Discussion

KM has 98.5% sensitivity and 91.4% specificity for AF detection in patients aged 65 years and more [32]. Most of the patients were comfortable with using the device, finding it easy to use, nonrestrictive, and that it did not cause anxiety [33,34]. We found no literature regarding AF monitoring using portable devices without medical staff involvement for AIS patients treated in a stroke unit. According to the present study's results, selected AIS patients can use the device properly. Feasibility was demonstrated with the achievement of all operationalization criteria, except for the withdrawal rate before starting the intervention. The recruitment rate among AIS patients was 26.3% (75/285), which was higher than the target set at $\geq$20%. Patients were not able to use the device due to aphasia (32/193, 16.6%), severe neurological condition (17/193, 8.8%), unstable general condition (8/193, 4.1%), visual disturbances (6/193, 3.1%), dementia (5/193, 2.6%), and tremors (2/193, 1%) (Figure 1). There was a group of patients (40/193, 20.7%) who were not able to use the device properly, but the reasons why were difficult to ascertain. Their neurological condition seemed to not be debilitating but they still did not manage to obtain a proper ECG record using the KM device. Patients in that group, as it could be expected, were elderly and in a worse neurological condition than that of the study group (Table 1). The average age of patients in the study group was 64.44 years. AF incidence continues to be much higher in the elderly population, and in adults older than 65 years it is estimated to be 8.6% [35].

The withdrawal rate before the start of the intervention was 26% (20/75), which was higher than the established $\leq$20%. Five patients were excluded from the study group as the device was not available. The reason was malfunction of the device or an insufficient number of KM pads, which were available for patients (two devices). Fifteen patients finally declined to participate, although they gave preconsent first. All those patients were afraid of accidentally damaging the devices (KM pad and the portable tablet computer), although there were no repercussions to fear.

The withdrawal rate after the start of the intervention was 6% (5/75), which met the established criteria of $\leq$10%. All five participants had standard training on how to use the KM card and how to record ECG on tablet. They were provided a tablet and KM pad for three days but no ECG record was registered, even though these patients claimed that they used the device regularly and properly. The reason was not a technical one and remains unexplained. It was most likely caused by the patients' noncompliance.

The technical problem of connection between KM pad and the portable tablet computer was solved using additional headphones attached to portable tablet computer. Patients' difficulties with touching the KM metal electrodes correctly to create the bipolar lead were the result of neurological deficits. This was solved by fixing the KM pad to the portable tablet computer. Such technical issues can be solved and should not be a reason to preclude the use of KM in selected AIS patients.

The varying population involvement, settings, and clinical scenarios make it difficult to compare results of the published studies on KM device. Further, the automated arrhythmia detection algorithm provided inconclusive results of between 0.8%–27.6% of KM ECGs [36].

In the presented study, across 72 h of screening, patients used the device 17.9 $\pm$ 7.98 times. AliveCor algorithm interpreted ECG as possible AF in 16% (8/50) of AIS study patients, 2.8% (8/285) of all AIS, and 4.1% (8/193) of non-AF AIS patients. Halcox et al. reported AF diagnosis in 3.8% participants in high-risk population (CHADS-VASc score $\geq$ 2) screening using KM device twice weekly over 12 months (plus additional use if symptomatic). Stroke in patients who are found to have clinical AF is one of the most powerful predictors of a recurrent stroke, and this was incorporated into the CHA2DS2-VASc (congestive heart failure, hypertension, age category, diabetes, stroke/TIA/systemic embolism history, sex, vascular disease history) score with two points [33,37]. Proportion of the newly identified AF after stroke are related not only to duration of registration, but also to the timing. The yield of AF detection is highest in the first weeks to months after the cerebral ischemic event, and more intensive monitoring detects more AF [9,38]. When cryptogenic AIS patients are not screened for AF using standard methods, the chance to register AF could be increased by using mobile devices, even at the time of hospitalization.

Although the recruitment rate in the AIS patients (26.3%) was relatively low compared to that of the other screening studies (76.1%), the AF detection rate (4.1% of non-AF patients) was similar to that of the other populations screened with mobile ECG (0.12–8%) [14,39,40]. The detection rate depends on the characteristics of the population being studied and was higher for the hospitalized patients with an increased risk of AF. The results of the presented study were also compared to the results of prolonged, continuous ECG using electrode belt in patients with a recent embolic stroke of unknown source [41]. Using this method in the pilot study, AF was diagnosed in 6.7% of patients. In the presented study, AIS patients able to use mobile ECG were younger and in better neurological condition than that of those excluded. AF diagnosis and proper treatment could help to prevent the next cerebrovascular event and avoid neurological deterioration in relatively independent patients. Diagnosis of patients who have a history of symptoms suggesting AF remains difficult. The only way to establish the underlying heart rhythm is to capture an ECG while the patient has symptoms. Many patients go for years without diagnosis due to the difficulty in capturing the underlying heart rhythm. The advantage of the KM is that patients could capture the heart rhythm at the time of AF symptoms [42,43]. In our study, we found no correlation between KM AF detection and the history of AF clinical symptoms in AIS patients. AF screening using KM could be useful for both symptomatic and asymptomatic AIS patients.

The AliveCor app rhythm analysis algorithm automatically reported any ECG recorded as normal, possible AF, or unclassified. The ECG should be reviewed by consultants, and the therapeutic decision should be made based on their opinion. In the present study, the cardiologist confirmed AF in one patient for every five ECG records. Ninety-five percent of these single-lead ECGs were of adequate quality for a rhythm decision. Pitman et al. reported 89.61% single-lead ECGs to be an adequate quality for a rhythm decision [44]. Although the correlations between extrasystoles and bigeminy were reported by the cardiologist and ECG interpreted by AliveCor algorithm as AF, the observed sensitivity and specificity of KM for AF in the selected high-risk group of AIS patients was 100% and 98.3%, respectively. This confirms data from other studies that reported its sensitivity (66.7–98.5%) and specificity (91.4–97%) [32,45–49]. The clinical outcome of the present study was AF diagnosis leading to therapeutic decisions and to a change in antithrombotic strategy from antiplatelets to anticoagulation in one AIS patient (1/50, 2% of study group; 1/285, 0.3% AIS patients; 1/193, 0.5% non-AF AIS patients). Considering the short duration of the observation time for each patient (three days), it seems to be a simple method to enhance diagnostic efficacy in selected AIS patients. Early diagnosis of AF provides the opportunity for early initiation of anticoagulation treatment to reduce stroke risk and other complications. Data from the literature also point to the role of statins in the treatment of AF-related stroke patients. Poststroke statin therapy reduced the risk of all-cause mortality [50]. In-hospital mortality is significantly higher in the AF-related stroke patients not taking statins before hospitalization than in those who did [51]. In the presented study, the AF screening was made within first nine days after admission to the stroke unit because of the AIS diagnosis (Table 1).

There are no data in the literature regarding AIS patients using smartphone electrographic screening tools while in the hospital. In Tu et al., the protocol for KM ECG monitoring is administered by nursing staff at the same frequency as the vital observations of pulse and blood pressure until discharge [52]. In our study, there was no limitation of KM usage, and selected patients used the device independently (without medical staff involvement). There was no correlation between each patient's total ECG records and AF registration, as we expected.

Regarding limitations of this study, the feasibility of implementing the KM device in AIS patients was relatively low compared to that of other studies. Patients screened in other studies were not neurologically related ones, nor were they stroke patients. Ischemic stroke is the third leading cause of disability globally [53,54]. Even patients with strokes that are considered mild in clinical assessment showed impaired concentration and memory and could have depression [55–57]. The prevalence of poststroke cognitive impairment ranges from 20–80%, which varies between countries, races, and diagnostic criteria [58]. AF increases risk of cognitive decline [59,60]. Those factors, which could affect ability to use KM device, were not investigated in the present study. It is difficult to assess cognitive function in early days of a stroke.

The duration of monitoring was short, as were the repeat intermittent screenings (three days). It was dictated by average length of AIS patient's hospital stay and another diagnostic method's schedule (for example, Holter ECG).

The selection bias was that the exclusion criteria for the study allowed only people who could use the KM device. There were 38.9% (75/193) of such participants in the high-risk group of non-AF AIS patients.

AIS patients included in the study were relatively young—64.4 years old, on average. Although some risk factors are the same for the embolic and thrombotic strokes, AF is leading cause of stroke among older patients (>75 years) [61,62]. In the present study, 40 elder patients (75 years old) were excluded from the study because they were unable to use KM device for unclear reasons. The age difference between those two groups was statistically significant ($p < 0.001$) (Table 1).

5. Conclusions

The results of this study suggest that it is feasible to use KardiaMobile (KM) device to detect atrial fibrillation (AF) in the selected acute ischemic stroke (AIS) patients (26.3% of AIS). These selected patients were younger (mean age 65 years, $p < 0.001$) and in better neurological condition (Scandinavian Stroke Scale, $p < 0.001$) than that of the remaining AIS patients, who were not able to use the device for unclear reasons (mean age: 75 years). KM detected AF in selected AIS patients with high specificity and sensitivity (100% and 98.3%, respectively).

Author Contributions: Conceptualization, M.L.-M. and U.F.; methodology, M.L.-M. and U.F.; software, M.K.; formal analysis, M.L.-M.; investigation, M.L.-M., A.K.-O., N.D., K.S. and A.O.; resources, M.L.-M.; data curation, M.L.-M.; writing—original draft preparation, M.L.-M.; writing—review and editing, M.L.-M. and U.F.; visualization, M.L.-M.; supervision, U.F.; project administration, M.L.-M.; funding acquisition, M.L.-M. and U.F. All authors have read and agreed to the published version of the manuscript.

Funding: This study was financially supported by the Centre of Postgraduate Medical Education, Warsaw, Poland (project number: 501-1-14-16-17). The funding body had no role in the design or conduct of the study, nor in the collection, management, analysis, or interpretation of the data. The funding body was not involved in the preparation, review, or approval of the manuscript in which the findings of the study were reported. There was no financial support from any company producing the evaluated devices.

Institutional Review Board Statement: The study was conducted in accordance with the Declaration of Helsinki, and approved by the local Ethics Committee of Centre of Postgraduate Medical Education (protocol 23/PB/2017 dated on 8th March 2017).

Informed Consent Statement: Informed consent was obtained from all subjects involved in the study.

Data Availability Statement: The data presented in this study are available on request from the corresponding author.

Conflicts of Interest: The authors declare no conflict of interest.

References

1. Yaghi, S.; Bernstein, R.A.; Passman, R.; Okin, P.M.; Furie, K.L. Cryptogenic Stroke. *Circ. Res.* **2017**, *120*, 527–540. [CrossRef] [PubMed]
2. Nah, H.-W.; Lee, J.-W.; Chung, C.-H.; Choo, S.-J.; Kwon, S.U.; Kim, J.S.; Warach, S.; Kang, D.-W. New brain infarcts on magnetic resonance imaging after coronary artery bypass graft surgery: Lesion patterns, mechanism, and predictors. *Ann. Neurol.* **2014**, *76*, 347–355. [CrossRef] [PubMed]
3. Dalen, J.E.; Alpert, J.S. Silent Atrial Fibrillation and Cryptogenic Strokes. *Am. J. Med.* **2017**, *130*, 264–267. [CrossRef] [PubMed]
4. Benjamin, E.J.; Muntner, P.; Alonso, A.; Bittencourt, M.S.; Callaway, C.W.; Carson, A.P.; Chamberlain, A.M.; Chang, A.R.; Cheng, S.; Das, S.R.; et al. Heart disease and stroke statistics—2019 update: A report from the American heart association. *Circulation* **2019**, *139*, e56–e528. [CrossRef] [PubMed]
5. Timmermans, C.; Lévy, S.; Ayers, G.M.; Jung, W.; Jordaens, L.; Rosenqvist, M.; Thibault, B.; Camm, J.; Rodriguez, L.-M.; Wellens, H.J. Spontaneous episodes of atrial fibrillation after implantation of the Metrix Atrioverter: Observations on treated and nontreated episodes. *J. Am. Coll. Cardiol.* **2000**, *35*, 1428–1433. [CrossRef]
6. Steinberg, J.S.; O'Connell, H.; Li, S.; Ziegler, P.D. Thirty-Second Gold Standard Definition of Atrial Fibrillation and Its Relationship with Subsequent Arrhythmia Patterns. *Circ. Arrhythmia Electrophysiol.* **2018**, *11*, e006274. [CrossRef] [PubMed]
7. Sgreccia, D.; Manicardi, M.; Malavasi, V.L.; Vitolo, M.; Valenti, A.C.; Proietti, M.; Lip, G.Y.H.; Boriani, G. Comparing Outcomes in Asymptomatic and Symptomatic Atrial Fibrillation: A Systematic Review and Meta-Analysis of 81,462 Patients. *J. Clin. Med.* **2021**, *10*, 3979. [CrossRef]
8. Sposato, L.A.; Cipriano, L.E.; Saposnik, G.; Vargas, E.R.; Riccio, P.M.; Hachinski, V. Diagnosis of atrial fibrillation after stroke and transient ischaemic attack: A systematic review and meta-analysis. *Lancet Neurol.* **2015**, *14*, 377–387. [CrossRef]
9. Sanna, T.; Diener, H.-C.; Passman, R.S.; Di Lazzaro, V.; Bernstein, R.A.; Morillo, C.; Rymer, M.M.; Thijs, V.; Rogers, T.; Beckers, F.; et al. Cryptogenic Stroke and Underlying Atrial Fibrillation. *N. Engl. J. Med.* **2014**, *370*, 2478–2486. [CrossRef]
10. Cerasuolo, J.O.; Cipriano, L.E.; Sposato, L.A. The complexity of atrial fibrillation newly diagnosed after ischemic stroke and transient ischemic attack: Advances and uncertainties. *Curr. Opin. Neurol.* **2017**, *30*, 28–37. [CrossRef]
11. Sposato, L.A.; Klein, F.R.; Jáuregui, A.; Ferrúa, M.; Klin, P.; Zamora, R.; Riccio, P.M.; Rabinstein, A. Newly Diagnosed Atrial Fibrillation after Acute Ischemic Stroke and Transient Ischemic Attack: Importance of Immediate and Prolonged Continuous Cardiac Monitoring. *J. Stroke Cerebrovasc. Dis.* **2012**, *21*, 210–216. [CrossRef] [PubMed]
12. Yang, X.; Li, S.; Zhao, X.; Liu, L.; Jiang, Y.; Li, Z.; Wang, Y.; Wang, Y. Atrial fibrillation is not uncommon among patients with ischemic stroke and transient ischemic stroke in China. *BMC Neurol.* **2017**, *17*, 207. [CrossRef] [PubMed]
13. Proesmans, T.; Mortelmans, C.; Van Haelst, R.; Verbrugge, F.; Vandervoort, P.; Vaes, B. Mobile Phone–Based Use of the Photoplethysmography Technique to Detect Atrial Fibrillation in Primary Care: Diagnostic Accuracy Study of the FibriCheck App. *JMIR mHealth uHealth* **2019**, *7*, e12284. [CrossRef] [PubMed]
14. Perales, C.R.L.; Van Spall, H.G.C.; Maeda, S.; Jimenez, A.; Laţcu, D.G.; Milman, A.; Kirakoya-Samadoulougou, F.; Mamas, M.A.; Muser, D.; Arroyo, R.C. Mobile health applications for the detection of atrial fibrillation: A systematic review. *Europace* **2020**, *23*, 11–28. [CrossRef]
15. Hindricks, G.; Potpara, T.; Dagres, N.; Arbelo, E.; Bax, J.J.; Blomström-Lundqvist, C.; Boriani, G.; Castella, M.; Dan, G.A.; ESC Scientific Document Group; et al. 2020 ESC Guidelines for the diagnosis and management of atrial fibrillation developed in collaboration with the European Asso-ciation for Cardio-Thoracic Surgery (EACTS). *Eur. Heart J.* **2021**, *42*, 373–498. [CrossRef] [PubMed]
16. Al-Alusi, M.A.; Ding, E.; McManus, D.D.; Lubitz, S.A. Wearing Your Heart on Your Sleeve: The Future of Cardiac Rhythm Monitoring. *Curr. Cardiol. Rep.* **2019**, *21*, 158. [CrossRef]
17. Chan, P.; Wong, C.; Poh, Y.C.; Pun, L.; Leung, W.W.; Wong, Y.; Wong, M.M.; Poh, M.; Chu, D.W.; Siu, C. Diagnostic Performance of a Smartphone-Based Photoplethysmographic Application for Atrial Fibrillation Screening in a Primary Care Setting. *J. Am. Heart Assoc.* **2016**, *5*, e003428. [CrossRef]
18. Treskes, R.W.; Gielen, W.; Wermer, M.J.; Grauss, R.W.; Van Alem, A.P.; Dehnavi, R.A.; Kirchhof, C.J.; Van Der Velde, E.T.; Maan, A.C.; Wolterbeek, R.; et al. Mobile phones in cryptogenic strOke patients Bringing sIngle Lead ECGs for Atrial Fibrillation detection (MOBILE-AF): Study protocol for a randomised controlled trial. *Trials* **2017**, *18*, 402. [CrossRef]
19. Ding, E.Y.; Svennberg, E.; Wurster, C.; Duncker, D.; Manninger, M.; Lubitz, S.A.; Dickson, E.; Fitzgibbons, T.P.; Akoum, N.; Al-Khatib, S.M.; et al. Survey of current perspectives on consumer-available digital health devices for detecting atrial fibrillation. *Cardiovasc. Digit. Heal. J.* **2020**, *1*, 21–29. [CrossRef]
20. Schnabel, R.B.; Haeusler, K.G.; Healey, J.S.; Freedman, B.; Boriani, G.; Brachmann, J.; Brandes, A.; Bustamante, A.; Casadei, B.; Crijns, H.J.; et al. Searching for Atrial Fibrillation Poststroke. *Circulation* **2019**, *140*, 1834–1850. [CrossRef]
21. Koh, K.T.; Law, W.C.; Zaw, W.M.; Foo, D.H.P.; Tan, C.T.; Steven, A.; Samuel, D.; Fam, T.L.; Chai, C.H.; Wong, Z.S.; et al. Smartphone electrocardiogram for detecting atrial fibrillation after a cerebral ischaemic event: A multicentre randomized controlled trial. *Europace* **2021**, *23*, 1016–1023. [CrossRef] [PubMed]
22. Tayal, A.H.; Tian, M.; Kelly, K.M.; Jones, S.C.; Wright, D.G.; Singh, D.; Jarouse, J.; Brillman, J.; Murali, S.; Gupta, R. Atrial fibrillation detected by mobile cardiac outpatient telemetry in cryptogenic TIA or stroke. *Neurology* **2008**, *71*, 1696–1701. [CrossRef] [PubMed]
23. Gladstone, D.; Spring, M.; Dorian, P.; Panzov, V.; Thorpe, K.; Hall, J.; Vaid, H.; O'Donnell, M.; Laupacis, A.; Côté, R.; et al. Atrial Fibrillation in Patients with Cryptogenic Stroke. *N. Engl. J. Med.* **2014**, *370*, 2467–2477. [CrossRef] [PubMed]

24. Miller, D.J.; Khan, M.A.; Schultz, L.R.; Simpson, J.R.; Katramados, A.M.; Russman, A.N.; Mitsias, P.D. Outpatient cardiac telemetry detects a high rate of atrial fibrillation in cryptogenic stroke. *J. Neurol. Sci.* **2013**, *324*, 57–61. [CrossRef] [PubMed]
25. Wang, Y.; Qian, Y.; Smerin, D.; Zhang, S.; Zhao, Q.; Xiong, X. Newly Detected Atrial Fibrillation after Acute Stroke: A Narrative Review of Causes and Implications. *Cardiology* **2019**, *144*, 112–121. [CrossRef]
26. Scandinavian Stroke Study Group. Multicenter trial of hemodilution in ischemic stroke—Background and study protocol. *Stroke* **1985**, *16*, 885–890. [CrossRef]
27. Persson, C.U.; Holmegaard, L.; Redfors, P.; Jern, C.; Blomstrand, C.; Jood, K. Increased muscle tone and contracture late after ischemic stroke. *Brain Behav.* **2020**, *10*, e01509. [CrossRef]
28. Vanbosse, H.; Fafara, A.; Gagnon, M.; Collins, J.; Elfassy, C.; Merlo, G.M.; Marsh, J.; Sawatzky, B.; Yap, R.; Hamdy, R.; et al. A Telerehabilitation Intervention for Youths with Arthrogryposis Multiplex Congenita: Protocol for a Pilot Study. *JMIR Res. Protoc.* **2020**, *9*, e18688. [CrossRef]
29. Gagnon, M.; Merlo, G.M.; Yap, R.; Collins, J.; Elfassy, C.; Sawatzky, B.; Marsh, J.; Hamdy, R.; Veilleux, L.-N.; Dahan-Oliel, N. Using Telerehabilitation to Deliver a Home Exercise Program to Youth with Arthrogryposis: Single Cohort Pilot Study. *J. Med. Internet Res.* **2021**, *23*, e27064. [CrossRef]
30. Garabelli, P.; Stavrakis, S.; Po, S. Smartphone-based arrhythmia monitoring. *Curr. Opin. Cardiol.* **2017**, *32*, 53–57. [CrossRef]
31. Goldenthal, I.L.; Sciacca, R.R.; Riga, T.; Bakken, S.; Baumeister, M.; Biviano, A.B.; Dizon, J.M.; Wang, D.; Wang, K.C.; Whang, W.; et al. Recurrent atrial fibrillation/flutter detection after ablation or cardioversion using the AliveCor KardiaMobile device: iHEART results. *J. Cardiovasc. Electrophysiol.* **2019**, *30*, 2220–2228. [CrossRef] [PubMed]
32. Lowres, N.; Neubeck, L.; Salkeld, G.; Krass, I.; McLachlan, A.J.; Redfern, J.; Bennett, A.A.; Briffa, T.; Bauman, A.; Martinez, C.; et al. Feasibility and cost-effectiveness of stroke prevention through community screening for atrial fibrillation using iPhone ECG in pharmacies. *Thromb. Haemost.* **2014**, *111*, 1167–1176. [CrossRef] [PubMed]
33. Halcox, J.P.; Wareham, K.; Cardew, A.; Gilmore, M.; Barry, J.P.; Phillips, C.; Gravenor, M.B. Assessment of Remote Heart Rhythm Sampling Using the AliveCor Heart Monitor to Screen for Atrial Fibrillation. *Circulation* **2017**, *136*, 1784–1794. [CrossRef] [PubMed]
34. Benezet-Mazuecos, J.; García-Talavera, C.S.; Rubio, J.M. Smart devices for a smart detection of atrial fibrillation. *J. Thorac. Dis.* **2018**, *10*, S3824–S3827. [CrossRef]
35. Atrial Fibrillation in the Old/Very Old: Prevalence and Burden, Predisposing Factors and Complications. Available online: https://www.escardio.org/Journals/E-Journal-of-Cardiology-Practice/Volume-17/Atrial-fibrillation-in-the-old-very-old-prevalence-and-burden-predisposing-factors-and-complications (accessed on 19 November 2021).
36. NICE Guidance. KardiaMobile for the Ambulatory Detection of Atrial Fibrillation. *Medtech Innovation Briefing [MIB232]*. Available online: https://www.nice.org.uk/advice/mib232 (accessed on 29 October 2020).
37. Lip, G.Y.; Nieuwlaat, R.; Pisters, R.; Lane, D.A.; Crijns, H.J. Refining Clinical Risk Stratification for Predicting Stroke and Thromboembolism in Atrial Fibrillation Using a Novel Risk Factor-Based Approach. *Chest* **2010**, *137*, 263–272. [CrossRef]
38. Thijs, V. Atrial Fibrillation Detection. *Stroke* **2017**, *48*, 2671–2677. [CrossRef]
39. Ghazal, F.; Theobald, H.; Rosenqvist, M.; Al-Khalili, F. Feasibility and outcomes of atrial fibrillation screening using intermittent electrocardiography in a primary healthcare setting: A cross-sectional study. *PLoS ONE* **2018**, *13*, e0198069. [CrossRef]
40. Rosenfeld, L.E.; Amin, A.N.; Hsu, J.; Oxner, A.; Hills, M.T.; Frankel, D.S. The Heart Rhythm Society/American College of Physicians Atrial Fibrillation Screening and Education Initiative. *Heart Rhythm.* **2019**, *16*, e59–e65. [CrossRef]
41. Lumikari, T.J.; Pirinen, J.; Putaala, J.; Sibolt, G.; Kerola, A.; Pakarinen, S.; Lehto, M.; Nieminen, T. Prolonged ECG with a novel recorder utilizing electrode belt and mobile device in patients with recent embolic stroke of undetermined source: A pilot study. *Ann. Noninvasive Electrocardiol.* **2020**, *25*, e12802. [CrossRef]
42. Jones, N.R.; Taylor, C.J.; Hobbs, F.D.R.; Bowman, L.; Casadei, B. Screening for atrial fibrillation: A call for evidence. *Eur. Heart J.* **2019**, *41*, 1075–1085. [CrossRef]
43. Reed, M.J.; Grubb, N.R.; Lang, C.C.; O'Brien, R.; Simpson, K.; Padarenga, M.; Grant, A.; Tuck, S.; Keating, L.; Coffey, F.; et al. Multi-centre Randomised Controlled Trial of a Smartphone-based Event Recorder Alongside Standard Care Versus Standard Care for Patients Presenting to the Emergency Department with Palpitations and Pre-syncope: The IPED (Investigation of Palpitations in the ED) study. *EClinicalMedicine* **2019**, *8*, 37–46. [CrossRef]
44. Pitman, B.M.; Chew, S.-H.; Wong, C.X.; Jaghoori, A.; Iwai, S.; Thomas, G.; Chew, A.; Sanders, P.; Lau, D.H. Performance of a Mobile Single-Lead Electrocardiogram Technology for Atrial Fibrillation Screening in a Semirural African Population: Insights From "The Heart of Ethiopia: Focus on Atrial Fibrillation" (TEFF-AF) Study. *JMIR mHealth uHealth* **2021**, *9*, e24470. [CrossRef] [PubMed]
45. Chan, P.-H.; Wong, C.-K.; Pun, L.; Wong, Y.-F.; Wong, M.M.-Y.; Chu, D.W.-S.; Siu, C.-W. Head-to-Head Comparison of the AliveCor Heart Monitor and Microlife WatchBP Office AFIB for Atrial Fibrillation Screening in a Primary Care Setting. *Circulation* **2017**, *135*, 110–112. [CrossRef] [PubMed]
46. Brasier, N.; Raichle, C.J.; Dörr, M.; Becke, A.; Nohturfft, V.; Weber, S.; Bulacher, F.; Salomon, L.; Noah, T.; Birkemeyer, R.; et al. Detection of atrial fibrillation with a smartphone camera: First prospective, international, two-centre, clinical validation study (DETECT AF PRO). *Europace* **2018**, *21*, 41–47. [CrossRef] [PubMed]

47. Koltowski, L.; Balsam, P.; Głowczynska, R.; Rokicki, J.K.; Peller, M.; Maksym, J.; Blicharz, L.; Maciejewski, K.; Niedziela, M.; Opolski, G.; et al. Kardia Mobile applicability in clinical practice: A comparison of Kardia Mobile and standard 12-lead electrocardiogram records in 100 consecutive patients of a tertiary cardiovascular care center. *Cardiol. J.* **2021**, *28*, 543–548. [CrossRef] [PubMed]

48. Wong, K.C.; Klimis, H.; Lowres, N.; Von Huben, A.; Marschner, S.; Chow, C.K. Diagnostic accuracy of handheld electrocardiogram devices in detecting atrial fibrillation in adults in community versus hospital settings: A systematic review and meta-analysis. *Heart* **2020**, *106*, 1211–1217. [CrossRef] [PubMed]

49. Duarte, R.; Stainthorpe, A.; Greenhalgh, J.; Richardson, M.; Nevitt, S.; Mahon, J.; Kotas, E.; Boland, A.; Thom, H.; Marshall, T.; et al. Lead-I ECG for detecting atrial fibrillation in patients with an irregular pulse using single time point testing: A systematic review and economic evaluation. *Health Technol. Assess.* **2020**, *24*, 1–164. [CrossRef]

50. Eun, M.-Y.; Jung, J.-M.; Choi, K.-H.; Seo, W.-K. Statin Effects in Atrial Fibrillation-Related Stroke: A Systematic Review and Meta-Analysis. *Front. Neurol.* **2020**, *11*, 589684. [CrossRef]

51. Wańkowicz, P.; Staszewski, J.; Dębiec, A.; Nowakowska-Kotas, M.; Szylińska, A.; Turoń-Skrzypińska, A.; Rotter, I. Pre-Stroke Statin Therapy Improves In-Hospital Prognosis Following Acute Ischemic Stroke Associated with Well-Controlled Nonvalvular Atrial Fibrillation. *J. Clin. Med.* **2021**, *10*, 3036. [CrossRef]

52. Tu, H.T.; Chen, Z.; Swift, C.; Churilov, L.; Guo, R.; Liu, X.; Jannes, J.; Mok, V.; Freedman, B.; Davis, S.M.; et al. Smartphone electrographic monitoring for atrial fibrillation in acute ischemic stroke and transient ischemic attack. *Int. J. Stroke* **2017**, *12*, 786–789. [CrossRef]

53. *Global Health Estimates 2020: Disease Burden by Cause, Age, Sex, by Country and by Region, 2000–2019*; World Health Organization: Geneva, Switzerland. 2020. Available online: https://www.who.int/data/gho/data/themes/mortality-and-global-health-estimates/global-health-estimates-leading-causes-of-dalys (accessed on 18 May 2021).

54. Panahi, Y.; Sahebkar, A.; Naderi, Y.; Barreto, G.E. Neuroprotective effects of minocycline on focal cerebral ischemia injury: A systematic review. *Neural Regen. Res.* **2020**, *15*, 773–782. [CrossRef] [PubMed]

55. Mahon, S.; Parmar, P.; Barker-Collo, S.; Krishnamurthi, R.; Jones, K.; Theadom, A.; Feigin, V. Determinants, Prevalence, and Trajectory of Long-Term Post-Stroke Cognitive Impairment: Results from a 4-Year Follow-Up of the Auckland Stroke Regional Outcomes Study-IV Study. *Neuroepidemiology* **2017**, *49*, 129–134. [CrossRef] [PubMed]

56. Kapoor, A.; Lanctôt, K.L.; Bayley, M.; Kiss, A.; Herrmann, N.; Murray, B.J.; Swartz, R.H. "Good Outcome" Isn't Good Enough. *Stroke* **2017**, *48*, 1688–1690. [CrossRef] [PubMed]

57. Halevi, D.R.; Bursaw, A.W.; Karamchandani, R.R.; Alderman, S.; Breier, J.I.; Vahidy, F.; Aden, J.K.; Cai, C.; Zhang, X.; Savitz, S.I. Cognitive deficits in acute mild ischemic stroke and TIA and effects of rt-PA. *Ann. Clin. Transl. Neurol.* **2019**, *6*, 466–474. [CrossRef] [PubMed]

58. Sun, J.-H.; Tan, L.; Yu, J.-T. Post-stroke cognitive impairment: Epidemiology, mechanisms and management. *Ann. Transl. Med.* **2014**, *2*. [CrossRef]

59. Chen, L.Y.; Agarwal, S.K.; Norby, F.; Gottesman, R.F.; Loehr, L.; Soliman, E.Z.; Mosley, T.H.; Folsom, A.R.; Coresh, J.; Alonso, A. Persistent but not Paroxysmal Atrial Fibrillation Is Independently Associated with Lower Cognitive Function. *J. Am. Coll. Cardiol.* **2016**, *67*, 1379–1380. [CrossRef]

60. Singh, N.; Chun, S.; Hadley, D.; Froelicher, V. Clinical Implications of Technological Advances in Screening for Atrial Fibrillation. *Prog. Cardiovasc. Dis.* **2018**, *60*, 550–559. [CrossRef]

61. Baturova, M.A.; Lindgren, A.; Carlson, J.; Shubik, Y.V.; Olsson, S.B.; Platonov, P.G. Predictors of new onset atrial fibrillation during 10-year follow-up after first-ever ischemic stroke. *Int. J. Cardiol.* **2015**, *199*, 248–252. [CrossRef]

62. Wańkowicz, P.; Nowacki, P.; Gołąb-Janowska, M. Atrial fibrillation risk factors in patients with ischemic stroke. *Arch. Med. Sci.* **2021**, *17*, 19–24. [CrossRef]

Journal of
Clinical Medicine

Article

Effect of Intravenous Alteplase on Functional Outcome and Secondary Injury Volumes in Stroke Patients with Complete Endovascular Recanalization

Gabriel Broocks [1,*], Lukas Meyer [1], Celine Ruppert [1], Wolfgang Haupt [1], Tobias D. Faizy [1], Noel Van Horn [1], Matthias Bechstein [1], Helge Kniep [1], Sarah Elsayed [1], Andre Kemmling [2,3], Ewgenia Barow [4], Jens Fiehler [1] and Uta Hanning [1]

[1]   Department of Neuroradiology, University Medical Center Hamburg-Eppendorf, 20251 Hamburg, Germany; lu.meyer@uke.de (L.M.); celine@ruppertinheidelberg.de (C.R.); wolfgang.haupt0@gmail.com (W.H.); t.faizy@uke.de (T.D.F.); no.vanhorn@uke.de (N.V.H.); m.bechstein@uke.de (M.B.); h.kniep@uke.de (H.K.); s.elsayed@uke.de (S.E.); fiehler@uke.de (J.F.); u.hanning@uke.de (U.H.)

[2]   Department of Neuroradiology, Philipps-University Marburg, 35037 Marburg, Germany; akemmling@gmail.com

[3]   Department of Neuroradiology, University Medical Center Schleswig-Holstein, Campus Lübeck, 23538 Lübeck, Germany

[4]   Department of Neurology, University Medical Center Hamburg-Eppendorf, 20251 Hamburg, Germany; e.barow@uke.de

*   Correspondence: g.broocks@uke.de

Citation: Broocks, G.; Meyer, L.; Ruppert, C.; Haupt, W.; Faizy, T.D.; Van Horn, N.; Bechstein, M.; Kniep, H.; Elsayed, S.; Kemmling, A.; et al. Effect of Intravenous Alteplase on Functional Outcome and Secondary Injury Volumes in Stroke Patients with Complete Endovascular Recanalization. *J. Clin. Med.* **2022**, *11*, 1565. https://doi.org/10.3390/jcm11061565

Academic Editor: Georgios Tsivgoulis

Received: 11 February 2022
Accepted: 8 March 2022
Published: 12 March 2022

Publisher's Note: MDPI stays neutral with regard to jurisdictional claims in published maps and institutional affiliations.

Abstract: Intravenous thrombolytic therapy with alteplase (IVT) is a standard of care in ischemic stroke, while recent trials investigating direct endovascular thrombectomy (EVT) approaches showed conflicting results. Yet, the effect of IVT on secondary injury volumes in patients with complete recanalization has not been analyzed. We hypothesized that IVT is associated with worse functional outcome and aggravated secondary injury volumes when administered to patients who subsequently attained complete reperfusion after EVT. Anterior circulation ischemic stroke patients with complete reperfusion after thrombectomy defined as thrombolysis in cerebral infarctions (TICI) scale 3 after thrombectomy admitted between January 2013–January 2021 were analyzed. Primary endpoints were the proportion of patients with functional independence defined as modified Rankin Scale (mRS) score 0–2 at day 90, and secondary injury volumes: Edema volume in follow-up imaging measured using quantitative net water uptake (NWU), and the rate of symptomatic intracerebral hemorrhage (sICH). A total of 219 patients were included and 128 (58%) patients received bridging IVT before thrombectomy. The proportion of patients with functional independence was 28% for patients with bridging IVT, and 34% for patients with direct thrombectomy ($p = 0.35$). The rate of sICH was significantly higher after bridging IVT (20% versus 7.7%, $p = 0.01$). Multivariable logistic and linear regression analysis confirmed the independent association of bridging IVT with sICH (aOR: 2.78, 95% CI: 1.02–7.56, $p = 0.046$), and edema volume (aOR: 8.70, 95% CI: 2.57–14.85, $p = 0.006$). Bridging IVT was associated with increased edema volume and risk for sICH as secondary injury volumes. The results of this study encourage direct EVT approaches, particularly in patients with higher likelihood of successful EVT.

Keywords: stroke; thrombectomy; thrombolysis

1. Introduction

The administration of intravenous recombinant tissue plasminogen activator (IVT) within the first hours after stroke onset is a standard of care in ischemic stroke, and is often used in combination with endovascular thrombectomy (EVT), which has proven to be beneficial to patients with ischemic stroke caused by a large vessel occlusion (LVO) in the anterior circulation [1–4]. The primary treatment target of IVT and EVT in LVO stroke is vessel recanalization, which has shown to be the main determinant of functional outcome [5–7]. Vessel recanalization after IVT alone, however, occurs rarely, depending on vessel size [5]. On the other side, reported recanalization rates in EVT trials increased continuously: 40% in IMS-III, 2013; 60% in MR CLEAN, 2015; 77% in DEFUSE-3, 2018; 86% in ESCAPE-NA1, 2020, respectively [8–10], mainly due to better devices and rising experience. IVT has been administered in the majority of the included patients, except DEFUSE-3 with only 10% of patients receiving IVT. Recently, it has been observed that direct EVT was non-inferior compared to patients undergoing EVT with prior IVT with regards to functional outcomes, however, other trials did not observe non-inferiority of direct EVT [11,12]. Consequently, it is discussed controversially to ascertain in which patients to apply IVT, and in particular to investigate factors associated with additional harm [13–15]. The increasing rates of vessel recanalization after EVT alone emphasize the need to investigate the effects of IVT on lesion pathophysiology and clinical outcome in patients with complete reperfusion. The purpose of this study was to compare clinical outcomes and secondary injury volumes in patients with versus without IVT and complete EVT. We hypothesized that IVT is associated with worse functional outcome at day 90 and increased secondary injury volumes at follow-up imaging (i.e., symptomatic intracranial hemorrhage and ischemic edema formation) when administered in patients with ischemic stroke who subsequently underwent complete recanalization by EVT with a thromboloysis in cerebral infarction (TICI) of 3.

2. Materials and Methods

2.1. Patients

All ischemic stroke patients with LVO in the anterior circulation admitted between January 2013–January 2021 at a high-volume tertiary stroke center were consecutively analyzed. Only anonymized data were analyzed after ethical review board approval, and informed consent was waived after review.

The a priori defined inclusion criteria for this study were: (1) ischemic anterior circulation LVO stroke with multimodal CT imaging on admission (non-enhanced CT (NECT), CT angiography (CTA) and CT perfusion (CTP)); (2) occlusion of the intracranial internal carotid artery or proximal middle cerebral artery (M1 segment); (3) known onset of symptoms; (4) complete reperfusion after EVT defined as a Thrombolysis in Cerebral Infarction (TICI) score of 3; (5) absence of intracranial hemorrhage; (6) absence of significant imaging artifacts. TICI rating was determined by the operating neurointerventionalist and validated by a further attending neuroradiologist. Good functional outcomes were defined as modified Rankin Scale (mRS) scores 0–2. The mRS scores were evaluated at the 90-day follow-up by a physician or a trained and certified mRS nurse. sICH was defined according to the second European–Australasian Acute Stroke Study (ECASS II) as presence of intracerebral hemorrhage and a 4-point neurological deterioration on the National Institute of Health Stroke Scale (NIHSS) [16].

2.2. Revascularization Protocol

IVT was administered to patients according to current guidelines [17]. Laboratory and conventional clinical inclusion and exclusion criteria for IVT applied. EVT was performed via a femoral artery approach under general anesthesia or conscious sedation. Endovascular procedures were performed by board-certified interventional neuroradiologists. The choice of thrombectomy device was left to the operator.

2.3. Image Analysis

The readers were blinded for all clinical data. Image analyses including volumetric and densitometric analysis were performed using commercially available software (Analyze 11.0, Biomedical Imaging Resource, Mayo Clinic, Rochester, MN, USA) by a board-certified experienced neuroradiologist using standardized procedures. Ischemic lesion net water uptake (NWU) was measured as an imaging biomarker to quantify cerebral ischemic edema before treatment on admission CT and on follow-up CT (FCT) 24 h after treatment with a standardized procedure, as previously described in detail [7,18–20]. NWU describes ischemic lesion water uptake per volume of brain infarct (i.e., relative proportion of edema) [7,20]. ΔNWU was defined as the additional NWU from admission to FCT for each patient (ΔNWU = NWU$_{fct}$ − NWU$_{admission}$) [7]. Total infarct volume has been defined as the volume of the whole visually evident lesion in FCT and was measured using semiautomatic volumetric segmentation. Edema volume (EV) as a measure of the absolute volume of edema was then determined as described before [21] as the product of the total infarct volume and NWU$_{fct}$ according to Equation (1) according to Nawabi et al. [21].

Equation (1) [19,21]:

$$\text{Edema volume} = \text{Total infarct volume} * \% - \text{NWU} \qquad (1)$$

2.4. Statistical Analysis

Univariable distribution of metric variables was described by median and interquartile range (IQR) or means and standard deviation (SD). Categorial variables were compared using χ^2-tests. Shapiro–Wilk tests were used to test for normal distribution. Patients who received bridging IVT were compared to patients who did not receive IVT and directly underwent EVT. Patient characteristics are shown in Table 1. As endpoints, we defined the proportion of patients with functional independence (mRS 0–2), the degree of edema formation (ΔNWU), the occurrence of sICH, and edema volume. To determine the treatment effect of IVT on the aforementioned endpoints, we used inverse-probability weighted regression adjustments using logit outcome and treatment models adjusted for baseline variables (age, NIHSS, ASPECTS, occlusion location, and time from onset to imaging). Secondly, we investigated the association of the independent variables on functional independence and sICH using uni- and multivariable logistic regression analysis, as well as ΔNWU and edema volume using linear regression analysis, respectively. Correlation between all independent variables was tested to exclude multicollinearity.

Table 1. Patient characteristics.

Baseline Characteristics	Bridgin IVT	Direct MT	*p* Value
Subjects, *n* (%)	128 (58)	91 (42)	
Baseline variables			
Age in years, median (IQR)	73 (62–80)	76 (66–83)	0.06
Female sex, *n* (%)	56 (44)	72 (56)	0.10
Admission NIHSS, median (IQR)	16 (12–20)	16 (10–20)	0.66
ASPECTS, median (IQR)	6 (6–8)	7 (5–8)	0.63
Time from onset to imaging in h, median (IQR)	3.1 (1.4–4.4)	2.0 (1.0–4–0)	0.38
Time imaging to reperfusion in h, median (IQR)	1.7 (1.3–2.0)	1.8 (1.6–2.8)	0.05
Endpoints			
Follow-up infarct volume in mL, median (IQR)	48 (18–92)	37 (11–65)	0.04
Follow-up NWU, mean % (SD)	14.7 (8.1)	11.9 (0.8)	0.02
Edema volume in mL, median (IQR)	6 (2–17)	4 (1–11)	0.04
ΔNWU, median % (IQR)	6.1 (0.5–11.9)	4.1 (1.6–7.7)	0.09
NIHSS at 24 h	10 (4–21)	13 (5–18)	0.75
Modified Rankin Scale, median (IQR)	4 (2–5)	3 (1–5)	0.61
mRS 0–2, *n* (%)	36 (28)	31 (34)	0.35
sICH, *n* (%)	26 (20.3)	7 (7.7)	0.01

NIHSS: National Institute of Health Stroke Scale, IQR: Interquartile Range, ASPECTS: Alberta Stroke Program Early CT Score, h: hours, mL: milliliters, NWU: net water uptake, sICH: symptomatic intracranial hemorrhage.

A statistically significant difference was accepted at a *p*-value of less than 0.05. Analyses were performed using Stata 17.0 (StataMP, StataCorp, College Station, TX, USA).

3. Results

3.1. Study Cohort

A total of 219 patients were included. The patient characteristics are displayed in Table 1. The median age was 74 (IQR: 63–81) and 106 patients were female (48%). The median NIHSS was 16 (IQR: 11–20) and the median ASPECTS was 7 (IQR: 6–8). The median time from symptom onset to imaging was 3 h (IQR: 1.3–4.0). A total of 128 patients received IVT (58%). The median time from imaging to recanalization was 1.7 h (IQR: 1.4–2.4). After 24 h, the median NIHSS was 11 (IQR: 4–20), and the median mRS at day 90 was 3 (IQR: 2–5).

Comparing patients who received IVT to patients who did not receive IVT, there were no significant differences in sex, admission NIHSS (median 16, $p = 0.66$), or ASPECTS (median 6 versus 7, $p = 0.63$). Tendentially, patients with IVT were younger (73 versus 76 years, $p = 0.06$). There were no differences regarding the time from symptom onset to imaging (3.1 versus 2.0 h, $p = 0.38$) and time from imaging to complete reperfusion (1.7 versus 1.8 h, $p = 0.05$) (Table 1). Figure 1 illustrates an example of a patient receiving IVT who develops significant edema formation. In total, 94 patients (43%) showed futile recanalization defined as mRS 4–6 at day 90 despite complete recanalization. Considering early neurological change, 92% of all patients showed a lower NIHSS score at discharge compared to the NIHSS on admission.

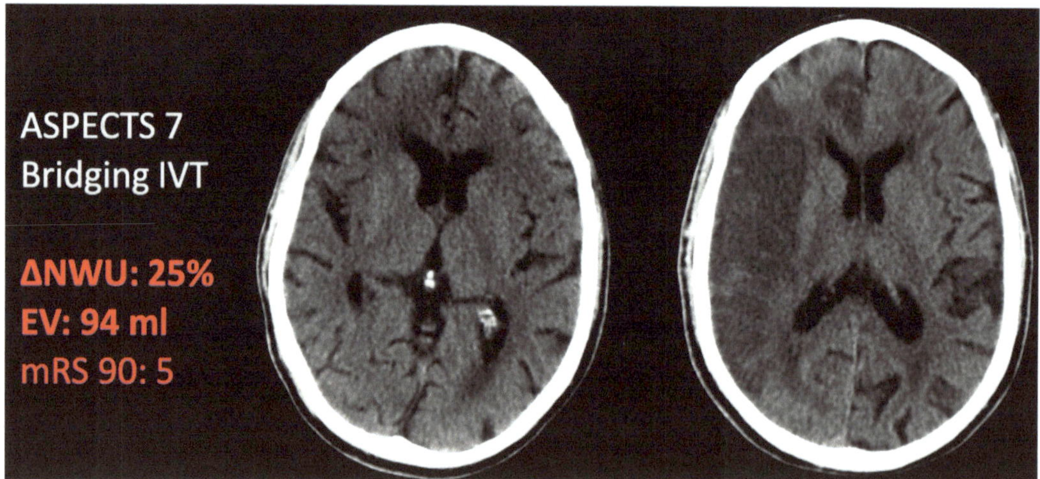

Figure 1. Illustration of a patient receiving bridging intravenous alteplase prior to complete endovascular recanalization. The patient showed significant edema progression with aggravated absolute edema volume in follow-up imaging. ΔNWU reflects the change of the relative edema proportion from admission to follow-up (ΔNWU = % NWU admission − % NWU follow-up) as previously defined [7].

3.2. Impact of IVT on Clinical Endpoints

Comparing patients with bridging IVT to patients with direct EVT, no differences were observed in the median NIHSS at 24 h (10 versus 13, $p = 0.75$). The median mRS at day 90 was similar: 4 (IQR: 2–5) in patients with bridging IVT, and 3 (IQR: 1–5) in patients with direct EVT ($p = 0.61$). The proportion of patients with functional independence was 28% for patients with bridging IVT and 34% for patients with direct EVT ($p = 0.35$) (Table 1). After regression adjustment, no significant treatment effect of IVT on functional independence as an endpoint was observed, adjusted for age, NIHSS, ASPECTS, occlusion location, and time from onset to imaging ($p = 0.11$), with an adjusted proportion of functional independence of 38% (95% CI: 28–48%) for patients with direct EVT, compared to 28% (95% CI: 19–37%) for patients with bridging IVT. In multivariable logistic regression analysis, IVT was not associated with functional independence, despite a small trend towards reduced probability for good outcome (OR: 0.48, $p = 0.09$). Independent significant predictors were age (OR: 0.93, $p < 0.001$), and NIHSS (OR: 0.85, $p < 0.001$) (Table 2, Figure 2).

Table 2. Binary logistic regression analysis for good clinical outcome (A; mRS 0–2) and for symptomatic intracranial hemorrhage (B; sICH).

(A) mRS 0–2	Univariable Analysis			Multivariable Analysis			
	OR	95% CI	*p* Value	OR	95% CI		*p* Value
Age	0.94	0.92–0.97	<0.001	0.93	0.89–0.97		<0.001
NIHSS	0.82	0.77–0.88	<0.001	0.85	0.78–0.91		<0.001
Time onset to imaging	1.04	0.81–1.33	0.77	–	–	–	–
ASPECTS	1.20	0.98–1.47	0.004	–	–	–	0.26
IVT	0.76	0.42–1.35	0.35	0.48	0.21–1.12		0.09
Time image to recan	0.88	0.64–1.20	0.41	–	–	–	–
(B) sICH							
Age	1.00	0.98–1.03	0.65	–	–		–
NIHSS	1.03	0.98–1.09	0.25	–	–		–
Time onset to imaging	1.38	1.00–1.90	0.046	1.34	0.97–1.84	–	0.07
ASPECTS	0.74	0.55–0.99	0.045	0.73	0.54–0.99	–	0.044
IVT	3.06	1.27–7.39	0.01	2.78	1.02–7.56		0.046
Time imaging to recan	0.85	0.48–1.49	0.57	–	–	–	–

IVT: Intravenous treatment with alteplase.

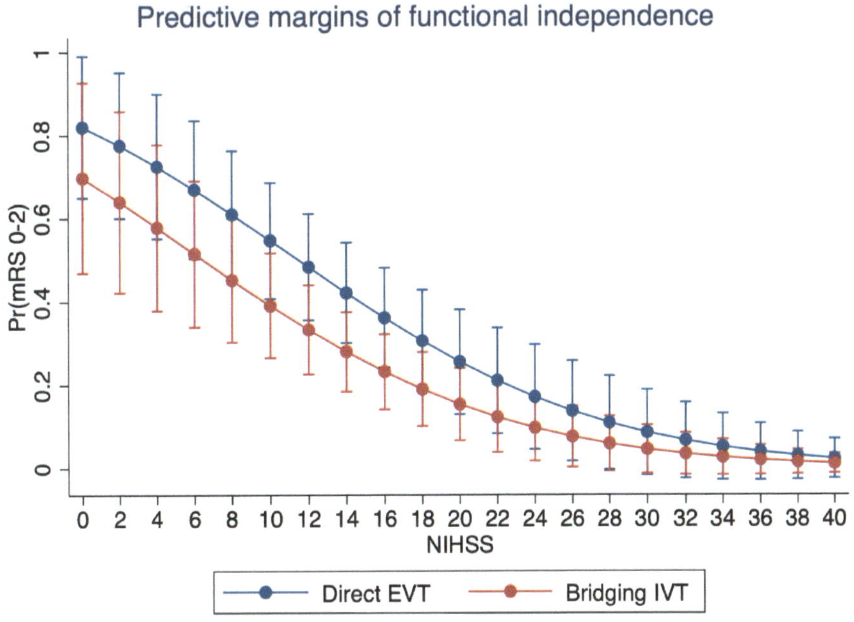

Figure 2. Prediction of functional independence (*y* axis) based on multivariable logistic regression analysis to show the impact of IVT application according to the baseline NIHSS adjusted for age and ASPECTS (Table 2).

3.3. Impact of IVT on Secondary Injury Volumes

Patients with bridging IVT had higher total infarct volumes (48 mL versus 37 mL, $p = 0.04$), and higher EV (6 mL versus 4 mL, $p = 0.039$). The rate of sICH was significantly higher in patients with bridging IVT (26% versus 7%, $p = 0.01$). After regression adjustment, IVT had a significant effect on the occurrence of sICH ($p = 0.029$) with an adjusted proportion of sICH of 7.8% (95% CI: 1.7–13.8%) in direct EVT patients versus 19.0% (95% CI: 10.9–27.1%), adjusted for age, NIHSS, ASPECTS, occlusion location, and time from onset to imaging. Second, IVT had a significant effect on EV ($p = 0.01$) after regression adjustment, with an adjusted EV of 6.9 mL (95% CI: 4.9–9.0 mL) for patients with direct EVT and 14.2 mL (95% CI: 9.0–19.5 mL) for patients with bridging EVT, adjusted for the aforementioned variables. In multivariable logistic regression analysis, IVT (OR: 2.78, $p = 0.046$), time from onset to imaging (OR: 1.34, $p = 0.046$), and ASPECTS (OR: 0.74, $p = 0.044$) were significantly and independently associated with sICH (Table 2, Figure 3). In multivariable linear regression analysis, IVT was significantly associated with higher EV (ß = 8.7, 95% CI: 2.6–14.8, $p < 0.01$) (Figure 4).

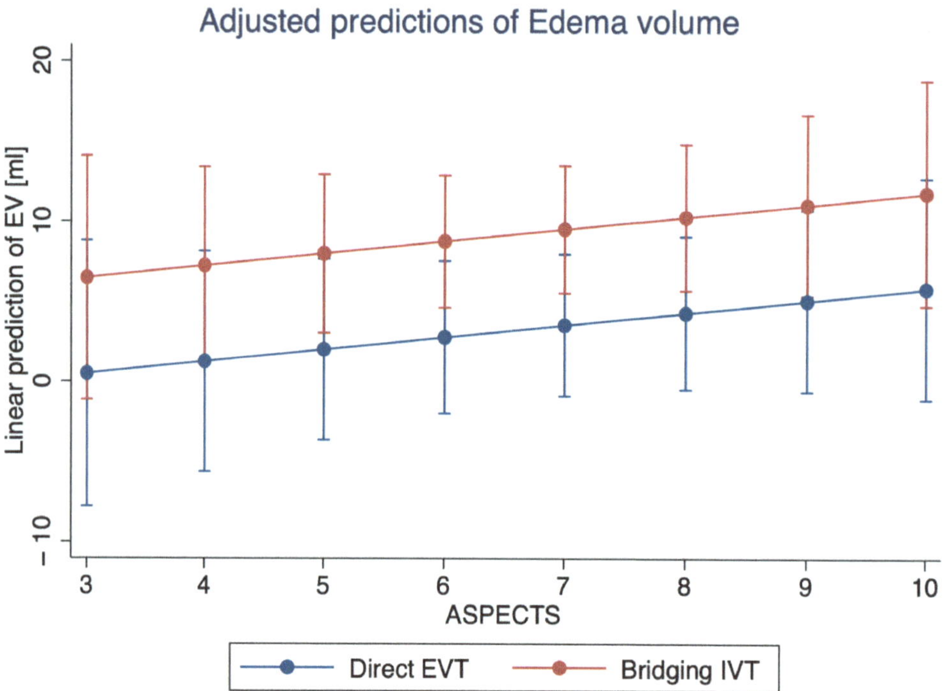

Figure 3. Prediction of edema volume (*y* axis) based on multivariable linear regression analysis to show the impact of IVT application according to the baseline ASPECTS. No further variable was independently associated with EV.

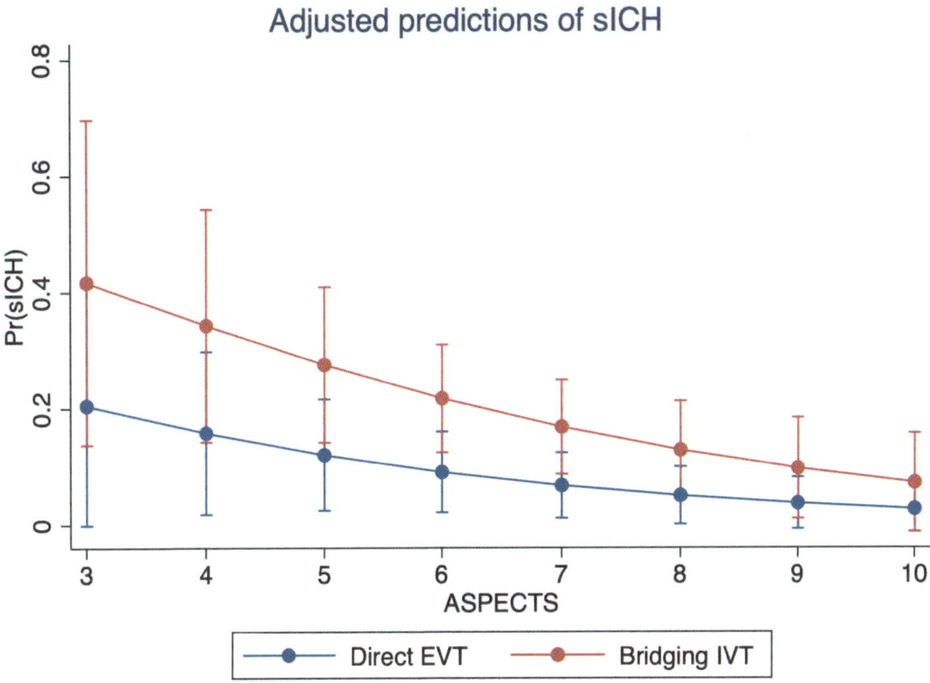

Figure 4. Prediction of symptomatic intracranial hemorrhage (*y* axis) based on multivariable logistic regression analysis to show the impact of IVT application according to the baseline ASPECTS, adjusted for time from onset to imaging.

4. Discussion

The aim of this study was to investigate the impact of IVT on clinical outcome and secondary injury volumes in patients with complete endovascular recanalization compared to direct EVT. The main findings of this study are: (1) IVT was not associated with better functional outcomes compared to direct EVT but was (2) significantly and independently associated with increased rates of sICH and EV as secondary injury volumes. (3) Important predictors of outcome in TICI 3 recanalized patients were age and NIHSS, while ASPECTS, time from onset to imaging, or time from imaging to recanalization were not independent predictors, which emphasizes the importance of complete recanalization as major determinant of outcome.

In the light of the currently ongoing debate on whether to treat ischemic stroke patients with bridging IVT or direct EVT, the results of our study indicate that IVT may not be associated with better outcome, when EVT results in complete reperfusion. In the past, IVT has been associated with reduced lesion volume, as well as increased risk of secondary hemorrhage as a marker of blood brain barrier breakdown by destructive effects on the extracellular matrix and endothelial basal lamina [1,22]. Furthermore, IVT may be associated with neurotoxic cell injury in the ischemic brain by activation of excitatory aminoacid receptors [22]. In line with this, we observed a significantly increased EV after EVT in patients who received IVT, alongside the markedly increased rates of sICH, suggesting that IVT might be associated with increasing blood-brain barrier disturbance in this subgroup of patients. Previously, it has been observed that reperfusion in anterior circulation LVO patients may be associated with increased edema formation [23,24]. On the other hand, several studies showed that reperfusion is associated with reduced edema formation even in patients with low ASPECTS [25,26]. These ambiguous findings hint toward different courses of edema formation after thrombolytic and endovascular therapy and emphasize two important problems: a lack of understanding of the relationship between treatment and lesion pathophysiology, and the resulting consequences for optimization of individual treatment strategies.

Since approval of IVT for the treatment of ischemic stroke by the FDA in 1996, there has been an ongoing debate about harm and benefit of IVT administration and about the optimal treatment selection [27–33]. In 2018, the WAKE-UP trial reported a beneficial treatment effect of IVT compared to placebo despite a slightly increased rate of sICH [34]. This observation, however, lacked statistical significance in patients with an NIHSS > 10, which might be of importance considering the higher median NIHSS in the typical EVT eligible patient cohort (i.e., median NIHSS of 17 in the HERMES metaanalysis [2]).

More recently, the results of the DIRECT-MT, DEVT, and SKIP trial demonstrated non-inferiority of direct EVT compared to EVT + IVT, with a 4% CI margin in meta-analysis [35]. More importantly, all trials did not demonstrate that IVT increases the rate of successful EVT compared to direct EVT, which has been used as an important argument in favor of IVT administration [36–38]. In the DEVT trial, the proportion of patients with successful EVT (TICI 2b-3) and the end of the procedure was 88.5% for patients with direct EVT, and 87.2% for patients with bridging IVT, respectively, also showing no significant differences in times from onset to randomization, and time from onset to puncture. Second, the rate of reperfusion at follow-up was not different: 97% in patients with direct EVT, and 94% for patients with bridging IVT [37]. A further argument in favor of IVT was a supposedly improved microcirculation, and less peripheral emboli in new territories [39]. A recent study, however, observed that IVT had no effect on the number and volume of peripheral emboli after EVT independent of the degree of recanalization using high-resolution diffusion-weighted imaging [40].

Yet, the published results of the aforementioned trials on direct EVT versus bridging IVT still leave room for interpretation, and may not be sufficient to change clinical practice. Ethnical aspects might also contribute to differing results of the recently published studies, with an odds ratio (>1 favoring direct EVT) for the Asian trials (DIRECT-MT, DEVT, SKIP) of 1.08 (95% CI: 0.85–1.38) compared to 0.82 (95%: 0.63–1.07) in non-Asian trials (MR CLEAN noIV, SWIFT DIRECT). These differences could be suggestive of a different stroke etiology (i.e., atherothrombotic versus cardioembolic), which might yield a potential relevance for treatment selection. Nevertheless, the potential implications of an EVT only practice should be considered and discussed. Decision-making and workflow times could be significantly reduced, also considering the decline in primary stroke center admissions prior to transfer to an EVT center ("drip-and-ship"). Furthermore, the application of IVT is associated with high economical impact, with an approximate base payment of >12,000$ in the US [41]. Additionally, the administration of neuroproctants in combination with reperfusion might lead to further improvements in functional outcomes and could be tested as an alternative to bridging IVT [42]. In the ESCAPE-NA1 trial, the application of nerinetide was associated with better outcomes in patients who did not receive IVT [8]. Likewise, further drugs, such as glyburide, are currently tested as adjuvant treatment options [43,44].

To our knowledge, this is the first study that investigated the effect of bridging IVT on functional outcome and secondary injury volumes using quantitative imaging biomarkers in a patient collective exclusively consisting of complete reperfusion cases. Considering the rising frequency of successful vessel recanalization over time (e.g., approximately 60% in MR CLEAN, 72% in ESCAPE, 77% in DEFUSE-3, and 86% in ESCAPE-NA1 [2,8]), the importance of optimizing treatment strategies for these patients is suggested by the high proportion of patients with poor outcomes despite successful EVT as described in a recent meta-analysis (i.e., 45% mRS 3–6 in TICI 2b-3) [45,46]. The early identification of stroke patients who might not benefit from IVT could therefore be important to further improve functional outcome. The present study might give further insights into the effect of IVT on lesion pathophysiology in the setting of complete vessel recanalization, and might help to tailor individual adjuvant treatment options.

Limitations include the retrospective single-center nature of this study and the lack of a control group containing patients with incomplete reperfusion. This includes a potential selection bias regarding the indication of IVT which is caused by the study design. Finally, quantitative edema analysis by densitometry requires the absence of a space-occupying hemorrhage, is subject of time-consuming post-processing, and cannot be applied in patients with significant artifacts. A further limitation is the missing status of the cerebral collateral circulation, which might have an impact on secondary injury volumes. Furthermore, details on blood-pressure management, which might also affect the degree of secondary injury volumes, are unknown for this study. Future studies should analyze the identification of baseline variables to guide IVT, particularly in cases with high probability of vessel recanalization (e.g., time window, vessel anatomy, thrombus properties). Moreover, future research is necessary to investigate the impact of IVT in relationship to the degree of reperfusion for patients with posterior circulation stroke.

5. Conclusions

Bridging IVT was not associated with better functional outcome in patients receiving complete recanalization. The significantly increased edema volume and risk for sICH as secondary injury volumes could be a major reason for the lack of a clinical benefit of IVT in this patient cohort. The results of the present study support direct EVT approaches, particularly in patients with a higher likelihood of successful EVT.

Author Contributions: Conceptualization, G.B., W.H., A.K., J.F. and U.H.; data curation, C.R., T.D.F., S.E. and U.H.; formal analysis, G.B., L.M., C.R., M.B. and H.K.; investigation, T.D.F., N.V.H., H.K., S.E. and E.B.; methodology, E.B.; project administration, M.B.; resources, N.V.H.; software, H.K.; supervision, W.H. and U.H.; validation, L.M., A.K., E.B., J.F. and U.H.; visualization, G.B., C.R. and H.K.; writing—original draft, G.B., W.H. and M.B.; writing—review & editing, G.B., L.M., T.D.F., N.V.H., M.B., S.E., A.K., E.B., J.F. and U.H. All authors have read and agreed to the published version of the manuscript.

Funding: This research received no specific grant from any funding agency in the public, commercial, or not-for-profit sectors.

Institutional Review Board Statement: The study was conducted in accordance with the Declaration of Helsinki, and approved by the Institutional Review Board (or Ethics Committee) of Ethikkommission der Ärztekammer Hamburg, WF04/13).

Informed Consent Statement: Patient consent was waived as data was fully anonymized.

Data Availability Statement: The data that support the findings of this study are available from the corresponding author upon reasonable request.

Conflicts of Interest: Fiehler: German Ministry of Science and Education (BMBF)/German Ministry of Economy and Innovation (BMWi)/German Research Foundation (DFG)/European Union (EU)/Hamburgische Investitions und Förderbank/Medtronic/Microvention/Philips/Stryker, Consultant: Acandis/Boehringer Ingelheim/Cerenovus/Covidien/Evasc/Neurovascular/MDClinicals/Medtronic/Medina/Microvention/Penumbra/Route92/Stryker/Transverse Medical. Kemmling: Siemens. Other authors: No disclosures.

References

1. Mair, G.; Von Kummer, R.; Morris, Z.; Von Heijne, A.; Bradey, N.; Cala, L.; Peeters, A.; Farrall, A.J.; Adami, A.; Potter, G.; et al. Effect of IV alteplase on the ischemic brain lesion at 24–48 h after ischemic stroke. *Neurology* **2018**, *91*, e2067–e2077. [CrossRef] [PubMed]
2. Goyal, M.; Menon, B.K.; Van Zwam, W.H.; Dippel, D.W.J.; Mitchell, P.J.; Demchuk, A.M.; Dávalos, A.; Majoie, C.B.L.M.; Van Der Lugt, A.; De Miquel, M.A.; et al. Endovascular thrombectomy after large-vessel ischaemic stroke: A meta-analysis of individual patient data from five randomised trials. *Lancet* **2016**, *387*, 1723–1731. [CrossRef]
3. Bhogal, P.; Andersson, T.; Maus, V.; Mpotsaris, A.; Yeo, L. Mechanical Thrombectomy—A Brief Review of a Revolutionary new Treatment for Thromboembolic Stroke. *Clin. Neuroradiol.* **2018**, *28*, 313–326. [CrossRef] [PubMed]

4. Turc, G.; Tsivgoulis, G.; Audebert, H.J.; Boogaarts, H.; Bhogal, P.; De Marchis, G.M.; Fonseca, A.C.; Khatri, P.; Mazighi, M.; de la Ossa, N.P.; et al. European Stroke Organisation (ESO)–European Society for Minimally Invasive Neurological Therapy (ESMINT) expedited recommendation on indication for intravenous thrombolysis before mechanical thrombectomy in patients with acute ischemic stroke and anterior circulation large vessel occlusion. *J. NeuroInterv. Surg.* **2022**, *14*, 209–227. [CrossRef] [PubMed]
5. Bhatia, R.; Hill, M.; Shobha, N.; Menon, B.; Bal, S.; Kochar, P.; Watson, T.; Goyal, M.; Demchuk, A.M. Low Rates of Acute Recanalization With Intravenous Recombinant Tissue Plasminogen Activator in Ischemic Stroke. *Stroke* **2010**, *41*, 2254–2258. [CrossRef] [PubMed]
6. Yang, D.; Geng, Y.; Zhang, M.; Lin, M.; Shi, Z.; Zi, W.; Hao, Y.; Wang, H.; Liu, W.; Wang, W.; et al. Complete Recanalization May Exert the Most Important Effect on Outcomes of Endovascular Treatment in Acute Ischemic Stroke with Small Infarct Core Beyond 6 Hours. *World Neurosurg.* **2019**, *125*, e544–e551. [CrossRef] [PubMed]
7. Broocks, G.; Flottmann, F.; Hanning, U.; Schön, G.; Sporns, P.; Minnerup, J.; Fiehler, J.; Kemmling, A. Impact of endovascular recanalization on quantitative lesion water uptake in ischemic anterior circulation strokes. *J. Cereb. Blood Flow Metab.* **2020**, *40*, 437–445. [CrossRef] [PubMed]
8. Hill, M.; Goyal, M.; Menon, B.K.; Nogueira, R.G.; McTaggart, R.A.; Demchuk, A.M.; Poppe, A.Y.; Buck, B.; Field, T.; Dowlatshahi, D.; et al. Efficacy and safety of nerinetide for the treatment of acute ischaemic stroke (ESCAPE-NA1): A multicentre, double-blind, randomised controlled trial. *Lancet* **2020**, *395*, 878–887. [CrossRef]
9. Albers, G.W.; Marks, M.P.; Kemp, S.; Christensen, S.; Tsai, J.P.; Ortega-Gutierrez, S.; McTaggart, R.A.; Torbey, M.T.; Kim-Tenser, M.; Leslie-Mazwi, T.; et al. Thrombectomy for Stroke at 6 to 16 Hours with Selection by Perfusion Imaging. *N. Engl. J. Med.* **2018**, *378*, 708–718. [CrossRef]
10. Berkhemer, O.A.; Fransen, P.S.S.; Beumer, D.; Berg, L.A.V.D.; Lingsma, H.F.; Yoo, A.J.; Schonewille, W.J.; Vos, J.A.; Nederkoorn, P.J.; Wermer, M.J.H.; et al. A Randomized Trial of Intraarterial Treatment for Acute Ischemic Stroke. *N. Engl. J. Med.* **2015**, *372*, 11–20. [CrossRef]
11. Yang, P.; Zhang, Y.; Zhang, L.; Zhang, Y.; Treurniet, K.M.; Chen, W.; Peng, Y.; Han, H.; Wang, J.; Wang, S.; et al. Endovascular Thrombectomy with or without Intravenous Alteplase in Acute Stroke. *N. Engl. J. Med.* **2020**, *382*, 1981–1993. [CrossRef] [PubMed]
12. Treurniet, K.M.; For the MR CLEAN-NO IV Investigators; LeCouffe, N.E.; Kappelhof, M.; Emmer, B.J.; van Es, A.C.G.M.; Boiten, J.; Lycklama, G.J.; Keizer, K.; Yo, L.S.F.; et al. MR CLEAN-NO IV: Intravenous treatment followed by endovascular treatment versus direct endovascular treatment for acute ischemic stroke caused by a proximal intracranial occlusion—Study protocol for a randomized clinical trial. *Trials* **2021**, *22*, 141. [CrossRef] [PubMed]
13. Podlasek, A.; Dhillon, P.S.; Butt, W.; Grunwald, I.Q.; England, T.J. To bridge or not to bridge: Summary of the new evidence in endovascular stroke treatment. *Stroke Vasc. Neurol.* **2022**. [CrossRef] [PubMed]
14. Chandra, R.V.; Leslie-Mazwi, T.M.; Mehta, B.P.; Derdeyn, C.P.; Demchuk, A.M.; Menon, B.K.; Goyal, M.; González, R.G.; Hirsch, J.A. Does the use of IV tPA in the current era of rapid and predictable recanalization by mechanical embolectomy represent good value? *J. NeuroInterv. Surg.* **2016**, *8*, 443–446. [CrossRef] [PubMed]
15. Hassan, A.E.; Ringheanu, V.M.; Preston, L.; Tekle, W.; Qureshi, A.I. IV tPA is associated with increase in rates of intracerebral hemorrhage and length of stay in patients with acute stroke treated with endovascular treatment within 4.5 hours: Should we bypass IV tPA in large vessel occlusion? *J. NeuroInterv. Surg.* **2020**, *13*, 114–118. [CrossRef] [PubMed]
16. Hacke, W.; Kaste, M.; Fieschi, C.; von Kummer, R.; Davalos, A.; Meier, D.; Larrue, V.; Bluhmki, E.; Davis, S.; Donnan, G.; et al. Randomised double-blind placebo-controlled trial of thrombolytic therapy with intravenous alteplase in acute ischaemic stroke (ECASS II). *Lancet* **1998**, *352*, 1245–1251. [CrossRef]
17. Powers, W.J.; Rabinstein, A.A.; Ackerson, T.; Adeoye, O.M.; Bambakidis, N.C.; Becker, K.; Biller, J.; Brown, M.; Demaerschalk, B.M.; Hoh, B.; et al. Guidelines for the Early Management of Patients With Acute Ischemic Stroke: 2019 Update to the 2018 Guidelines for the Early Management of Acute Ischemic Stroke: A Guideline for Healthcare Professionals From the American Heart Association/American Stroke Association. *Stroke* **2019**, *50*, e344–e418.
18. Minnerup, J.; Broocks, G.; Kalkoffen, J.; Langner, S.; Knauth, M.; Psychogios, M.N.; Wersching, H.; Teuber, A.; Heindel, W.; Eckert, B.; et al. Computed tomography-based quantification of lesion water uptake identifies patients within 4.5 h of stroke onset: A multicenter observational study. *Ann. Neurol.* **2016**, *80*, 924–934. [CrossRef] [PubMed]
19. Broocks, G.; Faizy, T.D.; Flottmann, F.; Schön, G.; Langner, S.; Fiehler, J.; Kemmling, A.; Gellissen, S. Subacute Infarct Volume With Edema Correction in Computed Tomography Is Equivalent to Final Infarct Volume After Ischemic Stroke. *Investig. Radiol.* **2018**, *53*, 472–476. [CrossRef]
20. Broocks, G.; Flottmann, F.; Ernst, M.; Faizy, T.D.; Minnerup, J.; Siemonsen, S.; Fiehler, J.; Kemmling, A. Computed Tomography–Based Imaging of Voxel-Wise Lesion Water Uptake in Ischemic Brain. *Investig. Radiol.* **2018**, *53*, 207–213. [CrossRef]
21. Nawabi, J.; Flottmann, F.; Hanning, U.; Bechstein, M.; Schön, G.; Kemmling, A.; Fiehler, J.; Broocks, G. Futile Recanalization With Poor Clinical Outcome is Associated with Increased Edema Volume After Ischemic Stroke. *Investig. Radiol.* **2019**, *54*, 282–287. [CrossRef] [PubMed]
22. Micieli, G. Safety and efficacy of alteplase in the treatment of acute ischemic stroke. *Vasc. Health Risk Manag.* **2009**, *5*, 397–409. [CrossRef] [PubMed]
23. Ng, F.C.; Yassi, N.; Sharma, G.; Brown, S.B.; Goyal, M.; Majoie, C.B.; Jovin, T.G.; Hill, M.D.; Muir, K.W.; Saver, J.L.; et al. Cerebral Edema in Patients With Large Hemispheric Infarct Undergoing Reperfusion Treatment: A HERMES Meta-Analysis. *Stroke* **2021**, *52*, 3450–3458. [CrossRef] [PubMed]

24. Irvine, H.J.; Ostwaldt, A.-C.; Bevers, M.B.; Dixon, S.; Battey, T.W.; Campbell, B.C.; Davis, S.M.; Donnan, G.A.; Sheth, K.N.; Jahan, R.; et al. Reperfusion after ischemic stroke is associated with reduced brain edema. *J. Cereb. Blood Flow Metab.* **2017**, *38*, 1807–1817. [CrossRef] [PubMed]

25. Thorén, M.; Azevedo, E.; Dawson, J.; Egido, J.A.; Falcou, A.; Ford, G.A.; Holmin, S.; Mikulik, R.; Ollikainen, J.; Wahlgren, N.; et al. Predictors for Cerebral Edema in Acute Ischemic Stroke Treated With Intravenous Thrombolysis. *Stroke* **2017**, *48*, 2464–2471. [CrossRef] [PubMed]

26. Broocks, G.; Hanning, U.; Flottmann, F.; Schönfeld, M.; Faizy, T.D.; Sporns, P.B.; Baumgart, M.; Leischner, H.; Schön, G.; Minnerup, J.; et al. Clinical benefit of thrombectomy in stroke patients with low ASPECTS is mediated by oedema reduction. *Brain* **2019**, *142*, 1399–1407. [CrossRef] [PubMed]

27. Jovin, T.G. MRI-Guided Intravenous Alteplase for Stroke—Still Stuck in Time. *N. Engl. J. Med.* **2018**, *379*, 682–683. [CrossRef]

28. Ospel, J.; Kashani, N.; Fischer, U.; Menon, B.; Almekhlafi, M.; Wilson, A.; Foss, M.; Saposnik, G.; Goyal, M.; Hill, M. How Do Physicians Approach Intravenous Alteplase Treatment in Patients with Acute Ischemic Stroke Who Are Eligible for Intravenous Alteplase and Endovascular Therapy? Insights from UNMASK-EVT. *Am. J. Neuroradiol.* **2020**, *41*, 262–267. [CrossRef]

29. Gilbert, B.W.; Huffman, J.B. Time to Stop Looking at Alteplase for Stroke Through a Prism. *J. Pharm. Pract.* **2019**, *33*, 127–128. [CrossRef] [PubMed]

30. Thomalla, G.; Boutitie, F.; Ma, H.; Koga, M.; Ringleb, P.; Schwamm, L.H.; Wu, O.; Bendszus, M.; Bladin, C.F.; Campbell, B.C.V.; et al. Intravenous alteplase for stroke with unknown time of onset guided by advanced imaging: Systematic review and meta-analysis of individual patient data. *Lancet* **2020**, *396*, 1574–1584. [CrossRef]

31. Whiteley, W.; Emberson, J.; Lees, K.R.; Blackwell, L.; Albers, G.; Bluhmki, E.; Brott, T.; Cohen, G.; Davis, S.; Donnan, G.; et al. Risk of intracerebral haemorrhage with alteplase after acute ischaemic stroke: A secondary analysis of an individual patient data meta-analysis. *Lancet Neurol.* **2016**, *15*, 925–933. [CrossRef]

32. Aoki, J.; Kimura, K. The question of alteplase dose for stroke is not resolved. *Nat. Rev. Neurol.* **2016**, *12*, 376–377. [CrossRef] [PubMed]

33. Goyal, M.; Ospel, J.M.; Hill, M.D. Will there be a rapid change towards an EVT-only paradigm? *Interv. Neuroradiol.* **2021**, *27*, 744–745. [CrossRef] [PubMed]

34. Thomalla, G.; Simonsen, C.Z.; Boutitie, F.; Andersen, G.; Berthezene, Y.; Cheng, B.; Cheripelli, B.; Cho, T.-H.; Fazekas, F.; Fiehler, J.; et al. MRI-Guided Thrombolysis for Stroke with Unknown Time of Onset. *N. Engl. J. Med.* **2018**, *379*, 611–622. [CrossRef] [PubMed]

35. Podlasek, A.; Dhillon, P.S.; Butt, W.; Grunwald, I.Q.; England, T.J. Direct mechanical thrombectomy without intravenous thrombolysis versus bridging therapy for acute ischemic stroke: A meta-analysis of randomized controlled trials. *Int. J. Stroke* **2021**, *16*, 621–631. [CrossRef] [PubMed]

36. Suzuki, K.; Matsumaru, Y.; Takeuchi, M.; Morimoto, M.; Kanazawa, R.; Takayama, Y.; Kamiya, Y.; Shigeta, K.; Okubo, S.; Hayakawa, M.; et al. Effect of Mechanical Thrombectomy Without vs With Intravenous Thrombolysis on Functional Outcome Among Patients With Acute Ischemic Stroke. *JAMA J. Am. Med Assoc.* **2021**, *325*, 244–253. [CrossRef] [PubMed]

37. Zi, W.; Qiu, Z.; Li, F.; Sang, H.; Wu, D.; Luo, W.; Liu, S.; Yuan, J.; Song, J.; Shi, Z.; et al. Effect of Endovascular Treatment Alone vs. Intravenous Alteplase Plus Endovascular Treatment on Functional Independence in Patients With Acute Ischemic Stroke. *JAMA* **2021**, *325*, 234–243. [CrossRef]

38. Flottmann, F.; Broocks, G.; Faizy, T.D.; McDonough, R.; Watermann, L.; Deb-Chatterji, M.; Thomalla, G.; Herzberg, M.; Nolte, C.H.; Fiehler, J.; et al. Factors Associated with Failure of Reperfusion in Endovascular Therapy for Acute Ischemic Stroke. *Clin. Neuroradiol.* **2020**, *31*, 197–205. [CrossRef] [PubMed]

39. Desilles, J.-P.; Loyau, S.; Syvannarath, V.; Gonzalez-Valcarcel, J.; Cantier, M.; Louedec, L.; Lapergue, B.; Amarenco, P.; Ajzenberg, N.; Jandrot-Perrus, M.; et al. Alteplase Reduces Downstream Microvascular Thrombosis and Improves the Benefit of Large Artery Recanalization in Stroke. *Stroke* **2015**, *46*, 3241–3248. [CrossRef]

40. Broocks, G.; Meyer, L.; Kabiri, R.; Kniep, H.C.; McDonough, R.; Bechstein, M.; van Horn, N.; Lindner, T.; Sedlacik, J.; Cheng, B.; et al. Impact of intravenous alteplase on sub-angiographic emboli in high-resolution diffusion-weighted imaging following successful thrombectomy. *Eur. Radiol.* **2021**, *31*, 8228–8235. [CrossRef] [PubMed]

41. Kleindorfer, D.; Broderick, J.; Demaerschalk, B.; Saver, J. Cost of Alteplase Has More Than Doubled Over the Past Decade. *Stroke* **2017**, *48*, 2000–2002. [CrossRef] [PubMed]

42. Shi, L.; Rocha, M.; Leak, R.K.; Zhao, J.; Bhatia, T.; Mu, H.; Wei, Z.; Yu, F.; Weiner, S.L.; Ma, F.; et al. A new era for stroke therapy: Integrating neurovascular protection with optimal reperfusion. *J. Cereb. Blood Flow Metab.* **2018**, *38*, 2073–2091. [CrossRef] [PubMed]

43. Sheth, K.N.; Kimberly, W.T.; Elm, J.J.; Kent, T.A.; Mandava, P.; Yoo, A.J.; Thomalla, G.; Campbell, B.; Donnan, G.A.; Davis, S.M.; et al. Pilot Study of Intravenous Glyburide in Patients With a Large Ischemic Stroke. *Stroke* **2014**, *45*, 281–283. [CrossRef] [PubMed]

44. Sheth, K.N.; Petersen, N.H.; Cheung, K.; Elm, J.J.; Hinson, H.E.; Molyneaux, B.J.; Beslow, L.A.; Sze, G.K.; Simard, J.M.; Kimberly, W.T. Long-Term Outcomes in Patients Aged ≤70 Years With Intravenous Glyburide From the Phase II GAMES-RP Study of Large Hemispheric Infarction. *Stroke* **2018**, *49*, 1457–1463. [CrossRef] [PubMed]

45. Van Horn, N.; Kniep, H.; Leischner, H.; McDonough, R.; Deb-Chatterji, M.; Broocks, G.; Thomalla, G.; Brekenfeld, C.; Fiehler, J.; Hanning, U.; et al. Predictors of poor clinical outcome despite complete reperfusion in acute ischemic stroke patients. *J. NeuroInterv. Surg.* **2020**, *13*, 14–18. [CrossRef] [PubMed]
46. Kaesmacher, J.; Dobrocky, T.; Heldner, M.R.; Bellwald, S.; Mosimann, P.J.; Mordasini, P.; Bigi, S.; Arnold, M.; Gralla, J.; Fischer, U. Systematic review and meta-analysis on outcome differences among patients with TICI2b versus TICI3 reperfusions: Success revisited. *J. Neurol. Neurosurg. Psychiatry* **2018**, *89*, 910–917. [CrossRef] [PubMed]

Journal of
Clinical Medicine

Article

Cerebral Hypoperfusion Intensity Ratio Is Linked to Progressive Early Edema Formation

Noel van Horn [1], Gabriel Broocks [1], Reza Kabiri [1], Michel C. Kraemer [1], Soren Christensen [2], Michael Mlynash [2], Lukas Meyer [1], Maarten G. Lansberg [2], Gregory W. Albers [2], Peter Sporns [3], Adrien Guenego [4], Jens Fiehler [1], Max Wintermark [5], Jeremy J. Heit [5,†] and Tobias D. Faizy [1,*,†]

[1] Department of Diagnostic and Interventional Neuroradiology, University Medical Center Hamburg-Eppendorf, 20246 Hamburg, Germany; no.vanhorn@uke.de (N.v.H.); g.broocks@uke.de (G.B.); reza.kabiri@live.de (R.K.); michel_1@hotmail.de (M.C.K.); lu.meyer@uke.de (L.M.); fiehler@uke.de (J.F.)

[2] Department of Neurology and Neurological Sciences, Stanford University School of Medicine, Stanford, CA 94305, USA; sorenc@stanford.edu (S.C.); mmlynash@stanford.edu (M.M.); lansberg@stanford.edu (M.G.L.); albers@stanford.edu (G.W.A.)

[3] Department of Diagnostic and Interventional Neuroradiology, University Hospital Basel, 4031 Basel, Switzerland; peter.sporns@usb.ch

[4] Department of Interventional Neuroradiology, Erasme University Hospital, 1070 Brussels, Belgium; adrien.guenego@erasme.ulb.ac.be

[5] Department of Radiology, Stanford University School of Medicine, Stanford, CA 94305, USA; max.wintermark@gmail.com (M.W.); jeremyheit@gmail.com (J.J.H.)

* Correspondence: t.faizy@uke.de; Tel.: +49-0-152-2283-5161

† These authors contributed equally to this work.

Citation: van Horn, N.; Broocks, G.; Kabiri, R.; Kraemer, M.C.; Christensen, S.; Mlynash, M.; Meyer, L.; Lansberg, M.G.; Albers, G.W.; Sporns, P.; et al. Cerebral Hypoperfusion Intensity Ratio Is Linked to Progressive Early Edema Formation. *J. Clin. Med.* **2022**, *11*, 2373. https://doi.org/10.3390/jcm11092373

Academic Editor: Hubertus Axer

Received: 12 March 2022
Accepted: 16 April 2022
Published: 23 April 2022

Publisher's Note: MDPI stays neutral with regard to jurisdictional claims in published maps and institutional affiliations.

Abstract: The hypoperfusion intensity ratio (HIR) is associated with collateral status and reflects the impaired microperfusion of brain tissue in patients with acute ischemic stroke and large vessel occlusion (AIS-LVO). As a deterioration in cerebral blood flow is associated with brain edema, we aimed to investigate whether HIR is correlated with the early edema progression rate (EPR) determined by the ischemic net water uptake (NWU) in a multicenter retrospective analysis of AIS-LVO patients anticipated for thrombectomy treatment. HIR was automatically calculated as the ratio of time-to-maximum (TMax) > 10 s/(TMax) > 6 s. HIRs < 0.4 were regarded as favorable (HIR+) and ≥0.4 as unfavorable (HIR−). Quantitative ischemic lesion NWU was delineated on baseline NCCT images and EPR was calculated as the ratio of NWU/time from symptom onset to imaging. Multivariable regression analysis was used to assess the association of HIR with EPR. This study included 731 patients. HIR+ patients exhibited a reduced median NWU upon admission CT (4% (IQR: 2.1–7.6) versus 8.2% (6–10.4); $p < 0.001$) and less median EPR (0.016%/h (IQR: 0.007–0.04) versus 0.044%/h (IQR: 0.021–0.089; $p < 0.001$) compared to HIR− patients. Multivariable regression showed that HIR+ (β: 0.53, SE: 0.02; $p = 0.003$) and presentation of the National Institutes of Health Stroke Scale (β: 0.2, SE: 0.0006; $p = 0.001$) were independently associated with EPR. In conclusion, favorable HIR was associated with lower early edema progression and decreased ischemic edema formation on baseline NCCT.

Keywords: brain edema; collateral circulation; ischemic stroke; perfusion imaging; thrombectomy

1. Introduction

Cerebral hypoperfusion below a critical threshold of 10 to 15 mL blood/100 g brain tissue per minute may result in ischemic tissue damage and subsequent brain edema formation [1–3]. Brain edema formation is one of the critical hallmarks for the prognostication of clinical and procedural outcomes of AIS-LVO patients [1,4]. Aggravated brain edema development is inherently linked to persistent critical hypoperfusion, and extensive edema formation was found to be associated with poor clinical outcomes and the occurrence of malignant infarction, despite successful endovascular treatment [1,3,5]. Thus, a proper and reliable assessment of tissue edema may have strong implications for treatment triage in stroke patients.

Computed tomography perfusion imaging offers an estimation of cerebral blood perfusion over time. The time when the residue function reaches its peak (TMax) is a parameter broadly utilized for infarct core volume and penumbra tissue estimation. The hypoperfusion intensity ratio (HIR) is automatically calculated from perfusion imaging source data as the ratio of the time-to-maximum of the tissue residue function: (TMax) > 10 s/(TMax) > 6 s [6]. Thereby, HIR also reflects the fraction of more severely infarcted brain tissue. Favorable HIR profiles are linked to robust tissue collateral status, lower baseline core volumes, higher penumbra volumes, and also to good functional outcomes after thrombectomy treatment [7–10]. However, whether HIR status is associated with time-dependent early edema progression and quantitative brain edema estimates on admission imaging has not been investigated.

The purpose of this study was to assess whether HIR may serve as a standardized and reliable imaging biomarker for the assessment of early brain edema progression in AIS-LVO patients anticipated for thrombectomy treatment. We hypothesized that favorable HIR profiles are associated with less temporal ischemic lesion growth from symptom onset to imaging, and decreased quantitative ischemic lesion net water uptake (NWU) on baseline non-contrast head computed tomography (NCCT) compared to patients who exhibited unfavorable HIR profiles.

2. Materials and Methods

2.1. Study Design

We conducted a multicenter cohort study of consecutive AIS-LVO patients undergoing thrombectomy triage at two comprehensive stroke centers (redacted). Patients were retrospectively screened for inclusion from a prospectively maintained stroke database between January 2013 and December 2020. The study protocol was approved by the institutional review boards of both study centers, and complied with the Health Insurance Portability and Accountability Act (HIPAA) and followed the guidelines of the Declaration of Helsinki. Informed patient consent was waived by the review boards for this retrospective study.

2.2. Patient Inclusion, Population, and Clinical Data

Clinical, imaging, and demographic data were obtained from the electronic medical records of each patient.

Inclusion criteria were as follows: (1) Patients with suspected AIS-LVO who underwent thrombectomy triage with the intention-to-treat by endovascular measures within 16 h after stroke onset; (2) available baseline multimodal head CT including NCCT, CT angiography, and CT perfusion imaging; (3) known time of symptom onset to imaging (in minutes); and (4) presence of anterior circulation large vessel occlusion of the internal carotid artery or first (M1) or second (M2) segment of the middle cerebral artery. Exclusion criteria were as follows: (1) poor admission CT image quality due to excessive patient motion or failed contrast bolus, (2) detection of parenchymal hemorrhage on baseline NCCT imaging that impedes delineation of ischemic lesion NWU, and (3) unknown time of symptom onset (e.g., in the scenario of a wake-up stroke).

2.3. Imaging Analysis

All CT perfusion studies were automatically analyzed with RAPID (iSchemaView, Menlo Park, CA, USA). The image analyzation software Horos (Horos Project©, v 3.3.6) was used to assess all of the NCCT and CT angiography images, including NWU determination.

The ischemic core was defined as the volume of tissue with ≥70% reduction in cerebral blood flow relative to the contralateral hemisphere on CT perfusion.

HIR was defined as the volume of ischemic brain tissue with a time-to-maximum of the residue function (TMax) delay of >10 s divided by the volume of brain tissue with a TMax delay of >6 s [11]. A favorable HIR was regarded as a ratio of ≤0.4, and unfavorable

HIR was defined as >0.4 based on previously published thresholds for this multicenter stroke cohort, as published in [12].

The Alberta Stroke Program Early CT Score (ASPECTS) [13] was determined on pretreatment NCCT images by two neuroradiologists (T.D.F. and J.J.H.) on a subset of studies to determine the inter-reader agreement, and then the remaining studies were assigned by a single neuroradiologist (T.D.F.) with 10 years of experience.

Pial arterial collateral status was determined on CT angiography in a manner identical to ASPECTS assessment using the modified Tan scale [14], and robust collaterals were defined as filling of $\geq$50% of the middle cerebral artery (MCA) territory, whereas poor collaterals filled <50% of the MCA territory.

Ischemic lesion NWU was determined by the approach reported by Broocks et al. and Minnerup et al. [1,15]. In brief, ischemic lesion hypodensity was evaluated on admission NCCT images and CT perfusion images were used to verify the area of tissue infarction. Next, ischemic lesion hypodensities were outlined defining a region of interest, which was mirrored to the contralateral unaffected hemisphere. Manual adjustments to the region of interest were made to exclude the sulci and cerebrospinal fluid, if applicable. A histogram was sampled from the region of interest for Houndsfield units between 20 and 80. NWU was calculated in %, as described by Broocks et al. [1].

The early edema progression rate (EPR) was calculated as the ratio of the time from symptom onset to imaging (per hour) divided by ischemic lesion NWU (in %) determined on admission NCCT [16].

2.4. Outcome Measures

The primary study outcome was EPR (in %/h). Functional outcomes were assessed by the modified Rankin scale at 90 days (mRS90), and good outcomes were regarded as mRS90 of 0–2.

2.5. Statistical Analysis

All data are presented as mean $\pm$ standard deviation or median and interquartile range (IQR). The Kolmogorov–Smirnov test was used to test for normal or non-normal distribution. Absolute and relative frequencies are given for categorical data. Univariate regression analysis was performed to compare the clinical, radiological, and outcome parameters for patients with favorable and unfavorable HIR profiles using the chi-square test for counts and the Mann–Whitney U test for measurements. For the logistic regression models, the proportional-odds assumption had to be met ($p < 0.05$) before further analysis. Tests for collinearity with regards to HIR were applied to the included variables. Multivariable linear logistic regression analysis was performed to estimate the independent impact of favorable HIR on EPR. This model was adjusted for variables showing significance in the univariate analysis, i.e., presentation NIHSS, baseline ASPECTS, favorable TAN collaterals, and proximal vessel occlusion, as well as for additional variables that did show a significant association with early edema development or progression in previous studies, such as age, sex, and baseline glucose, although these variables did not show a significant difference between the HIR groups in the univariate analyses (Table 1). Statistical significance was set at the level of $p = 0.05$. A statistical analysis of all data was performed using SPSS (IBM Corp. Released 2016. IBM SPSS Statistics for Windows, Version 24.0. Armonk, NY, USA) and the StataCorp. (2019. Stata Statistical Software: Release 16. StataCorp LLC: College Station, TX, USA). Any unknown data are indicated in the given section below the tables.

Table 1. Patient characteristics and stroke presentation details dichotomized by the hypoperfusion intensity ratio [a].

	HIR+ (*n* = 381)	HIR− (*n* = 350)	*p*-Value
Age (years), median (IQR)	75 (64–82)	76 (64–84)	0.562
Female	200 (52.2)	176 (50.3)	0.551
Medical History			
Arterial Fibrillation	146 (38.1)	141 (40.3)	0.645
Hypertension	248 (64.8)	256 (73.1)	0.024
Blood Glucose, median (IQR)	120 (105–146)	123 (104–148)	0.657
Diabetes Mellitus	66 (18.9)	85 (22.2)	0.235
Smoking			
Current Smoker	42 (11)	41 (11.7)	0.997
Never Smoked	245 (64)	246 (70.3)	0.571
Prior Smoker	61 (15.9)	53 15.1)	0.494
Unknown Smoking Status	33 (8.7)	10 (2.9)	
Stroke Presentation Details			
Presentation NIHSS, median (IQR)	12 (7–17)	17 (13–20)	<0.001
Time from Symptom Onset to i.v. tPA in min, median (IQR)	95 (73–164)	108 (66–150)	0.838
Time from Symptom Onset to imaging, min, median (IQR)	197 (100–396)	160 (86–335)	0.61
Treatment Details			
Intravenous tPA Details			
i.v. tPA Administration	198 (51.7)	168 (48)	0.343
Endovascular Treatment Details			
Treated by endovascular thrombectomy	342 (89.8)	348 (99.4)	0.58
Recanalization after thrombectomy (TICI 2b-3)	287 (83.9)	270 (77.5)	0.01
Vessel Recanalization by tPA, Thrombectomy or Spontaneous	305 (79.6)	248 (70.9)	0.001

Patient characteristics and stroke presentation details of all 731 patients dichotomized by favorable (HIR+) and unfavorable (HIR−) hypoperfusion intensity ratios. Values are displayed as absolute numbers and frequencies, mean ± SD or median and interquartile range (IQR). The univariate logistic regression analysis between HIR+ and HIR− for all parameters is displayed on the very right column with corresponding *p*-values. NIHSS = National Institutes of Health Stroke Scale; i.v. tPA = intravenous tissue plasminogen activator; TICI = thrombolysis in cerebral infarction. [a] Data are *n* (%), unless otherwise indicated.

3. Results

In this study, 731 patients met inclusion criteria (Figure 1). Based on their HIR profiles, 381 patients (52%) were considered to have favorable HIR profiles (HIR+), whereas 350 patients were found to have unfavorable HIR profiles (HIR−).

Fewer patients in the HIR+ group had a history of hypertension compared to the HIR- patients (*n* = 248 (65%) versus *n* = 256 (73%); *p* = 0.024). There were no significant differences with respect to other demographical or medical history data (Table 1).

Compared to patients with unfavorable HIR, HIR+ patients likely had significantly lower median admission NIHSS scores (12 (IQR 7–17) versus 17 (IQR 13–20); *p* < 0.001) and more patients exhibited vessel recanalization either spontaneously or after thrombectomy or intravenous alteplase treatment (*n* = 305 (80%) versus *n* = 248 (71%); *p* < 0.001).

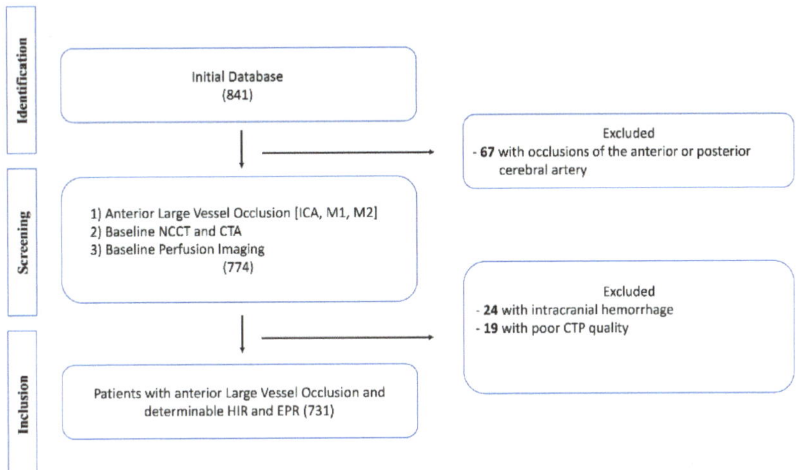

Figure 1. Patient inclusion and exclusion flow chart.

With respect to the admission imaging findings, patients in the HIR+ group had higher median ASPECTS (8 (IQR 7–10) versus 7 (IQR 6–8); $p < 0.001$), lower median baseline infarct core volumes (0 (IQR 0–10) versus 31 (IQR 13–64); $p < 0.001$) and more frequently exhibited favorable arterial CTA collateral profiles ($n = 304$ (79%) versus $n = 195$ [56%]; $p < 0.001$).

Patients with a favorable HIR showed lower median ischemic lesion NWU on admission imaging (4% (IQR 2.1–7.6) versus 8.2% (IQR 6–10.4); $p < 0.001$) and reduced EPR (0.016%/h (IQR 0.007–0.04) versus 0.044%/hour (IQR 0.021–0.089); $p < 0.001$) compared to those with unfavorable HIR profiles (Figure 2). Patients in the HIR+ group had lower median scores on the mRS90 compared to HIR- patients (median 2 (IQR 1–5) versus median 4 (IQR 3–6); $p < 0.001$) (Table 2).

For the assessment of the primary outcome (EPR), a multivariable linear regression model was performed. HIR was found to be non-normally distributed. However, as our sample size was large enough, a multivariable linear regression model was deemed suitable for the assessment of the primary study outcome. Furthermore, we found a weak but significant correlation between HIR and NIHSS ($r = 0.35$; $p = <0.001$), ASPECTS ($r = -0.33$; $p < 0.001$), CTA collaterals ($r = -0.33$; $p < 0.001$), and occlusion location ($r = -0.198$, $p < 0.001$) when accounting for the collinearity of the included variables in the model. These findings were similar to other previous studies on HIR, and as the variance of the inflation factor was found to be low (<2), the variables were included into the multivariable linear model. We found that favorable HIR (β: 0.53, SE: 0.02; $p = 0.003$) and presentation NIHSS scores (β: 0.2, SE: 0.0006; $p = 0.001$) were independently associated with EPR, regardless of the ASPECTS score on admission NCCT (β: -0.11, SE: 0.002; $p = 0.551$), arterial CTA collaterals (β: -0.46, SE: 0.008; $p = 0.579$), proximal vessel occlusion localization (β: 0.03, SE: 0.0034; $p = 0.349$), blood glucose (β: 0.0004, SE: 0; $p = 0.62$), age (β: 0.0013, SE: 0.0002; $p = 0.604$), and sex (β: -0.0084, SE: 0.007; $p = 0.229$) (Table 3 and Figure 3).

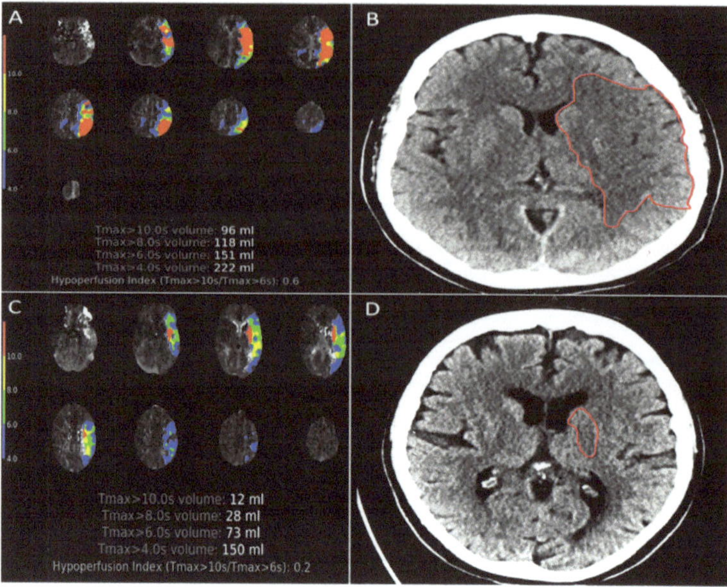

Figure 2. Imaging example of early edema progression in two patients with poor and favorable HIR. Two patients with comparable large vessel occlusion (proximal M1 occlusion) were both treated by endovascular thrombectomy. A 67-year old male patient (**A,B**) with symptom onset 2 h before admission imaging showing unfavorable HIR profiles (HIR = 0.6) and a considerable large baseline infarct hypodensity on non-contrast head CT (marked by the red field in (**B**)). The early edema progression rate was calculated at 5.18%/hour for this patient. A 71-year old female patient (**C,D**) who was last seen well 3.2 h before admission imaging exhibited only small infarct hypoattenuation in the basal ganglia (marked by the red circle in (**D**)), showing a favorable HIR profiles of 0.2 on admission perfusion imaging. The related early edema progression rate for this patient was calculated at 0.48% per hour.

Table 2. Presentation imaging details and clinical outcome dichotomized by the hypoperfusion intensity ratio [a].

	HIR+ (*n* = 381)	HIR− (*n* = 350)	*p* Value
ASPECTS, median (IQR)	8 (7–10)	7 (6–8)	<0.001
Baseline Infarct Core Volume [CBF < 30%] (mL), median (IQR)	0 (0–10)	31 (13–64)	<0.001
Penumbra Tmax > 6 s volume (mL), median (IQR)	90.5 (54.5–140)	145 (88–197)	<0.001
Penumbra Tmax > 10 s volume (mL), median (IQR)	20.8 (8–45.2)	89.6 (49.7–126)	<0.001
NWU on admission (%), median (IQR)	4 (2.1–7.6)	8.2 (6–10.4)	<0.001
Early Edema Progression Rate, (%/h), median (IQR)	0.96 (0.42–2.4)	2.64 (1.26–5.34)	<0.001
Hypoperfusion intensity ratio (HIR), median (IQR)	0.2 (0.1–0.3)	0.6 (0.5–0.7)	<0.001
Favorable CTA collaterals (TAN)	304 (79.4)	195 (55.7)	<0.001
Vessel occlusion localization on CTA			
Internal carotid artery	60 (15.7)	87 (24.9)	0.004
Proximal MCA 1 segment occlusion	119 (31.1)	153 (43.7)	<0.001

Table 2. *Cont.*

	HIR+ (*n* = 381)	HIR− (*n* = 350)	*p* Value
Distal MCA 1 segment occlusion	103 (26.9)	59 (16.9)	0.001
MCA 2 segment occlusion	96 (25.1)	51 (14.6)	0.001
Long-term clinical outcome			
Modified Ranking Scale after 90 days, median (IQR)	2 (1–5)	4 (3–6)	<0.001
Unknown	19 (5)	8 (2.3)	

Details of baseline imaging and functional outcome measures of all 731 patients dichotomized by favorable (HIR+) and unfavorable (HIR−) hypoperfusion intensity ratios. Values are displayed as absolute numbers and frequencies, mean ± SD or median and interquartile range (IQR). Univariate logistic regression analysis between HIR+ and HIR− for all parameters is displayed on the very right column with corresponding *p*-values. ASPECTS = Alberta Stroke Program Early CT Score; CBF = cerebral blood flow; Tmax > 6 = time-to-maximum of the tissue residue function with a delay of >6 s; Tmax > 10 = time-to-maximum of the tissue residue function with a delay of >10 s; NWU = Net Water Uptake; MCA = Middle Cerebral Artery. Favorable CTA collaterals were defined as TAN > 50%, assessed on CT angiography. [a] Data are *n* (%), unless otherwise indicated.

Table 3. Linear multivariable regression analysis for primary outcome (EPR).

	Early Edema Progression Rate		
Predictors	β	*SE*	*p-Value*
Favorable HIR	0.53	0.017	0.003
Presentation NIHSS	0.2	0.0006	0.001
Baseline ASPECTS	−0.11	0.002	0.551
Favorable TAN collaterals	−0.46	0.0083	0.579
Proximal vessel occlusion	0.03	0.0034	0.349
Blood glucose	0.00036	0	0.62
Age	0.0013	0.0002	0.604
Sex	−0.0084	0.007	0.229
Observations *n* = 701			

Calculated Beta (β), Standard error (SE) and *p*-value of all available (*n* = 701) patients screened via multivariable linear regression analysis. *n* = 30 were excluded from this analysis due to unknown blood glucose values. EPR = early edema progression rate; HIR = hypoperfusion intensity ratio; NIHSS = National Institute and Health Stroke Scale; ASPECTS = Alberta Stroke Program Early CT Score; NCCT = non-contrast head computed tomography. Proximal vessel occlusion is regarded as presence of anterior circulation large vessel occlusion of the internal carotid artery or first (M1) or second (M2) segment of the middle cerebral artery. Favorable HIR is defined as HIR < 0.4; favorable CTA collaterals were defined as TAN > 50%, assessed on CT angiography.

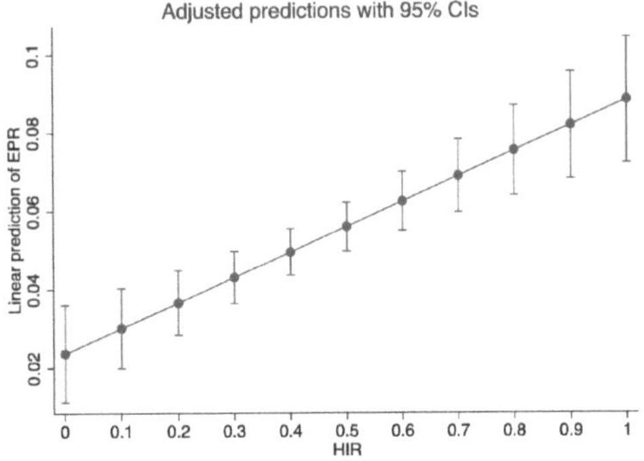

Figure 3. Adjusted prediction of early edema progression by HIR. A multivariable logistic regression plot was used to demonstrate the linear prediction of early edema progression rate (EPR, *y* axis) based on the hypoperfusion intensity ratio (HIR, *x* axis). EPR = Early Edema Progression Rate; HIR = Hypoperfusion Intensity Ratio; 95% CI = 95% Confidence Interval.

4. Discussion

In this study of AIS-LVO patients anticipated for thrombectomy treatment, we aimed to investigate whether imaging biomarkers of cerebral tissue perfusion are associated with early brain edema formation. We found that favorable HIR profiles, defined as a HIR of 0.4, were independently associated with decreased ischemic lesion NWU on admission imaging and a reduced EPR, defined as relative NWU per hour from symptom onset to imaging. Our findings underline the clinical potential of perfusion imaging biomarkers as surrogates for the assessment of infarct growth and brain edema development prior to treatment.

On a larger scale, our observations that favorable HIR profiles were found to be associated with less edema formation and reduced EPR at admission support the assumption that favorable collateral blood flow preserves tissue viability in ischemic brains after the occlusion of a large vessel. This hypothesis is further strengthened by our finding that favorable HIR profiles were correlated to favorable baseline clinical and imaging characteristics, i.e., presentation NIHSS and ASPECTS on admission. HIR is a standardized measure, which is automatically calculated from perfusion imaging [6]. Favorable HIR profiles were reported to correlate with robust arterial collateral status and robust microvascular blood transit through ischemic tissue, both of which are related to edema formation within ischemic brains [7–9,16–18].

These observations are similar to those of Broocks et al. [16], who also found higher admission ASPECTS, lower admission NIHSS, better median mRS scores after three months, and higher vessel recanalization rates in patients with good arterial collateral scores and less edema formation. Interestingly, HIR, but not pial arterial collateral status, was independently associated with reduced brain edema in our study in the multivariable regression analysis. In contrast, while most HIR+ patients also exhibited robust pial arterial collaterals, approximately 20% of patients were found to have favorable HIR profiles but poor arterial collateralization (Table 2). To date, it is not completely understood why some AIS-LVO patients may exhibit robust cerebral tissue perfusion, despite the lack of pial arterial collaterals and vice versa. One explanation for that may be that both parameters, although somewhat interrelated, assess different levels of microvascular blood flow in ischemic brains. While the arterial collateral status is associated with blood transit to the ischemic tissue, HIR is believed to be more likely related to microvascular blood transit on a tissue level (so called tissue-level collaterals), which differs substantially from the findings of previous studies [12,16]. Another explanation may be that the temporal recruitment of tissue-level collaterals differs from pial collaterals during the time-course of AIS-LVO, e.g., due to local autoregulation of the brain or aggravated progressive brain edema, thus leading to divergent imaging findings [12,19]. This assumption may further be supported by the finding of another study, which found that favorable HIR, but not arterial collaterals, was correlated to delayed edema formation 48–72 h after thrombectomy treatment [7]. Brain edema formation follows a non-linear pattern [16] and the robustness of "collateral-governed" microvascular perfusion in ischemic brains is known to vary over time, which may lead to the assumption that some collateral parameters may influence the tissue fate in earlier stages of ischemia, while other collateral biomarkers more strongly influence the long-term perseverance of ischemic brain tissue [20–25]. However, as demonstrated in previous studies, HIR seems to be strongly associated with edema formation, infarct progression, and long-term clinical outcomes after thrombectomy treatment [6,7,11]. Thus, it is conceivable that HIR may have strong prognostic implications regarding the procedural and clinical outcomes of AIS-LVO patients anticipated for thrombectomy treatment. Further studies are needed to better understand the timely mechanisms that alter microvascular collateral perfusion with respect to cerebral brain edema development over time and the specific mechanisms that influence clinical recovery after treatment.

Our study has several limitations. First, the retrospective design of the study may introduce bias. Next, the generalizability of our results may be limited, due to the specific utilized study protocol. To date, NWU determination is still time-consuming and laborious, thus limiting the broader applicability of this approach. There is still no standardized or automated approach for the determination of pial arterial collaterals on CT angiography images, which may introduce bias. Lastly, CT perfusion is still not available at many primary and secondary care centers, which limits the practicability of our approach.

Lastly, different thresholds for the dichotomization of HIR exist in the literature, and further research is needed to identify the optimal HIR threshold for the prediction of early edema development. Multicollinearity remains a potential problem in studies examining the impact of biomarkers that share a common pathophysiology. We tested for multicollinearity and found the intercorrelation between HIR and other co-variables to be low; however, we cannot fully exclude potential bias from multicollinearity.

5. Conclusions

Favorable HIR profiles were associated with less ischemic edema formation and reduced temporal edema development from symptom onset to imaging in AIS-LVO patients.

Author Contributions: Conceptualization, G.B., R.K., M.C.K., S.C., G.W.A., J.F., M.W., J.J.H. and T.D.F.; data curation, N.v.H., G.B., R.K., M.C.K., S.C., M.M., L.M., M.G.L., P.S., A.G., J.F., M.W. and T.D.F.; formal analysis, N.v.H. and M.M.; investigation, N.v.H., J.J.H. and T.D.F.; methodology, G.B.; software, J.J.H.; supervision, M.G.L., G.W.A., J.F., M.W. and J.J.H.; validation, G.W.A.; visualization, T.D.F.; writing—original draft, N.v.H., R.K., M.C.K., S.C., M.M., L.M., M.G.L., G.W.A., P.S., A.G., J.F., M.W., J.J.H. and T.D.F.; Writing–review and editing, N.v.H., G.B., R.K., M.C.K., S.C., M.M., L.M., M.G.L., G.W.A., P.S., A.G., J.F., M.W., J.J.H. and T.D.F. All authors have read and agreed to the published version of the manuscript.

Funding: Tobias Djamsched Faizy was funded by the German Research Foundation (DFG) for his work as a postdoctoral research scholar at Stanford University, Department of Neuroradiology (project number: 411621970).

Institutional Review Board Statement: The study protocol of this multicenter cohort study was approved by the institutional review boards of both study centers (eRef. Number: 37209 and GSR 689-15), and complied with the Health Insurance Portability and Accountability Act (HIPAA) and followed the guidelines of the Declaration of Helsinki. Patient informed consent was waived by the review boards for this retrospective study.

Informed Consent Statement: Patient consent was waived due to the retrospective design of this study and the comprehensive anonymization.

Data Availability Statement: The data that support the findings of this study are available from the corresponding author upon reasonable request.

Conflicts of Interest: Gregory W. ALBERS reports equity and consulting for iSchemaView and consulting from Medtronic. Jens FIEHLER reports unrelated-consultancy for Acandis, Boehringer Ingelheim, Cerenovus, Covidien, Evasc Neurovascular, MD-Clinicals, Medtronic, Medina, MicroVention, Penumbra, Route 92 Medical, Stryker, Transverse Medical; Grants/Grants Pending: MicroVention, Medtronic, Stryker, Cerenovus. CEO Eppdata. The other authors report no disclosures.

References

1. Broocks, G.; Hanning, U.; Flottmann, F.; Schönfeld, M.; Faizy, T.D.; Sporns, P.B.; Baumgart, M.; Leischner, H.; Schön, G.; Minnerup, J.; et al. Clinical benefit of thrombectomy in stroke patients with low ASPECTS is mediated by oedema reduction. *Brain* **2019**, *142*, 1399–1407. [CrossRef] [PubMed]
2. Broocks, G.; Flottmann, F.; Ernst, M.; Faizy, T.D.; Minnerup, J.; Siemonsen, S.; Fiehler, J.; Kemmling, A. Computed Tomography-Based Imaging of Voxel-Wise Lesion Water Uptake in Ischemic Brain: Relationship between Density and Direct Volumetry. *Investig. Radiol.* **2018**, *53*, 207–213. [CrossRef] [PubMed]
3. Broocks, G.; Flottmann, F.; Scheibel, A.; Aigner, A.; Faizy, T.D.; Hanning, U.; Leischner, H.; Broocks, S.I.; Fiehler, J.; Gellissen, S.; et al. Quantitative Lesion Water Uptake in Acute Stroke Computed Tomography Is a Predictor of Malignant Infarction. *Stroke* **2018**, *49*, 1906–1912. [CrossRef] [PubMed]
4. Broocks, G.; Kniep, H.; Schramm, P.; Hanning, U.; Flottmann, F.; Faizy, T.; Schönfeld, M.; Meyer, L.; Schön, G.; Aulmann, L.; et al. Patients with low Alberta Stroke Program Early CT Score (ASPECTS) but good collaterals benefit from endovascular recanalization. *J. Neurointerv. Surg.* **2020**, *12*, 747–752. [CrossRef] [PubMed]

5. Nawabi, J.; Flottmann, F.; Hanning, U.; Bechstein, M.; Schön, G.; Kemmling, A.; Fiehler, J.; Broocks, G. Futile Recanalization with Poor Clinical Outcome Is Associated with Increased Edema Volume after Ischemic Stroke. *Investig. Radiol.* **2019**, *54*, 282–287. [CrossRef]
6. Olivot, J.M.; Mlynash, M.; Inoue, M.; Marks, M.P.; Wheeler, H.M.; Kemp, S.; Straka, M.; Zaharchuk, G.; Bammer, R.; Lansberg, M.; et al. Hypoperfusion Intensity Ratio Predicts Infarct Progression and Functional Outcome in the DEFUSE 2 Cohort. *Stroke* **2014**, *45*, 1018–1023. [CrossRef]
7. Faizy, T.D.; Kabiri, R.; Christensen, S.; Mlynash, M.; Kuraitis, G.; Broocks, G.; Hanning, U.; Nawabi, J.; Lansberg, M.G.; Marks, M.P.; et al. Perfusion imaging-based tissue-level collaterals predict ischemic lesion net water uptake in patients with acute ischemic stroke and large vessel occlusion. *J. Cereb. Blood Flow Metab.* **2021**, *41*, 2067–2075. [CrossRef]
8. Guenego, A.; Fahed, R.; Albers, G.W.; Kuraitis, G.; Sussman, E.S.; Martin, B.W.; Marcellus, D.G.; Olivot, J.; Marks, M.P.; Lansberg, M.G.; et al. Hypoperfusion intensity ratio correlates with angiographic collaterals in acute ischaemic stroke with M1 occlusion. *Eur. J. Neurol.* **2020**, *27*, 864–870. [CrossRef]
9. Guenego, A.; Marcellus, D.G.; Martin, B.W.; Christensen, S.; Albers, G.W.; Lansberg, M.G.; Marks, M.P.; Wintermark, M.; Heit, J.J. Hypoperfusion Intensity Ratio Is Correlated with Patient Eligibility for Thrombectomy. *Stroke* **2019**, *50*, 917–922. [CrossRef]
10. Heit, J.J.; Mlynash, M.; Christensen, S.; Kemp, S.M.; Lansberg, M.G.; Marks, M.P.; Olivot, J.M.; Gregory, A.W. What predicts poor outcome after successful thrombectomy in late time windows? *J. Neurointerv. Surg.* **2021**, *13*, 421–425. [CrossRef]
11. Guenego, A.; Mlynash, M.; Christensen, S.; Bs, S.K.; Heit, J.J.; Lansberg, M.G.; Albers, G.W. Hypoperfusion ratio predicts infarct growth during transfer for thrombectomy. *Ann. Neurol.* **2018**, *84*, 616–620. [CrossRef] [PubMed]
12. Faizy, T.D.; Kabiri, R.; Christensen, S.; Mlynash, M.; Kuraitis, G.M.; Broocks, G.; Flottmann, F.; Marks, M.P.; Lansberg, M.G.; Albers, G.W.; et al. Favorable Venous Outflow Profiles Correlate with Favorable Tissue-Level Collaterals and Clinical Outcome. *Stroke* **2021**, *52*, 1761–1767. [CrossRef] [PubMed]
13. Pexman, J.H.W.; Barber, P.A.; Hill, M.D.; Sevick, R.J.; Demchuk, A.M.; Hudon, M.E.; Hu, W.Y.; Buchan, A.M. Use of the Alberta Stroke Program Early CT Score (ASPECTS) for Assessing CT Scans in Patients with Acute Stroke. *Am. J. Neuroradiol.* **2001**, *22*, 1534–1542. [PubMed]
14. Yeo, L.; Paliwal, P.; Teoh, H.; Seet, R.; Chan, B.; Ting, E.; Venketasubramanian, N.; Leow, W.; Wakerley, B.; Kusama, Y.; et al. Assessment of Intracranial Collaterals on CT Angiography in Anterior Circulation Acute Ischemic Stroke. *Am. J. Neuroradiol.* **2015**, *36*, 289–294. [CrossRef]
15. Minnerup, J.; Broocks, G.; Kalkoffen, J.; Langner, S.; Knauth, M.; Psychogios, M.N.; Wersching, H.; Teuber, A.; Heindel, W.; Eckert, B.; et al. Computed tomography-based quantification of lesion water uptake identifies patients within 4.5 h of stroke onset: A multicenter observational study. *Ann. Neurol.* **2016**, *80*, 924–934. [CrossRef]
16. Broocks, G.; Kemmling, A.; Meyer, L.; Nawabi, J.; Schön, G.; Fiehler, J.; Kniep, H.; Hanning, U. Computed Tomography Angiography Collateral Profile Is Directly Linked to Early Edema Progression Rate in Acute Ischemic Stroke. *Stroke* **2019**, *50*, 3424–3430. [CrossRef]
17. Tan, J.C.; Dillon, W.P.; Liu, S.; Adler, F.; Smith, W.S.; Wintermark, M. Systematic comparison of perfusion-CT and CT-angiography in acute stroke patients. *Ann. Neurol.* **2007**, *61*, 533–543. [CrossRef]
18. Faizy, T.D.; Heit, J.J. Rethinking the Collateral Vasculature Assessment in Acute Ischemic Stroke: The Comprehensive Collateral Cascade. *Top. Magn. Reson. Imaging* **2021**, *30*, 181–186. [CrossRef]
19. Jansen, I.G.H.; van Vuuren, A.B.; van Zwam, W.H.; van den Wijngaard, I.R.; Berkhemer, O.A.; Lingsma, H.F.; Slump, C.H.; van Oostenbrugge, R.J.; Treurniet, K.M.; Dippel, D.W.J.; et al. Absence of Cortical Vein Opacification is Associated with Lack of Intra-arterial Therapy Benefit in Stroke. *Radiology* **2018**, *286*, 731. [CrossRef]
20. Liebeskind, D.S. Imaging the collaterome: A stroke renaissance. *Curr. Opin. Neurol.* **2015**, *28*, 1–3. [CrossRef]
21. Sheth, S.A.; Liebeskind, D.S. Collaterals in endovascular therapy for stroke. *Curr. Opin. Neurol.* **2015**, *28*, 10–15. [CrossRef] [PubMed]
22. Bang, O.Y.; Saver, J.L.; Buck, B.H.; Alger, J.R.; Starkman, S.; Ovbiagele, B.; Kim, D.; Jahan, R.; Duckwiler, G.R.; Yoon, S.R.; et al. Impact of collateral flow on tissue fate in acute ischaemic stroke. *J. Neurol. Neurosurg. Psychiatry* **2008**, *79*, 625–629. [CrossRef] [PubMed]
23. Liebeskind, D.S. Collaterals in Acute Stroke: Beyond the Clot. *Neuroimaging Clin. N. Am.* **2005**, *15*, 553–573. [CrossRef] [PubMed]
24. Liebeskind, D.S. Collateral circulation. *Stroke* **2003**, *34*, 2279–2284. [CrossRef]
25. Baek, J.H. Low Hypoperfusion Intensity Ratio Is Associated with a Favorable Outcome even in Large Ischemic Core and Delayed Recanalization Time. *J. Clin. Med.* **2021**, *10*, 1869. [CrossRef]

Journal of
Clinical Medicine

Article

Potential of Stroke Imaging Using a New Prototype of Low-Field MRI: A Prospective Direct 0.55 T/1.5 T Scanner Comparison

Thilo Rusche [1,*], Hanns-Christian Breit [1], Michael Bach [1], Jakob Wasserthal [1], Julian Gehweiler [1], Sebastian Manneck [1], Johanna Maria Lieb [1], Gian Marco De Marchis [2], Marios Nikos Psychogios [1,†] and Peter B. Sporns [1,3,†]

[1] Department of Radiology, Clinic of Radiology & Nuclear Medicine, University Hospital Basel, University Basel, 4031 Basel, Switzerland; hanns-christian.breit@usb.ch (H.-C.B.); michael.bach@usb.ch (M.B.); jakob.wasserthal@usb.ch (J.W.); julian.gehweiler@usb.ch (J.G.); smanneck@gmail.com (S.M.); johanna.lieb@usb.ch (J.M.L.); marios.psychogios@usb.ch (M.N.P.); peter.sporns@usb.ch (P.B.S.)
[2] Department of Neurology, University Hospital Basel, University Basel, 4031 Basel, Switzerland; gian.demarchis@usb.ch
[3] Department of Diagnostic and Interventional Neuroradiology, University Medical Center Hamburg-Eppendorf, 20246 Hamburg, Germany
* Correspondence: rusche.thilo@googlemail.com; Tel.: +41-(0)762957963; Fax: +41-(0)612654354
† These authors contributed equally to this work.

Citation: Rusche, T.; Breit, H.-C.; Bach, M.; Wasserthal, J.; Gehweiler, J.; Manneck, S.; Lieb, J.M.; De Marchis, G.M.; Psychogios, M.N.; Sporns, P.B. Potential of Stroke Imaging Using a New Prototype of Low-Field MRI: A Prospective Direct 0.55 T/1.5 T Scanner Comparison. *J. Clin. Med.* **2022**, *11*, 2798. https://doi.org/10.3390/jcm11102798

Academic Editor: Hugues Chabriat

Received: 23 March 2022
Accepted: 13 May 2022
Published: 16 May 2022

Publisher's Note: MDPI stays neutral with regard to jurisdictional claims in published maps and institutional affiliations.

Abstract: Objectives: Ischemic stroke is a leading cause of mortality and acquired disability worldwide and thus plays an enormous health-economic role. Imaging of choice is computed-tomographic (CT) or magnetic resonance imaging (MRI), especially diffusion-weighted (DW) sequences. However, MR imaging is associated with high costs and therefore has a limited availability leading to low-field-MRI techniques increasingly coming into focus. Thus, the aim of our study was to assess the potential of stroke imaging with low-field MRI. Material and Methods: A scanner comparison was performed including 27 patients (17 stroke cohort, 10 control group). For each patient, a brain scan was performed first with a 1.5T scanner and afterwards with a 0.55T scanner. Scan protocols were as identical as possible and optimized. Data analysis was performed in three steps: All DWI/ADC (apparent diffusion coefficient) and FLAIR (fluid attenuated inversion recovery) sequences underwent Likert rating with respect to image impression, resolution, noise, contrast, and diagnostic quality and were evaluated by two radiologists regarding number and localization of DWI and FLAIR lesions in a blinded fashion. Then segmentation of lesion volumes was performed by two other radiologists on DWI/ADC and FLAIR. Results: DWI/ADC lesions could be diagnosed with the same reliability by the most experienced reader in the 0.55T and 1.5T sequences (specificity 100% and sensitivity 92.9%, respectively). False positive findings did not occur. Detection of number/location of FLAIR lesions was mostly equivalent between 0.55T and 1.5T sequences. No significant difference ($p = 0.789$–0.104) for FLAIR resolution and contrast was observed regarding Likert scaling. For DWI/ADC noise, the 0.55T sequences were significantly superior ($p < 0.026$). Otherwise, the 1.5T sequences were significantly superior ($p < 0.029$). There was no significant difference in infarct volume and volume of infarct demarcation between the 0.55T and 1.5T sequences, when detectable. Conclusions: Low-field MRI stroke imaging at 0.55T may not be inferior to scanners with higher field strengths and thus has great potential as a low-cost alternative in future stroke diagnostics. However, there are limitations in the detection of very small infarcts. Further technical developments with follow-up studies must show whether this problem can be solved.

Keywords: stroke imaging; low-field MRI; reading study; scanner comparison

1. Introduction

Stroke is the second leading cause of death and a major cause of disability worldwide. Its incidence is increasing because the population is aging [1]. Thus, stroke has an enormous health-economic impact with total costs of approximately EUR 26.6 billion in the European Union (EU) [2] and USD 71.55 billion in the USA [1]. The global incidence in the <65 years age group has increased by approximately 25% [3], particularly affecting younger age groups in low- and middle-income countries [2]. Causes for this increase can partly be attributed to inadequate prevention behavior to reduce risk factors and an insufficient infrastructure [4], especially regarding the availability of stroke imaging and further etiological work-up [5,6]. However, even in high-income countries, the demand for stroke imaging will continue to increase with rising stroke incidences [2]. In particular, MR imaging with diffusion-weighted (DW) sequences will become more important as MRI is superior to CT with a higher sensitivity, especially in the acute phase and for smaller lacunar strokes [6]. On the other hand, costs of MRI are significantly higher than those of CT [7] and metallic implants/cardiac pacemakers may pose contraindications. With this in mind, low-field MRI is increasingly coming into focus, offering MR imaging at a much lower cost but also reducing possible complications with metallic implants [5]. Therefore, the aim of our study was to evaluate the performance of a new prototype of low-field MRI in a prospective cohort of suspected stroke patients and to directly compare the diagnostic value to an established 1.5T MRI.

2. Materials and Methods

This prospective study was reviewed and approved by the cantonal (Basel Stadt, Switzerland) ethics committee (*BASEC2021 00166*). All included patients signed an informed consent form.

2.1. Patient Selection and Data Acquisition

Patient selection and data acquisition was performed from 1 May 2021 to 30 June 2021 at the Department of Radiology and Nuclear Medicine, University Hospital Basel, Switzerland, with the following steps (Figure 1): First, all patients who underwent MRI using a 1.5T scanner (Siemens MAGNETOM Avanto FIT 1.5T, Siemens Healthcare, Erlangen, Germany) as part of the diagnostic stroke workup for suspected stroke or transient ischemic attack (TIA) were preselected. Afterwards, MRI scans were reviewed for completeness and quality and regarding diagnosis of ischemic stroke or other acute pathologies. If no other acute pathology was detected, consent was obtained and patients were additionally examined using a new prototype of 0.55T scanner (Siemens MAGNETOM FreeMax 0.55T, Siemens Healthcare, Erlangen, Germany) directly after the 1.5T examination. If 0.55T datasets were complete and of good quality the patient was included in the study. Included patients without ischemic stroke on 1.5T imaging or any other acute pathology were defined as the control group. Patients were excluded if they fulfilled the following exclusion criteria:

(a)   Incomplete dataset 0.55T or 1.5T examination
(b)   Insufficient image quality 0.55T or 1.5T examination
(c)   No stroke but other acute pathology within initial 1.5T scan
(d)   No informed consent for additional 0.55T examination
(e)   Too large time difference between 1.5T and 0.55T examination (cut off 2 h)
(f)   Foreign materials not authorized for 0.55T scanners (i.e., cardiac pacemakers)

The 1.5T scanning protocol was in accordance with the hospital's internal standard protocol for emergency stroke diagnostics including axial DWI/ADC, FLAIR, and SWI (susceptibility-weighted imaging) sequences (Table 1). The 0.55T protocol (Table 1) corresponded to the standard protocol regarding the sequences used and was maximally adapted to the 1.5T protocol as far as technically possible (same slice thickness (ST) and slice spacing (SP); comparable in-plane resolution) to ensure the most sufficient scanner comparison. After subsequent verification with respect to data completeness (scan protocols with complete image acquisition) and image quality (artifacts, image contrast), the datasets were transferred to the Picture Archiving and Communication System (PACS, General Electric (GE), Boston, Massachusetts, MA, USA) for further analysis.

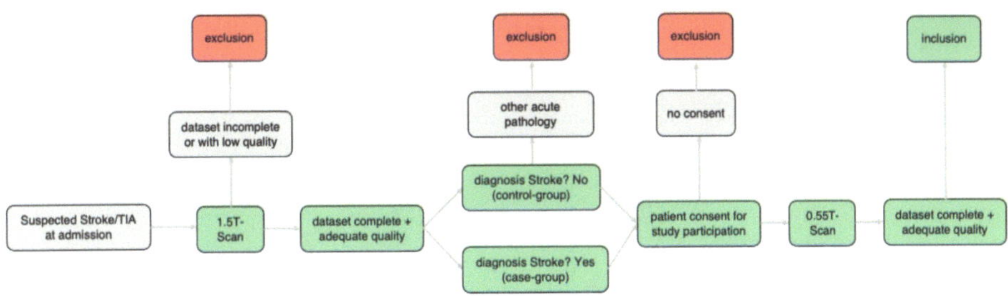

Figure 1. Workflow of patient selection and data acquisition.

Table 1. Scan protocols, Siemens MAGNETOM Avanto FIT 1.5 T and Siemens MAGNETOM FreeMax 0.55 T.

	Siemens MAGNETOM FreeMax 0.55 T	Siemens MAGNETOM Avanto Fit 1.5 T
FLAIR (fluid attenuated inversion recovery) transversal		
Field strength in T	0.55	1.5
FOV (field of view) in mm^2	209 × 230	187 × 230
ST (slice thickness) in mm	3	3
SS (slice spacing)	3.6	3.6
Number of slices	40	40
PS (pixel spacing) in mm^2	1.28 × 1.03	0.9 × 0.9
TR (repetition time) in msec	7780	8510
TE (echo time) in msec	96	112
TI (inversion delay) in msec	2368.8	2460
Turbo Factor	15	19
TA (time of acquisition) in min	5:28	3:26
BW (Bandwidth)	150	130
3D SWI (susceptibility weighted imaging) transversal		
Field strength in T	0.55	1.5
Sequence type	Multi shot 3D EPI	3D FLASH
FOV (field of view) in mm^2	201 × 230	194 × 230
ST (slice thickness) in mm	3	3
Number of slices	40	48
PS (pixel spacing) in mm^2	0.94 × 0.8	1.12 × 0.9
TR (repetition time) in msec	172	48
TE (echo time) in msec	100	40
Parallel imaging	-	GRAPPA factor 2
TA (time of acquisition) in min	2:23	2:17
BW (Bandwidth)	276	80
Single shot diffusion EPI (echo-planar imaging) transversal		
Field strength in T	0.55	1.5
FOV (field of view) in mm^2	220 × 220	230 × 230
ST (slice thickness) in mm	3	3
SS (slice spacing)	3.6	3.6
Number of slices	40	40
PS (pixel spacing) in mm^2	1.67 × 1.67	1.44 × 1.44
b values in s/mm^2	0, 1000	0, 1000
TR (repetition time) in msec	7400	6200
TE (echo time) in msec	102	103
Parallel imaging	GRAPPA factor 2	GRAPPA factor 2
TA (time of acquisition) in min	4:35	2:04
BW (Bandwidth)	842	1490

2.2. Data Analysis

Data analysis was performed in a three-step procedure (Figure 2): First, the 0.55 T and 1.5 T datasets were evaluated using Likert rating. Second, a reading study was performed regarding identification and localization of DWI and FLAIR lesions. Finally, a correlation analysis of DWI and FLAIR lesion volumes was performed.

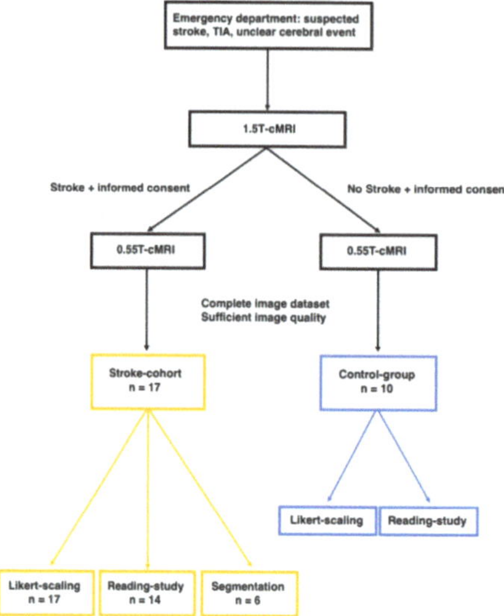

Figure 2. Workflow of data processing.

2.2.1. Likert Rating

Likert rating was performed by a neuroradiologist and a neuroradiologist in training with experience of 9 and 5 years. Each acquired 0.55 T or 1.5 T DWI/ADC and FLAIR dataset was rated with respect to the following criteria with a numerical value between 1 and 10 (1 maximum poor, 10 maximum good):

(a)   Overall image quality;
(b)   Resolution;
(c)   Noise;
(d)   Contrast;
(e)   Diagnostic quality.

Sample DWI/ADC and FLAIR sequences from a 3T scanner (Siemens MAGNETOM Skyra 3T, Siemens Healthcare, Erlangen, Germany) were set as gold standard (numerical value = 10). Dataset assessment was PACS-based using a standardized bookmark. Both raters were blinded to the results of the other rater.

2.2.2. Reading Study

Reading of 0.55 T and 1.5 T datasets was performed PACS-based and blinded (no clinical information, no image information) by two neuroradiologists with 8 and 13 years of professional experience, the latter defined as the most experienced rater. PACS-based post-processing as part of image analysis was allowed. For each dataset, the following reading tasks were conducted:

(a)   Evaluation stroke yes/no;
(b)   Number of DWI lesions: 0, 1, 2–10, >10;
(c)   DWI lesion main localization (especially in the case of multiple DWI lesions);
(d)   Number of FLAIR lesions: 0, 1, 2–10, >10;
(e)   FLAIR lesion main localization (especially in the case of multiple FLAIR lesions).

Reading was performed for only 24 of the 27 patient datasets (Figure 2, Table 2) because the time gap between the 1.5 T and 0.55 T examination was >1 h in three cases and possible data distortions caused by this time delay (e.g., infarct demarcation increasing in the meantime and therefore easier detectability of the lesions) should be excluded.

Table 2. Detailed patient data.

Patient	Patient Age	Neurological Symptoms at Admission	NIHSS	Control-Group (C), Stroke-Cohort (S)	Final Radiological Report	Time Gap between Scans in min	Time Gap between Onset and Scan 1 in min
Patient 1	87	facial droop, dysarthria, hemiparesis right side	3	S	acute to subacute infarct corpus nuclei caudati and cranial parts of the nucleus lentiformis left side	46	1166
Patient 2	88	ataxic gait and standing	no data	S	acute to subacute infarct lateral pontin left side	37	unknown
Patient 3	82	visual deficit	no data	S	acute to subacute punctiform infarcts parietal left and cerebellar right	93	unknown
Patient 4	84	intermittent dysarthria	1	S	acute to subacute infarct thalamus left side	33	unknown
Patient 5	58	facial droop, hemiparesis right side	10	S	multiple subacute infarcts posterior circulation on both sides	40	1050
Patient 6	65	low-grade facial paresis left side	1	S	acute to subacute infarcts of the thalamus and occipital/occipitotemporal right side	20	1175
Patient 7	65	dysdiadochokinesis right side	no data	S	subacute punctiform infarcts frontal and parietal left side	24	1704
Patient 8	75	facial droop right side	0	S	punctiform infarct gyrus postcentralis left side	22	1135
Patient 9	82	confusion	no data	S	acute to subacute infarcts frontal and parietal left side	32	unknown
Patient 10	79	hemiataxia left side	2	S	acute to subacute infarct thalamus right side	42	1492
Patient 11	86	dysarthria, ataxia right leg	3	S	acute infarct posterolateral pons left side	25	unknown
Patient 12	83	leg-emphasized hemiparesis left side	5	S	subacute infarct gyrus precentralis right side	31	unknown
Patient 13	89	visual deficit	2	S	acute cortical infarcts parietooccipital and frontal right side	38	2198
Patient 14	69	no data	no data	S	subacute infarct central left side	42	unknown
Patient 15	73	amaurosis fugax	0	C	no stroke	25	1197
Patient 16	29	strong nystagmus to left side, headache right frontal, vertigo	no data	C	no stroke	48	unknown
Patient 17	70	atypical transient global amnesia	no data	C	no stroke	44	unknown
Patient 18	87	aphasia	no data	C	no stroke	32	unknown
Patient 19	74	vertigo	0	C	no stroke	25	826
Patient 20	60	aphasia	0	C	no stroke	21	unknown
Patient 21	44	vertigo	0	C	no stroke	49	unknown
Patient 22	80	short-term loss of vision left side	no data	C	no stroke	35	2092
Patient 23	84	vertigo, gait instability	no data	C	no stroke	48	425
Patient 24	84	vertigo, gait instability	no data	C	no stroke	32	unknown
Patient 25	53	aphasia	1	excluded	subacute punctiform infarcts frontal and parietal left side	916	1197
Patient 26	59	transient global amnesia	0	excluded	bilateral punctiform diffusion-restrictions of the hippocampus head	2936	1394
Patient 27	46	facial droop, descending arm left side	2	excluded	acute to subacute infarct lenticostriatal right side	2812	1406

The clinical neuroradiological report and final neurological diagnosis (Table 2) were defined as underlying gold standard for the accuracy of the reading study.

2.2.3. Segmentation of DWI/ADC and FLAIR Lesions

For segmentation, the 0.55 T and 1.5 T datasets were transferred to a research server and post-processing software (NORA Medical Imaging Platform Project, University Medical Center Freiburg, Freiburg, Germany). Then, segmentation of stroke lesions was performed separately in each of the 0.55 T and 1.5 T DWI, ADC, and FLAIR sequences by two radiologists with experience of 4 and 5 years. Stroke lesions with multifocal and inhomogeneous confluent distribution or punctate configuration (too small for segmentation) were excluded for the segmentation process because in these cases sufficient segmentation was limited, and the bias of the results possibly caused by this should be avoided (Figure 2). In total, segmentation was performed for six 0.55 T and 1.5 T datasets.

2.3. Statistical Analysis

For statistical evaluation of Likert rating, a mean of the ratings of readers 1 and 2 was first calculated for each 0.55 T and 1.5 T patient dataset and evaluation point (a)–(e). Subsequently, a Wilcoxon signed rank test was used to evaluate significant or non-significant differences in Likert rating between the 0.55 T and 1.5 T sequences. Then, inter-reader comparisons were performed to determine intraclass correlation coefficients (ICC).

Calculation of sensitivity and specificity of readers 1 and 2 in the reading study was performed in relation to the gold standard.

Volume correlation of infarct volume between the 0.55 T and 1.5 T DWI and ADC sequences and volume correlation of infarct demarcation between the 0.55 T and 1.5 T FLAIR sequences were performed by first obtaining a mean volume for each of the six 0.55 T and 1.5 T DWI, ADC, and FLAIR datasets from readers 1 and 2. Afterwards, the Wilcoxon signed rank test was used to examine significant differences in infarct volume and volume of infarct demarcation between the 0.55 T and 1.5 T datasets ($p < 0.05$). In addition, an inter-reader correlation was performed for the segmented volumes of the DWI, ADC, and FLAIR datasets by calculating an intraclass correlation coefficient (ICC).

3. Results

A total of 27 complete and artifact-free datasets (17 stroke cohort; 10 control group) were acquired (mean age $\pm$ standard deviation, 71 years $\pm$ 15; 12 women [44%]).

Most included patients had mild neurologic symptoms (for details see Table 2) at admission and a low NIHSS score (1.88 ± 2.52).

3.1. Likert Rating

3.1.1. DWI/ADC Datasets

Regarding overall image quality (a), resolution (b), contrast (d), and diagnostic quality (e), average Likert ratings of readers 1 and 2 (Figure 3) were significantly better for the 1.5 T sequences than the 0.55 T sequences: (a) $p < 0.001$; (b) $p < 0.001$; (d) $p = 0.001$; (e) $p < 0.001$. Regarding noise (c) 0.55 T sequences were significantly superior ($p < 0.026$) to the 1.5 T sequences. Inter-reader comparisons showed high levels of agreement between readers 1 and 2 (ICC: (a) 0.77; (b) 0.78; (c) 0.84; (d) 0.71; (e) 0.88).

3.1.2. FLAIR Datasets

Regarding overall image quality (a), noise (c), and diagnostic quality (e), average Likert ratings of both readers (Figure 4) were significantly better for the 1.5 T sequences than the 0.55 T sequences: (a) $p < 0.0027$; (c) $p < 0.001$; (e) $p < 0.0292$). There was no significant difference for the criteria resolution (b) and contrast (d) ((b) $p = 0.1039$ and (d) $p = 0.7890$).

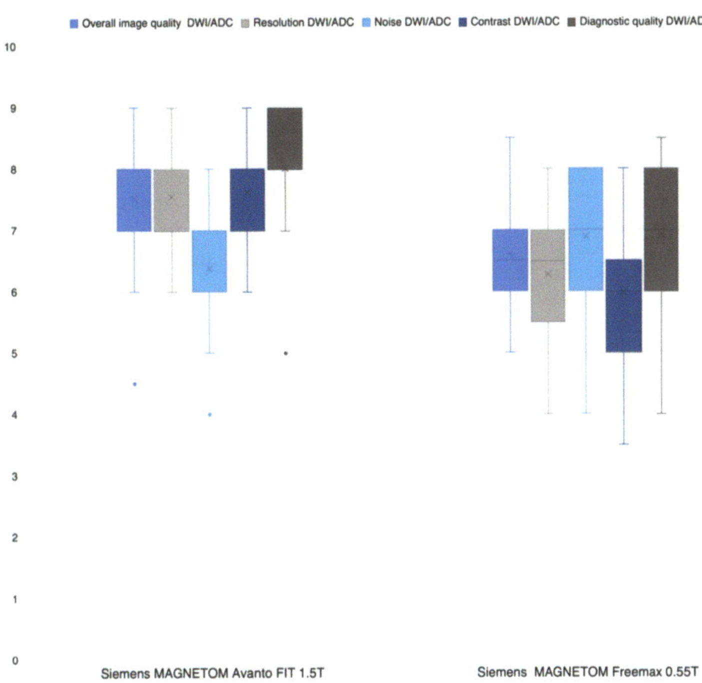

Figure 3. Average Likert-scoring reader 1 and 2 DWI/ADC sequences.

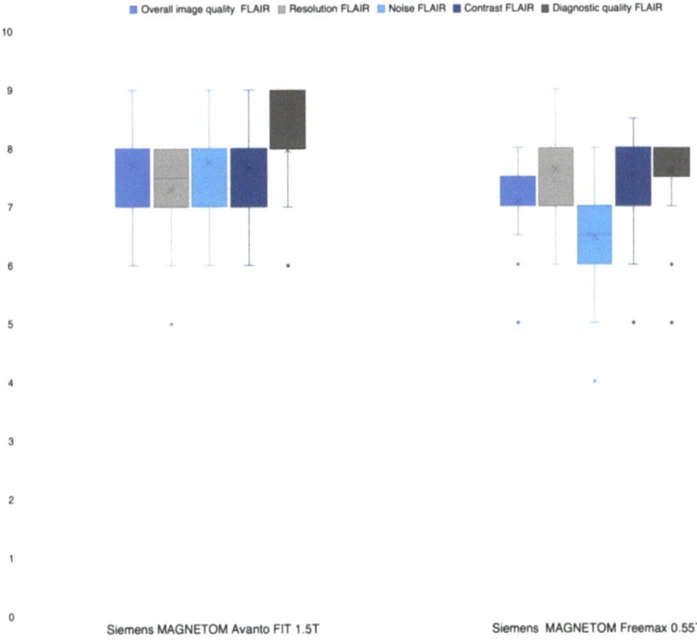

Figure 4. Average Likert-scoring reader 1 and 2 FLAIR sequences.

In the inter-reader comparison, there was medium to high agreement (ICC: (a) 0.64; (b) 0.87; (c) 0.64; (d) 0.73; (e) 0.71).

3.2. Reading Study

The reading study was performed with a total of 24 of the initially acquired 27 patient datasets (14 stroke cohort and 10 control group; mean age ± standard deviation, 74 years ± 14; 46% women) because in three datasets from the stroke cohort the time gap between the 1.5 T scan and 0.55 T scan was >1 h and possible result distortions caused by this should be excluded (Table 2).

The average time gap between the 1.5 T scans and 0.55 T scans was 36.8 ± 14.7 min.

3.2.1. DWI/ADC Datasets

There were no false positive findings in the 0.55 T sequences by readers 1 and 2, meaning no lesions were detected in the control group datasets (specificity at 0.55 T of readers 1 and 2: 100%). Reader 1 failed to detect a stroke in one case in both the 0.55 T and 1.5 T datasets (sensitivity of reader 1 at 0.55 T compared to gold standard: 92.9%).

Reader 2 did not detect two DWI lesions in the 0.55 T datasets but detected all lesions in the 1.5 T datasets (sensitivity of reader 2 at 0.55 T compared to gold standard: 85.6%). Both missed lesions had a point-like pattern, one being located subcortical/cortical and one infratentorial (Figures 5 and 6). Regarding the number of stroke lesions and lesion localization, there was complete agreement between readers 1 and 2 at the 1.5 T and 0.55 T datasets. Assessment of the extent of stroke as well as safe anatomic stroke localization was thus equivalent in the 0.55 T sequences compared with the 1.5 T sequences (Figure 7).

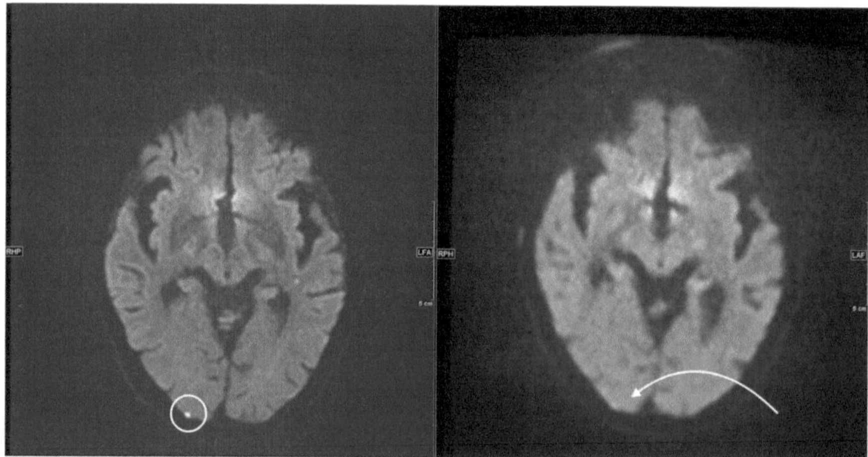

Figure 5. Left side, axial 1.5 T DWI sequence with sharply hyperintense delineable, punctiform DWI lesion located cortico-subcortical occipital right. On the right side, corresponding 0.55 T axial DWI sequence with the same slice localization without sufficiently detectable lesion.

3.2.2. FLAIR Datasets

In only one case did an inter-reader difference occur with respect to the assessment for a singular pontine lesion detected by reader 1 and not detected by reader 2. In the remaining 23 0.55 T and 1.5 T datasets, no inter-reader differences occurred. Reader 1 failed to detect single (1–2) FLAIR lesions with localization in the corona radiata in the 0.55 T FLAIR sequences compared with the 1.5 T FLAIR sequences in three cases. In contrast, there was complete agreement in the remaining 21 datasets. Reader 2 failed to detect single (1–2) FLAIR lesions with localization in the corona radiata in two cases in the 0.55 T FLAIR sequences compared with the 1.5 T FLAIR sequences. These cases overlapped with the

FLAIR lesions from reader 1 that were not detected in the 0.55 T sequences. In contrast, there was complete agreement in the remaining 22 datasets.

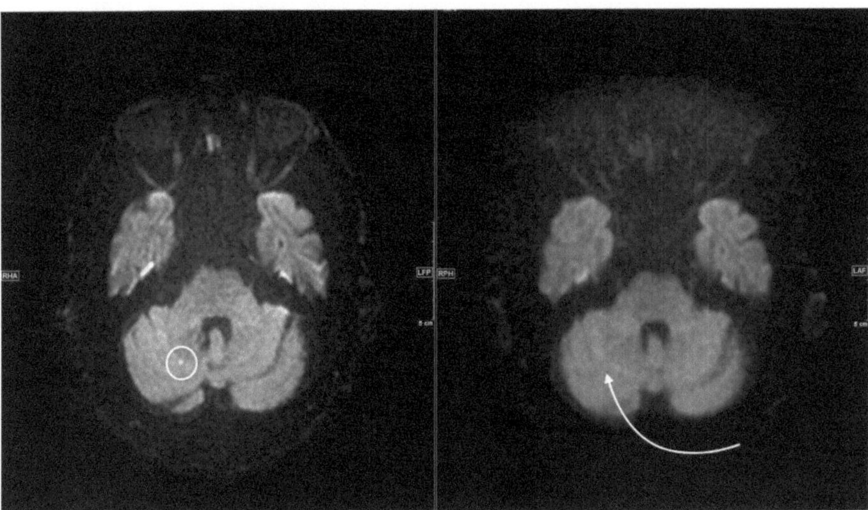

Figure 6. Left side, axial 1.5 T DWI sequence with sharply hyperintense delineable, punctiform DWI lesion located cerebellar right. On the right side, corresponding 0.55 T axial DWI sequence with the same slice localization without sufficiently detectable lesion.

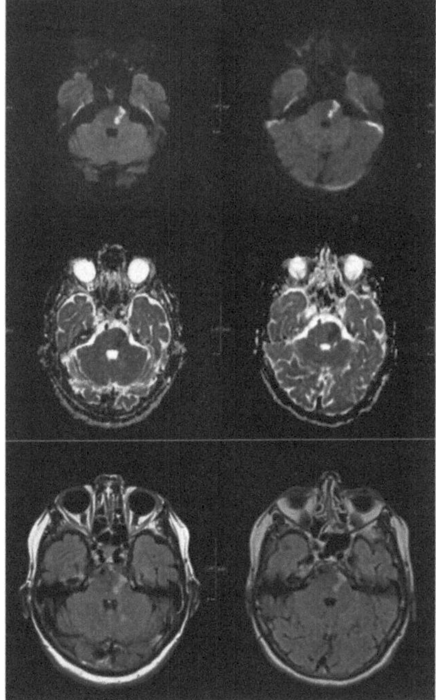

Figure 7. Left images from top to bottom axial 1.5 T DWI, ADC and FLAIR sequence with sharply delineated, demarcated, subacute pons infarct on the left. Right side, corresponding axial 0.55 T DWI, ADC and FLAIR sequence at the same slice localization with equivalent infarct morphology and thus identical diagnostic quality.

3.3. Segmentation of DWI/ADC and FLAIR Lesions

There was no significant difference regarding segmented lesion volumes between the 0.55 T and 1.5 T datasets (DWI $p = 0.375$, ADC $p = 0.63$, FLAIR $p = 0.38$). Inter-reader comparisons showed high levels of agreement for the volume segmentations of readers 1 and 2 for the individual DWI, ADC and FLAIR sequences (ICC DWI: 0.811, $p = < 0.0014$; ICC ADC: 0.89, $p = < 0.0001$; ICC FLAIR: 0.909, $p < 0.0001$).

4. Discussion

In our study, no significant differences were observed between the 0.55 T and 1.5 T sequences for FLAIR resolution and contrast regarding the Likert ratings. For DWI/ADC noise, the 0.55 T sequences were even significantly superior. For the diagnostic accuracy readings, the 0.55 T sequences were non-inferior for one of the two readers, meaning that DWI/ADC lesions were detected with the same specificity and sensitivity by the more experienced reader (100% and 92.9%, respectively). Regarding FLAIR lesions, there was almost complete agreement between the 0.55 T and 1.5 T sequences. In addition, we demonstrated that there were no significant differences between the 0.55 T and 1.5 T sequences in infarct volume and volume of infarct demarcation.

To the best of our knowledge, no comparable study has performed a 0.55 T versus 1.5 T scanner comparison regarding stroke imaging before. However, Mehdizade et al. [8] demonstrated, in a collective of a total of 18 patients, that an open low-field 0.23T scanner (Outlook, Marconi Medical Systems, Cleveland, OH, USA) can detect stroke lesions with equal confidence compared with a 1.5 T scanner (Eclipse, Marconi Medical Systems, Cleveland, OH, USA) using the acquired DWI sequences. In contrast, the diagnostic performance of the associated ADC maps was inferior for the 0.23T system. Terada et al. [9] investigated the same issue in a total of 24 patients with acute ischemic stroke using a 0.3T scanner (AIRIS II, Hitachi Medical Corporation, Tokyo, Japan) in comparison with a 1.5 T scanner (Gyroscan ACS NT, Philips Medical Systems, Hamburg, Germany). Similarly, the performance of the low-field MRI system was non-inferior, only more vulnerable to motion artifacts. The above-mentioned results are basically in agreement with the results of our study.

4.1. Likert Rating

The largest differences between DWI/ADC sequences at 0.55 T and 1.5 T were observed for resolution (6.2 $\pm$ 0.9 versus 7.6 $\pm$ 0.8) and contrast (7.6 $\pm$ 0.7 versus 6.0 $\pm$ 1.0). There may be several reasons for this: The resolution of the 0.55 T DWI/ADC sequences is nominally (about 14%) worse (pixel spacing 1.67 mm $\times$ 1.67 mm versus 1.44 mm $\times$ 1.44 mm; see Table 1). In addition, the bandwidth (BW) of the 0.55 T DWI/ADC sequences is significantly lower (BW 842 versus 1490, see Table 1). Therefore, the pass-through speed of the K space is slower compared to the 1.5 T DWI/ADC sequences. As a result, the data acquisition time on the 0.55 T scanner is longer and hence the time span of T2 * relaxation. According to the laws of the Fourier transformation, this is reflected in a signal reduction in K space and therefore a loss of resolution in spatial space (smoothing). The relatively high inferiority of the 0.55 T DWI/ADC sequences regarding contrast can be explained by the lower resolution (see above). This results in larger partial volume effects, which lead to lower contrast. Moreover, the 0.55 T scanner provides a priori a lower resonance signal (fewer protons are polarized) due to its lower magnetic field. The inferior head coil with only eight channels (1.5 T 32 channel head coil) also contributes to this effect. Interestingly, 0.55 T was superior to 1.5 T regarding DWI/ADC noise. Since the signal from the 0.55 T scanner is fundamentally inferior to the 1.5 T scanner due to the lower field strength and the inferior gradient and coil system, the latest post-processing applications for noise reduction were integrated into the 0.55 T system (deep resolve gain; Deep Resolve—Mobilizing the power of networks, Siemens Healthineers White paper, Behl et al.). This artificially reduces the visible and thus subjective noise noticeably without influencing or improving the signal and most likely explains the 0.55 T superiority regarding noise. With respect to FLAIR sequences, 0.55 T was non-inferior in resolution and contrast. This is because the resolution of the FLAIR sequences is already nominally better than the DWI/ADC 0.55 T sequences (Table 1) and the difference to the 1.5 T FLAIR sequences is therefore less detectable visually. In addition, the signal loss during K space acquisition is significantly lower for the FLAIR sequences than for the single-shot epi sequences (DWI/ADC), inducing less smoothing and not significantly reducing contrast. However, the decisive criterion for the radiologist is the diagnostic quality of the sequences. In this regard, there were only minor differences between the 0.55 T and 1.5 T DWI/ADC and FLAIR sequences (Figures 3 and 4) to the disadvantage of the 0.55 T system. Nevertheless, the advantages of the low-field MRI are lower costs for installation and maintenance (lower weight, smaller device size, no quench pipe, smaller MRI cabins) and thus for image acquisition, which in turn could increase future availability in developed economies, but especially in undeveloped areas [10,11].

Reading Study

Although the 0.55 T DWI/ADC sequences were maximally optimized (variation of b values, resolution and scan duration, application of artificial intelligence-based algorithms to increase visible resolution, as well as noise reduction), acute DWI lesions with punctuated configuration and cortical/subcortical or infratentorial localization could not be reliably detected in individual cases. In our opinion, possible causes are a too low contrast-to-noise ratio (CNR) and signal-to-noise ratio (SNR) or a too low resolution of the 0.55 T sequences. In conclusion, a too weak signal is the most likely cause. Future developments in coil design and the usability of techniques such as simultaneous multi-slice acquisition or stronger gradient systems may improve the SNR available per unit time and possibly the detection of small lesions.

In principle, the measurement time could simply be extended to generate more signal. However, this would lead to an increase in motion artifacts and long measurement times are counterproductive, especially in stroke patients, so a tradeoff between acquisition time and signal must be found. Overall, the impact of non-detection of a stroke lesion on further patient care is crucial. From a meta-analysis of 12 studies by Edlow et al. [12], it is known that, regardless of field strength, the prevalence of DWI negative acute ischemic strokes is approximately 6.8%. Factors associated with false negative DWI include stroke lesions of the posterior circulation, small volume, in which MRI was performed within the first 6 h of onset, or the NIHSS was <4 [12,13]. Some of these factors also apply to DWI lesions not detected in our study. Confirmed data on morbidity and mortality due to non-detection of stroke lesions in this context are not available. However, it is known that patients with missed strokes have a higher risk of recurrent strokes because of failure of stroke evaluation or secondary stroke prevention [13,14]. Related to the results of our study, we therefore recommend triage of patients with suspected stroke or TIA in the emergency setting: patients with mild, non-specific neurological symptoms or compatible with a posterior circulation stroke should thus be primarily examined on 1.5 T–3T devices to reliably detect even the smallest DWI lesions. Patients with clear or severe neurological symptoms, in the context of a diagnosis of exclusion or status after stroke for the evaluation of infarct demarcation, infarct size or complications after lysis (e.g., hemorrhage), could be examined sufficiently, safely, and equally on the 0.55 T system. This is also underlined by the results of our segmentation analysis, in which no significant differences in infarct volumes and volumes of infarct demarcation could be seen between the 0.55 T and 1.5 T sequences.

4.2. Limitations

There are several limitations that need to be addressed: First, a 1.5 T device of routine clinical use is the gold standard in this study. The question arises whether, for example, more lesions would have been detected at 3T or at 7T. Second, both scanners differ in respect to their gradient and coil system as well as the field strength. Therefore, it ultimately remains unclear whether the reason for a lack of delineation of the smallest lesions at 0.55 T is due to the field strength, the weaker gradient system, or even technical specifications such as the coils used. Third, the study cohort—while prospective—is still relatively small. Larger-scale studies to further define indications for stroke imaging at 0.55 T are needed and should assess whether scanner choice has an impact on patient outcomes. Fourth, the patients included in the study (control group and stroke cohort) were a convenient sample with limited representativeness in relation to the overall population (patients with suspected stroke or TIA). However, from our point of view this limitation plays only a minor role, because our study was primarily concerned with the basic evaluation of low-field MRI imaging in the context of a scanner comparison and less with the definition of, for example, possible selection criteria in the context of MRI-based stroke diagnosis, for example.

Fifth, there may be a lack of external validity, as the interpretation of the 0.55 T sequences may be difficult, especially for inexperienced radiologists. This may be aggravated in regions with low MRI availability and therefore a lack of neuroradiological specialists. On the other hand, this study presents an initial experience with 0.55 T MRI stroke imaging and could serve as a help for the interpretation.

In conclusion, low-field MRI stroke imaging at 0.55 T may not be inferior to scanners with higher field strength and thus has great potential as a low-cost alternative in future stroke diagnostics. However, there are limitations in the detection of very small infarcts. Further technical developments with follow-up studies must show whether these questions can be solved.

Author Contributions: Conceptualization, T.R.; Data curation, T.R., H.-C.B., M.B. and J.W.; Formal analysis, T.R., H.-C.B., J.G., S.M. and J.M.L.; Funding acquisition, T.R. and P.B.S.; Investigation, T.R.; Methodology, T.R., M.B. and P.B.S.; Project administration, T.R.; Resources, T.R. and P.B.S.; Software, T.R. and J.W.; Supervision, T.R.; Validation, T.R. and P.B.S.; Visualization, T.R.; Writing—original draft, T.R.; Writing—review & editing, T.R., H.-C.B., M.B., G.M.D.M., M.N.P. and P.B.S. All authors have read and agreed to the published version of the manuscript.

Funding: This research received no external funding.

Institutional Review Board Statement: The study was conducted in accordance with the Declaration of Helsinki, and approved by the Institutional Review Board (or Ethics Committee) of Basel Stadt, Switzerland (*BASEC2021 00166*; 21 January 2021).

Informed Consent Statement: Informed consent was obtained from all subjects involved in the study.

Data Availability Statement: The data presented in this study are available on request from the corresponding author.

Conflicts of Interest: The authors declare no conflict of interest.

Abbreviations

ADC	apparent diffusion coefficient
bSSFP	balanced steady-state free precession
BW	bandwidth
CNR	contrast-to-noise ratio
CT	computed tomography
DWI	diffusion-weighted imaging
Epi	echo-planar imaging
FLAIR	fluid attenuated inversion recovery
ICC	intraclass correlation coefficient
MRI	magnetic resonance imaging
NIHSS	National Institutes of Health Stroke Scale
PACS	Picture Archiving and Communication System
QALY	quality-adjusted life years
SNR	signal-to-noise ratio
SP	slice spacing

ST   slice thickness
SWI   susceptibility-weighted imaging
TIA   transient ischemic attack

References

1. Gorelick, P.B. The global burden of stroke: Persistent and disabling. *Lancet N*
2. Katan, M.; Luft, A. Global Burden of Stroke. *Semin. Neurol.* **2018**, *38*, 208–21
3. Krishnamurthi, R.V.; Moran, A.E.; Feigin, V.L.; Barker-Collo, S.; Norrving, B. M.H.; Nguyen, G.; et al. Stroke Prevalence, Mortality and Disability-Adjusted Data from the Global Burden of Disease 2013 Study. *Neuroepidemiology* **2015**
4. Kalkonde, Y.V.; Alladi, S.; Kaul, S.; Hachinski, V. Stroke Prevention Strategies [CrossRef] [PubMed]
5. Bhat, S.S.; Fernandes, T.T.; Poojar, P.; da Silva Ferreira, M.; Rao, P.C.; Hanum: S. Low-Field MRI of Stroke: Challenges and Opportunities. *J. Magn. Reson.*
6. Campbell, B.C.; Parsons, M.W. Imaging selection for acute stroke interv [PubMed]
7. Martinez, G.; Katz, J.M.; Pandya, A.; Wang, J.J.; Boltyenkov, A.; Malhotra, A of Initial Imaging Selection in Acute Ischemic Stroke Care. *J. Am. Coll. Rad*
8. Mehdizade, A.; Somon, T.; Wetzel, S.; Kelekis, A.; Martin, J.B.; Scheidegg Delavelle, J. Diffusion weighted MR imaging on a low-field open magnet. cerebral ischemia. *J. Neuroradiol.* **2003**, *30*, 25–30. [PubMed]
9. Terada, H.; Gomi, T.; Harada, H.; Chiba, T.; Nakamura, T.; Iwabuchi, S.; N et al. Development of diffusion-weighted image using a 0.3T open MRI. *J*
10. Heiss, R.; Nagel, A.M.; Laun, F.B.; Uder, M.; Bickelhaupt, S. Low-Field Breakthrough Technology in Clinical Imaging. *Investig. Radiol.* **2021**, *56*, 7
11. Runge, V.M.; Heverhagen, J.T. The Clinical Utility of Magnetic Resonan Addressing the Breadth of Current State-of-the-Art Systems, Which Incl 1–12. [CrossRef] [PubMed]
12. Edlow, B.L.; Hurwitz, S.; Edlow, J.A. Diagnosis of DWI-negative acute 256–262. [CrossRef] [PubMed]
13. Nagaraja, N. Diffusion weighted imaging in acute ischemic stroke: A rev: imaging application. *J. Neurol. Sci.* **2021**, *425*, 117435. [CrossRef] [PubMe
14. Makin, S.D.; Doubal, F.N.; Dennis, M.S.; Wardlaw, J.M. Clinically Confir Magnetic Resonance Imaging: Longitudinal Study of Clinical Outcomes, *46*, 3142–3148. [CrossRef] [PubMed]

Journal of
Clinical

Article

A Focus o
in Diagno

Yingwei Guo [1,2], **Y
Xueqiang Zeng [2,5]

Citation: Guo, Y.; Yang, Y.; C Wang, M.; Luo, Y.; Guo, J.; Liu Zeng, X.; Miu, X.; Zaman, A.; A Focus on the Role of DSC-F Dynamic Radiomics Features Diagnosis and Outcome Pred Ischemic Stroke. *J. Clin. Med.* 5364. https://doi.org/10.339 jcm11185364

Academic Editors: Gabriel Bro and Lukas Meyer

Received: 11 August 2022
Accepted: 8 September 2022
Published: 13 September 2022

Publisher's Note: MDPI stays with regard to jurisdictional cl published maps and institutior iations.

the Role of DSC-PWI Dynamic Radiomics Features is and Outcome Prediction of Ischemic Stroke

ıgjian Yang [1,2], Fengqiu Cao [1,2], Mingming Wang [3], Yu Luo [3,*], Jia Guo [4], Yang Liu [2],
ᵢiaoqiang Miu [1,2], Asim Zaman [2,5], Jiaxi Lu [2,5] and Yan Kang [1,2,6,*]

1 College of Medicine and Biological Information Engineering, Northeastern University,
  Shenyang 110169, China
2 College of Health Science and Environmental Engineering, Shenzhen Technology University,
  Shenzhen 518118, China
3 Department of Radiology, Shanghai Fourth People's Hospital Affiliated to Tongji University School of Medicine,
  Shanghai 200434, China
4 Department of Psychiatry, Columbia University, New York, NY 10027, USA
5 School of Applied Technology, Shenzhen University, Shenzhen 518060, China
6 Engineering Research Centre of Medical Imaging and Intelligent Analysis, Ministry of Education,
  Shenyang 110169, China
* Correspondence: duolan@hotmail.com (Y.L.); kangyan@sztu.edu.cn (Y.K.); Tel.: +86-13-94-047-2926 (Y.K.)

Abstract: Background: The ability to accurately detect ischemic stroke and predict its neurological recovery is of great clinical value. This study intended to evaluate the performance of whole-brain dynamic radiomics features (DRF) for ischemic stroke detection, neurological impairment assessment, and outcome prediction. Methods: The supervised feature selection (Lasso) and unsupervised feature-selection methods (five-feature dimension-reduction algorithms) were used to generate four experimental groups with DRF in different combinations. Ten machine learning models were used to evaluate their performance by ten-fold cross-validation. Results: In experimental group_A, the best AUCs (0.873 for stroke detection, 0.795 for NIHSS assessment, and 0.818 for outcome prediction) were obtained by outstanding DRF selected by Lasso, and the performance of significant DRF was better than the five-feature dimension-reduction algorithms. The selected outstanding dimension-reduction DRF in experimental group_C obtained a better AUC than dimension-reduction DRF in experimental group_A but were inferior to the outstanding DRF in experimental group_A. When combining the outstanding DRF with each dimension-reduction DRF (experimental group_B), the performance can be improved in ischemic stroke detection (best AUC = 0.899) and NIHSS assessment (best AUC = 0.835) but failed in outcome prediction (best AUC = 0.806). The performance can be further improved when combining outstanding DRF with outstanding dimension-reduction DRF (experimental group_D), achieving the highest AUC scores in all three evaluation items (0.925 for stroke detection, 0.853 for NIHSS assessment, and 0.828 for outcome prediction). By the method in this study, comparing the best AUC of $F_{t\text{-test}}$ in experimental group_A and the best_AUC in experimental group_D, the AUC in stroke detection increased by 19.4% (from 0.731 to 0.925), the AUC in NIHSS assessment increased by 20.1% (from 0.652 to 0.853), and the AUC in prognosis prediction increased by 14.9% (from 0.679 to 0.828). This study provided a potential clinical tool for detailed clinical diagnosis and outcome prediction before treatment.

Keywords: DSC-PWI; dynamic radiomics features; Lasso; dimension reduction; stroke detection; NIHSS assessment; outcome prediction

1. Introduction

Ischemic stroke is the primary reason for disability and the second-leading cause of death worldwide [1]. The surviving patients are usually accompanied by varying neurological deficits, resulting in impaired living quality and burdened families and society.

Good clinical outcomes were proven to be correlated with early vessel recanalization [2,3]. Early warning of stroke and accurate assessment of neurological recovery after treatment will facilitate the early prevention of stroke [4], the selection of individualized treatment plans [5], and the recovery of patients [6], thereby reducing the risk of stroke. Therefore, abnormal brain tissue detection and accurate prognostic status prediction are critical factors in stroke treatment.

Cerebral blood flow (CBF) is an essential physiological parameter to evaluate the state of brain tissue in the clinic. Normal blood flow transmission can provide blood oxygen for brain tissue and maintain a stable brain [7,8]. However, once a cerebral vascular occlusion results in ischemia in a region of brain tissue, the blood flow parameters in that ischemic region differ from those in normal tissue [9,10]. Therefore, CBF parameters have been widely used in the diagnosis and treatment of brain tumors [11,12], stroke [13–15], and Alzheimer's disease [16,17]. Clinically, the clinical images to observe blood flow mainly include ultrasonic doppler (UD), digital subtraction angiography (DSA), computed tomography angiography (CTA), computed tomography perfusion (CTP), perfusion-weighted imaging (PWI), and arterial spin labeling (ASL). Among them, due to the influence of the skull, images from UD are rarely used directly for studying brain diseases. Instead, they are primarily used for blood flow velocity detection in the carotid artery, heart, and other organs. The other images are widely used in stroke diagnosis and the prediction of functional recovery. In perfusion imaging, since the contrast agent is difficult to propagate effectively in the damaged tissue, the signal intensity in this area hardly changes. Therefore, the maximum tissue residual function (Tmax) extracted from PWI or CTA images can be used to detect ischemic stroke lesions (Tmax > 6 s) [18], and the region with a reduction of the relative CBF (rCBF) by 30% compared to that of the symmetric side defined as the core infarct area. In addition, the mismatch between ASL or PWI and diffusion-weighted imaging (DWI) can be used to detect the ischemic penumbra [19,20]. It has been proved that hemodynamic parameters (such as Tmax, CBF, etc.) obtained from medical images have been widely used to detect ischemic stroke lesions. With dozens of consecutively scanned three-dimensional (3D) images, there is more information in perfusion images than in other images. However, due to the difficulty in data processing brought by the vast data amount, the existing algorithms usually used the intermediate parameters calculated from the dynamic susceptibility contrast PWI (DSC-PWI) for clinical analysis rather than directly processing them.

In addition, two factors affecting the rehabilitation of stroke patients are the neurological impairment degree of the patients and the used treatment strategy, and the degree of neurological impairment is one of the influencing factors in deciding the treatment strategy. Therefore, the accurate assessment of neurological impairment in stroke patients is significant for treating and rehabilitating stroke. The main parameter for evaluating the degree of neurological impairment in stroke patients is the National Institutes of Health Stroke Scale (NIHSS) [21]. The NIHSS is obtained through questionnaires and usually includes the following domains: level of consciousness, eye movements, the integrity of visual fields, facial movements, arm and leg muscle strength, sensation, coordination, language, speech, and neglect. Each impairment is scored on an ordinal scale ranging from 0 to 2, 0 to 3, or 0 to 4. Item scores are summed to a total score ranging from 0 to 42 (the higher the score, the more severe the stroke) [22]. Previous studies have explored the association between medical images and NIHSS scores. Generally, stroke patients without vascular occlusion or peripheral occlusion in medical imaging have lower NIHSS and better prognoses. However, Ref. [23] reported that one patient with zero NIHSS might have a stroke. Therefore, NIHSS assessment alone or stroke detection alone may be misdiagnosed. If NIHSS-related information can be obtained based on images and combined with the results of stroke detection, it will be beneficial to assess the severity of patients.

Accurate outcome prediction will assist in customizing personalized treatment plans, reducing the situation of poor recovery, and objectively and accurately evaluating the treatment effect [24]. Several studies have shown that stroke outcomes correlate with clinical text information (CTI) and the parameters computed from medical images. For example, lea-Pereira et al. [25] predicted mortality risk scores during admission for ischemic stroke with CTI, such as age, sex, readmission, and neurological symptoms. Xie et al. [26] used patient information, clinical scores, and volumes of lesion tissue to predict the modified Rankin scale (mRS) in three months. Moreover, Brugnara et al. [27] combined location information for lesions, hypertension, diabetes, dizziness, and physical symptoms to perform the prediction. In addition, Ref. [28] used the neutrophil-lymphocyte ratio to predict the mRS Although diverse clinical information has consistently been associated with outcomes after ischemic stroke, the usefulness of neuroimaging in predicting outcomes has not been definitively established [29]. In previous studies, the characteristics of the lesion tissue were usually used to predict the outcome of patients, while the overall characteristics of the brain tissue were missing. However, the recovery of neurological function is a reflection of the brain's overall function, so it is necessary to explore the relationship between the overall characteristics of the brain and prognosis.

The information in medical images is crucial for the prevention, detection, treatment, and outcome prediction of ischemic stroke. Thus, extracting valuable information from medical images is an effective technique in clinical practice. Nowadays, radiomics, an innovative method to quantify high-dimensional features from medical images, is widely used in medical image processing. For example, it is used to investigate tumor heterogeneity [30,31] and in clinical decision support systems to improve treatment decision-making and accelerate advancements of clinical decision support systems in cancer medicine [32–37]. However, in the field of stroke, only a few studies have explored the role of radiomics in diagnosing ischemic stroke [38], penumbra-based prognosis assessment [39,40], and functional prediction [41]. Prior studies compared prognostic predictions between different diseases based on lesion characteristics. Few studies have used whole-brain features for clinical analysis. However, the appearance of local lesions will inevitably affect the whole-brain features. Therefore, the role of whole-brain features in diagnosing and treating stroke is of great value.

This study aims to explore the role of the whole-brain dynamic radiomics features (DRF) of DSC-PWI in the diagnosis of ischemic stroke, the assessment of neurological impairments, and outcome prediction. The main contributions lie in the following three aspects.

(1) This study explored the role of DRF in ischemic stroke. First, the radiomics features of 3D images in the time series of DSC-PWI were used to obtain the DRF of the whole brain. Then feature selection and dimensionality reduction methods were used to generate various combinations. Finally, by comparing the effects of multiple features in stroke diagnosis, NIHSS evaluation, and outcome prediction, the clinical value of DRF in stroke treatment and outcome prediction can be proved, providing a potential tool for clinical application.
(2) In this study, the DRF of the whole brain were extracted instead of lesion features, which reduced the process of lesion segmentation and saved time for clinical treatment.
(3) This study can extract useful features related to the target using feature analysis, reducing the problem of enormous computation costs caused by the direct analysis of a four-dimensional (4D) DSC-PWI image.

2. Materials and Methods

Detailed materials and methods are introduced in the following subsections. The materials are described in Section 2.1, and the methods are shown in Section 2.2.

2.1. Materials

The Institutional Review Boards approved this retrospective study of Shanghai Fourth People's Hospital, affiliated with the Tongji University School of Medicine and exempted from informed consent. The datasets in our study were collected by the neurology department of the Shanghai Fourth People's Hospital, affiliated with the Tongji University School of Medicine, China, from 2013 to 2016. A total of 156 DSC-PWI images from 88 patients were retrospectively reviewed and included. All patients were imaged within 24 h of symptom onset, and 22 patients were screened at least twice during pretreatment and post-treatment. After clinical examination, 78 (50%) DSC-PWI images were diagnosed as ischemic stroke. The primary clinical information includes income NHISS, outcome NHISS, and 90-day mRS. The DSC-PWI image for each patient was scanned on a 1.5T MR scanner (Siemens, Munich, Germany), and Table 1 shows the details.

Table 1. Patient information and scanning parameters of DSC-PWI datasets.

Patient Information		**Scanning Parameters of DSC-PWI Images**	
Numbers of patients	88	TE/TR	32/1590 ms
Datasets (sets)	156	Matrix	256×256
Image with ischemic stroke (%)	78 (50%)	FOV	230×230 mm^2
image of outcome patient (%)	73 (46.8%)	Thickness	5 mm
Female (%)	39 (25%)	Number of measurements	50
Age (Mean $\pm$ Std)	9.919 ± 6.747	Spacing between slices	6.5 mm
NIHSS (Mean $\pm$ Std)	6.275 ± 6.875	Pixel bandwidth	1347 Hz/pixel
90-day mRS	38 (47.5%)	Number of slices	20

2.2. Methods

The proposed method in this study includes four steps: preprocessing DSC-PWI datasets and computing DRF, feature selection and combination strategy, and evaluating the performance of the four combinations of DRF.

2.2.1. Preprocessing DSC-PWI Datasets and Computing Dynamic Radiomics Features

(A)   Registration and smoothing of DSC-PWI datasets

The preprocessing is intended to reduce noise and position deviation impacts. First, it includes registering all of the volumes in the time series, smoothing the voxel in the time series, and splitting the skull and brain tissue. This study corrected the DSC-PWI datasets for potential patient motion by registering all of the volumes in the time series. Then, a triple moving average filter was selected to smooth the data voxel-by-voxel with a 1×3 filtering kernel. The registration method was introduced in Refs. [42,43], and the filtering method was used in Ref. [40]. Next, the average 3D image was computed from the first ten 3D images and the last ten 3D images in the registered and smoothed DSC-PWI image. Finally, this study used neuroimaging software package FSL [44] to segment the skull from the average 3D image, and then the mask of brain tissue was obtained for each DSC-PWI image (seen in Figure 1a).

(B)   Making ground truth for three evaluation items

This study carried out three evaluation items to evaluate the role of DRF in the diagnosis and outcome prediction of ischemic stroke patients. The three evaluation items were ischemic stroke detection, NIHSS assessment, and outcome prediction. Ischemic stroke detection is used to recognize the presence of ischemic stroke lesions. The NIHSS can reflect the degree of neurological impairment, and the 90-day mRS can assess the recovery of neurological function in patients. In this study, we used the fully automated Rapid Processing of Perfusion and Diffusion (RAPID) software (iSchemaView, Menlo Park, CA, USA) [45] to detect ischemic stroke lesions in the brain tissue. According to the detection results, the ground truth (1—ischemic stroke, 0—normal) for the ischemic stroke of each DSC-PWI image can be obtained. For NIHSS assessment, depending on the NIHSS evaluated by two experienced neurologists, we redefined a score of zero as a normal state without neurological impairment and a score greater than zero as a patient with neurological impairment (neurological impairment: NIHSS > 0, normal: NIHSS = 0). For outcome prediction, we set a poor outcome as that with a 90-day mRS greater than 2 and a good outcome as that with a 90-day mRS less than 2 (good outcome: 90-day mRS $\leq$ 2, poor outcome: 90-day mRS > 2).

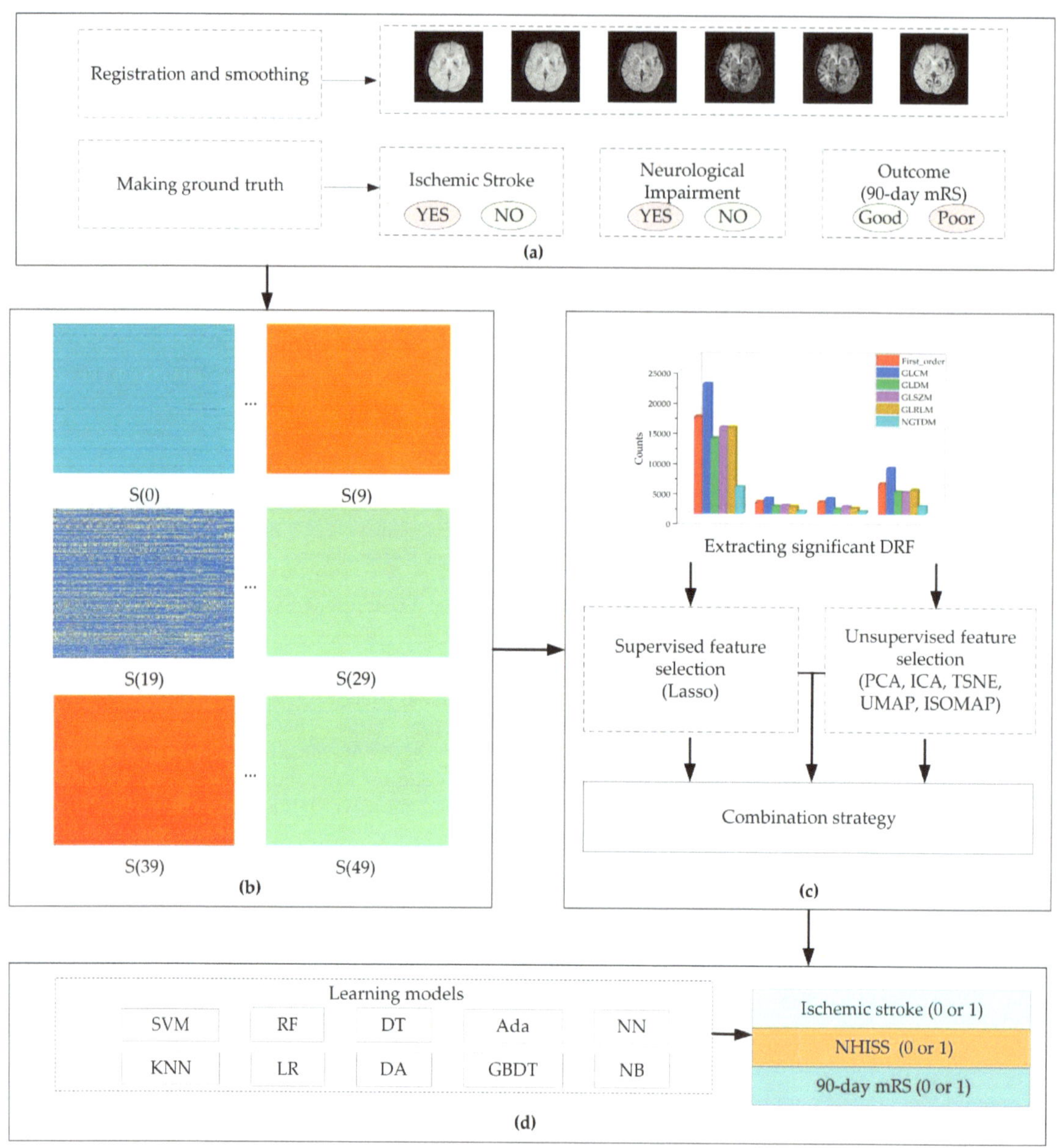

Figure 1. Flowchart of this study. (**a**) preprocessing of dynamic susceptibility contrast perfusion-weighted imaging (DSC-PWI) datasets; (**b**) computing whole-brain dynamic radiomics features (DRF); (**c**) feature selection and combination strategy; (**d**) evaluating performance with ten learning models. The DRF in (**b**) are combined with the radiomics features of 3D images in the time series of DSC-PWI image; the five unsupervised feature selection are principal component analysis (PCA), independent component correlation algorithm (ICA), t-distributed stochastic neighbor embedding (TSNE), uniform manifold approximation and projection (UMAP), and isometric feature mapping (ISOMAP); the ten models are support vector machine (SVM), decision tree (DT), Adaboost classifier (Ada), neural network (NN), random forest (RF), k-nearest neighbors (KNN), logistic regression (LR), linear discriminant analysis (DA), gradient boosting classifier (GBDT), and GaussianNB (NB).

(C)   Computing DRF

The DSC-PWI datasets are 4D images composed of N 3D images with a size of S × H × W, wherein N is the total number of 3D images in each DSC-PWI image and S, H, and W represent the slice, height, and width of the 3D images, respectively. This study used radiomics technology to compute the DRF of the brain tissue in the DSC-PWI image by splitting the DSC-PWI image into N 3D images. First, by decomposing the 4D images into N (50 in this study) single 3D images, the radiomics features of the brain tissue in each 3D image could be computed separately. Then, the DRF can be obtained by combining the radiomics features of all of the 3D images at the time order in the DSC-PWI image (seen in Figure 1b). This study calculated six original feature groups and used six filters to process the original feature groups. The original feature groups were the first-order statistics (First_order), gray-level co-occurrence matrix (GLCM), gray-level run-length matrix (GLRLM), gray-level size-zone matrix (GLSZM), gray-level dependency matrix (GLDM), and neighboring gray-tone difference matrix (NGTDM). The six filters included log sigma with scale {1.0, 2.0, 3.0, 4.0, 5.0}, wavelet, square, square root, logarithm, and exponential. To categorize the features, we summarized the filtering results into the original feature group. For example, the filtered GLCM features can be summarized into the group of GLCM. Thus, the final six feature sets were obtained, including First_order, GLCM, GLRLM, GLSZM, GLDM, and NGTDM. In this study, radiomics feature calculation was automatically performed using the PyRadiomics package implemented in Python [46,47]. Each 3D image in the DSC-PWI data was defined as S(n), wherein n was from zero to 49, and the DSC-PWI image was represented as set {S(0), S(1), . . . , S(49)}. Moreover, the calculated DRF were renamed by connecting their original name and the n-value of the 3D image S(n). For example, "log-sigma-1-0-mm-3D_firstorder_Skewness_17" represents the radiomics feature "log-sigma-1-0-mm-3D_firstorder_Skewness" of S(17), which is the 17th 3D image in DSC-PWI data, and this feature belongs to the First_order group.

2.2.2. Feature Selection and Combination Strategy

This study combined the feature selection method (least absolute shrinkage and selection operator, Lasso) and various feature dimension-reduction algorithms to explore the role of different combinations of DRF in ischemic stroke. The details are introduced in the following and shown in Figure 1c.

(A)   Extracting significant DRF

Before feature selection, feature normalization is necessary to eliminate the influence of dimension and value-range differences between features. This study used each feature's mean and standard deviation to normalize the feature vector. The transformation is given in Equation (1).

$$F_i^* = (F_i - \overline{F_i}) / (F_{imax} - F_{imin}) \tag{1}$$

where F_i^* is the normalized feature of the *ith* feature F_i, and the variables $\overline{F_i}$, F_{imax}, and F_{imin} are the mean, maximum, and minimum of F_i, respectively.

Then, this study used the *t*-test algorithm to extract the significant DRF from all DRF obtained in Section 2.2.1 (C). First, the homogeneity of variance test was performed to detect whether the feature has the homogeneity of variance. When the feature had homogeneity of variance, the *t*-test was performed directly. However, if the feature did not have homogeneity of variance, the parameter *equal_val = False* needed to be added during the *t*-test analysis. This study used the Levene test to realize the homogeneity of the variance test. Finally, the significant DRF with values of $p < 0.05$ in the *t*-test analysis remained to complete subsequent feature-selection processing. Therefore, according to the three evaluation items, three sets of significant DRF were obtained, and the significant DRF were defined as $F_{t\text{-test}}$ in this study.

(B)   Supervised feature selection

Feature selection aims to find the most compelling feature representing the target variable and compress the feature space. Lasso has been recognized as one of the most effective feature selection methods for selecting relevant features to the target variable [40,48,49]. This study used Lasso to select outstanding DRF depending on the three sets of ground truth, and the DRF with a non-zero coefficient were selected as the outstanding DRF. The Lasso was implemented by the *LassoCV* function imported from the *sklearn.linear_model* package in Python 3.6, and the *cv* was set as 10 in the function (seen in Table 2). The mathematical principle of Lasso is shown in Equation (2). By the supervised feature selection, three groups of outstanding DRF can be obtained according to three sets of ground truth. For each evaluation item, the selected outstanding features were defined as Lasso($F_{t\text{-test}}$, item) in this study.

$$Lasso(F_{t-\text{test}}, item) = argmin\left\{ \sum_{i=1}^{M} (y_i - \beta_0 - \sum_{j=1}^{q} \beta_j x_{ij})^2 + l\sum_{j=0}^{q} |\beta_j| \right\} \tag{2}$$

wherein Lasso ($F_{t\text{-test}}$, item) represents the selected outstanding DRF for the evaluation item from the significant DRF $F_{t\text{-test}}$; x_{ij} is the independent DRF in $F_{t\text{-test}}$; y_i is the ground truth of the ith case; λ is the penalty parameter greater than zero; β_j is the regression coefficient; M is the number of cases; q is the number of selected outstanding DRF; and i $\in$ [1, M], and j $\in$ [0, q].

(C)   Unsupervised feature selection

This study adopted unsupervised feature selection methods to extract various additional DRF. As one of the unsupervised feature selection methods, the feature dimension-reduction algorithm (DRA) has been applied. In detail, the DRAs used in this study include principal component analysis (PCA), independent component correlation algorithm (ICA), t-distributed stochastic neighbor embedding (TSNE), uniform manifold approximation and projection (UMAP), and isometric feature mapping (ISOMAP). All of the feature dimension-reduction methods were implemented by importing the corresponding package in python 3.6 (seen in Table 2). By setting the parameter *n_components* to 10, each dimensionality-reduction method reduced the significant DRF to 10 features. The obtained dimension-reduction DRF were defined as the name of the dimension-reduction method. For example, the dimension-reduction features obtained from PCA can be defined as *PCA*, and the dimension-reduction DRF in PCA were defined as {*PCA*0, *PCA*1, . . . , *PCA*9}.

Table 2. The implementation of feature selection and feature dimension reduction.

Method	Implementation in Python 3.6
Lasso	LassoCV (alphas = alphas, cv = 10, max_iter = 100,000, normalize = False). fit (features, targets)
PCA	sklearn.decomposition.PCA (svd_solver = 'auto', n_components = num_fea)
ICA	sklearn.decomposition.FastICA (n_components = num_fea, random_state = 12, max_iter = 1,000,000)
tSNE	sklearn.manifold.TSNE (n_components = num_fea, init = 'pca', random_state = 12, method = 'exact')
UMAP	umap.UMAP (n_neighbors = 5, min_dist = 0.3, n_components = num_fea). fit_transform(features)
ISOMAP	sklearn.manifold.Isomap (n_neighbors = 5, n_components = num_fea, n_jobs = −1). fit_transform(features)

(D)   Feature combination strategy

The feature combination strategy in this study was used to generate four experimental groups for each evaluation item (seen in Figure 2). First, experimental group_A {$F_{t\text{-test}}$, Lasso($F_{t\text{-test}}$, item) and DRA ($F_{t\text{-test}}$, item)} included significant DRF $F_{t\text{-test}}$, outstanding DRF Lasso($F_{t\text{-test}}$, item), and dimension-reduction feature DRA ($F_{t\text{-test}}$, item). Second, experimental group_B was the combination of Lasso($F_{t\text{-test}}$, item) and each DRA ($F_{t\text{-test}}$, item), which can be defined as group_B {Lasso($F_{t\text{-test}}$, item) + DRA ($F_{t\text{-test}}$, item)}. The signal '+' means combination. Third, the experimental group_ C was the collection of the selected outstanding DRF from DRA ($F_{t\text{-test}}$, item) by Lasso, which was defined as group_C {Lasso(DRA($F_{t\text{-test}}$, item), item)}. Finally, experimental group D was the combination of Lasso($F_{t\text{-test}}$, item) and each group_C, and the set was group D {Lasso($F_{t\text{-test}}$, item) + Lasso(DRA($F_{t\text{-test}}$, item), item)}. Based on the above, four experimental groups can be obtained for each evaluation item in this study.

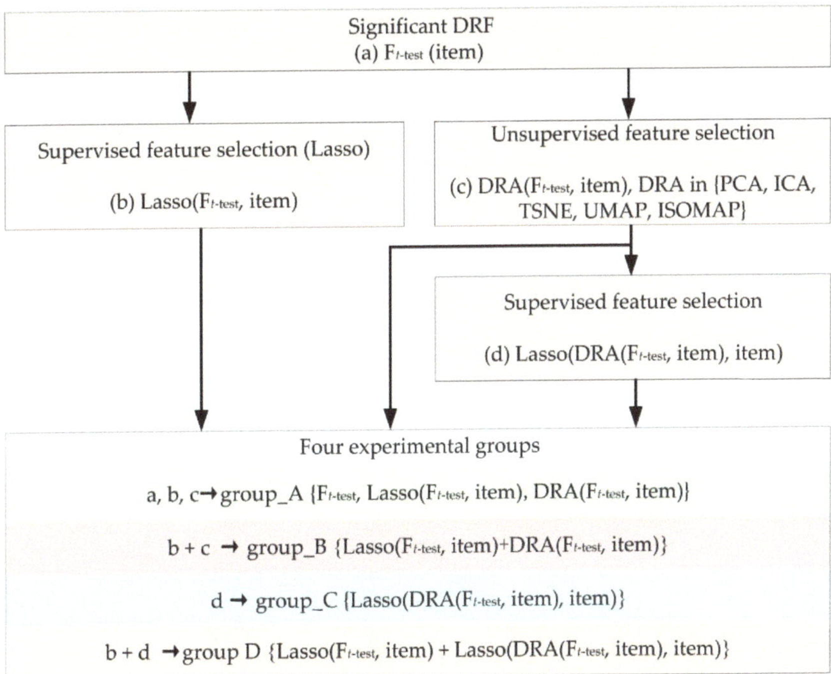

Figure 2. The flowchart of the feature combination strategy in our study.

2.2.3. Performance Evaluation

This study used ten supervised machine learning models to fully evaluate the effectiveness of the feature sets in the four experimental groups described above (seen in Figure 1d). The area under the curve score (AUC) was applied to evaluate the classification ability of each feature set in the four experimental groups. In detail, ten-fold cross-validation was performed to compute the AUC, and the ten machine learning models include support vector machine (SVM), decision tree (DT), Adaboost classifier (Ada), neural network (NN), random forest (RF), k-nearest neighbors (KNN), logistic regression (LR), linear discriminant analysis (DA), gradient boosting classifier (GBDT), and GaussianNB (NB) (seen in Table 3).

Table 3. Descriptions of the 10 models in this study.

Model	Definition in Python 3.6
SVM	sklearn.svm.SVC (kernel = 'rbf', probability = True)
DT	sklearn.tree. DecisionTreeClassifier ()
Ada	sklearn.ensemble.AdaBoostClassifier ()
NN	sklearn.neural_network. MLPClassifier (hidden_layer_sizes = (400, 100), alpha = 0.01, max_iter = 10,000)
RF	sklearn.ensemble.RandomForestClassifier (n_estimators = 200)
KNN	sklearn.neighbors. sklearn.neighbors ()
LR	sklearn.linear_model.logisticRegressionCV(max_iter = 100,000, solver = "liblinear")
DA	sklearn.discriminant_analysis ()
GBDT	sklearn.ensemble.GradientBoostingClassifier ()
NB	sklearn.naive_bayes. GaussianNB ()

3. Results

The results are divided into four sections, including preprocessing results, generated four experimental groups, and the performance of four experimental groups. The details are shown in the following.

3.1. Preprocessing Results

3.1.1. Ground Truth Distribution for Three Evaluation Items

According to the ischemic stroke detection by RAPID, this study detected 78 (50%) ischemic stroke cases in 156 DSC-PWI images. The number of NIHSS equal to zero was 95 (60.9%), and the 90-day mRS less than 2 (good outcome) was 101 (66%) (seen in Table 4). The ground truth equal to 1 means patients with ischemic stroke lesions, neurological impairment, or poor outcome, and the ground truth equal to zero means no ischemic stroke lesions, normal neurological function, or good outcome.

Table 4. Ground truth distribution for three evaluation items.

Ground Truth	Ischemic Stroke	NIHSS	90-Day mRS
1	78	61	55
0	78	95	101

3.1.2. Computed DRF

For each 4D DSC-PWI image, 83700 DRF (50 3D images $\times$ 1674 features) could be calculated (seen in Figure 3a and Table 5). These DRF were divided into six groups: (1) First-order (324 features $\times$ 50 = 16,200 features), (2) GLCM (432 features $\times$ 50 = 21,600 features), (3) GLRLM (288 features $\times$ 50 = 14,400 features), (4) GLSZM (288 features $\times$ 50 = 14,400 features), (5) NGTDM (90 features $\times$ 50 = 4500 features), and (6) GLDM (252 features $\times$ 50 = 12,600 features).

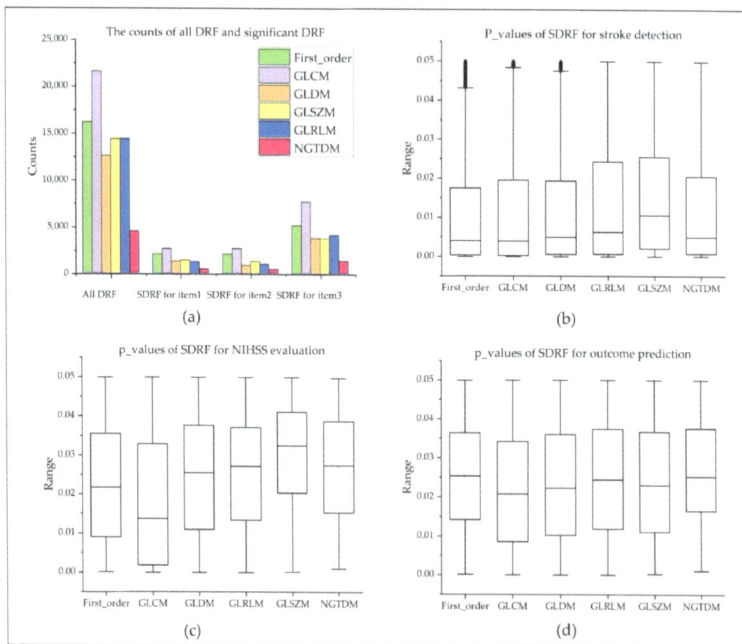

Figure 3. Statistics of all DRF and box plots of DRF for three evaluation items. (**a**) shows the distribution of DRF; (**b**–**d**) are the box plots of the p-values of significant DRF in each feature group for stroke detection, NIHSS evaluation, and outcome prediction, wherein item 1, item 2, and item 3 are stroke detection, NIHSS evaluation, and outcome prediction, respectively.

Table 5. Significant DRF statistics for the three evaluated items.

Item	Feature Group	Significant DRF	Mean	Std	Min	Medium	Max
	First_order	5118	0.0109	0.0139	<0.0001	0.0041	0.0500
	GLCM	7698	0.0114	0.0143	<0.0001	0.0040	0.0500
Stroke detection	GLDM	3800	0.0118	0.0142	<0.0001	0.0050	0.0500
	GLRLM	4117	0.0133	0.0149	<0.0001	0.0063	0.0500
	GLSZM	3737	0.0153	0.0148	<0.0001	0.0107	0.0500
	NGTDM	1352	0.0121	0.0145	<0.0001	0.0050	0.0499
	First_order	2061	0.0227	0.0146	0.0001	0.0217	0.0500
	GLCM	2655	0.0183	0.0166	<0.0001	0.0138	0.0500
NIHSS evaluation	GLDM	866	0.0243	0.0154	<0.0001	0.0255	0.0500
	GLRLM	1016	0.0256	0.0144	<0.0001	0.0272	0.0499
	GLSZM	1289	0.0300	0.0135	0.0001	0.0325	0.0500
	NGTDM	437	0.0269	0.0136	0.0009	0.0274	0.0498
	First_order	2089	0.0255	0.0137	0.0001	0.0254	0.0499
	GLCM	2650	0.0220	0.0147	<0.0001	0.0208	0.0500
Outcome prediction	GLDM	1304	0.0232	0.0148	<0.0001	0.0224	0.0500
	GLRLM	1254	0.0244	0.0147	<0.0001	0.0244	0.0500
	GLSZM	1439	0.0239	0.0143	0.0002	0.0231	0.0500
	NGTDM	467	0.0264	0.0127	0.0011	0.0253	0.0500

3.2. Selected Outstanding DRF and Dimension-Reduction DRF

3.2.1. Significant DRF for Three Evaluation Items

By the *t*-test analysis, this study extracted 25,822 significant DRF for ischemic stroke detection, and there were 5118 DRF in First_order, 7698 in GLCM, 3800 in GLDM, 4117 in GLRLM, 3737 in GLSZM, and 1352 in NGTDM. Their *p*-values ranged from 0.0123 ± 0.0144. Furthermore, 8324 significant DRF with p-values 0.0232 ± 0.0156 were extracted for NIHSS assessment. Among them, 2061 DRF were in First_order, 2655 in GLCM, 866 in GLDM, 1016 in GLRLM, 1289 in GLSZM, and 437 in NGTDM. Furthermore, 9203 significant DRF with *p*-values 0.0238 ± 0.0144 were extracted for outcome prediction, and there were 2089 DRF in First_order, 2650 in GLCM, 1304 in GLDM, 1254 in GLRLM, 1439 GLSZM, and 467 in NGTDM. The detailed statistics of each significant DRF were introduced in Table 5 and Figures 3b–d and 4a).

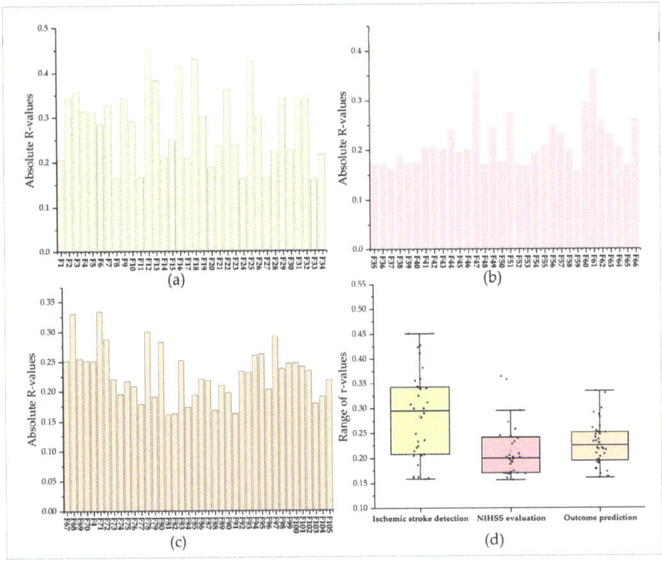

Figure 4. Correlation between outstanding DRF and the ground truths of three evaluation items. (**a–c**) are the Pearson correlation coefficients between outstanding DRF with ground truth for ischemic stroke detection, NIHSS evaluation, and outcome prediction; (**d**) is a box plot of the Pearson correlation coefficients for the three evaluation items.

3.2.2. Selected Outstanding DRF for Three Evaluation Items

This study used the Lasso algorithm to select outstanding DRF from significant DRF for each evaluation item. As a result, 34 outstanding DRF were selected for ischemic stroke detection and defined as Lasso ($F_{t\text{-test}}$, ischemic stroke detection). Among them, there were only 1 DRF in First_order, 9 in GLCM, 13 in GLDM, 5 in GLRLM, 4 in GLSZM, and 2 in NGTDM. Besides, 32 and 40 outstanding DRF were selected for NIHSS evaluation and outcome prediction, respectively. In the Lasso ($F_{t\text{-test}}$, NIHSS evaluation), there were 7 outstanding DRF in First_order, 5 in GLCM, 5 in GLDM, 2 GLRLM, 4 in GLSZM, and 8 in NGTDM. In the Lasso ($F_{t\text{-test}}$, Outcome prediction), there were 6 outstanding DRF in First_order, 21 in GLCM, 9 in GLDM, 2 in GLSZM, 2 in NGTDM, and none in GLRLM. All of the outstanding DRF for the three evaluation items were completely different, with only one DRF (F4) selected for both ischemic stroke detection and outcome prediction. Thus, a total of 105 outstanding DRF were selected. The outstanding DRF were renamed as F_k, and k represents the order of outstanding DRF.

Besides, the absolute values of Pearson correlation coefficients (R-values) between outstanding DRF and ground truths for ischemic stroke detection ranged from 0.158 to 0.451. For NIHSS evaluation, the absolute R-values between outstanding DRF and ground truths for NIHSS evaluation ranged from 0.156 to 0.365, and the absolute R-values between outstanding DRF and ground truths for outcome prediction ranged from 0.161 to 0.334. The results showed that the selected outstanding DRF had weak or moderate correlations with ischemic stroke, NIHSS, and outcome (seen in Figure 4b–d).

3.2.3. Dimension-Reduction DRF Obtained from Five Dimension-Reduction Algorithms

This study's feature dimension-reduction algorithms are unsupervised feature selection methods, and each feature dimension-reduction algorithm can obtain the same set of dimensionality-reduction DRF for the three evaluation items. The R-value of each dimension-reduction DRF with the corresponding ground truth was calculated and is shown in Figure 5. For example, for ischemic stroke detection, the dimension-reduction DRF obtained from PCA had R-values of 0.110 ± 0.121 and that from ICA, TSNE, IOSMAP, and UMAP had R-values of 0.140 ± 0.079, 0.110 ± 0.121, 0.294 ± 0.139, and 0.098 ± 0.133, respectively. For NIHSS evaluation, the dimension-reduction DRF obtained from PCA, ICA, TSNE, IOSMAP, and UMAP had R-values of 0.097 ± 0.089, 0.107 ± 0.075, 0.088 ± 0.069, 0.077 ± 0.052, and 0.155 ± 0.113. For outcome prediction, the dimension-reduction features obtained from PCA, ICA, TSNE, IOSMAP, and UMAP had R-values of 0.093 ± 0.092, 0.097 ± 0.087, 0.093 ± 0.092, 0.100 ± 0.088, and 0.157 ± 0.113.

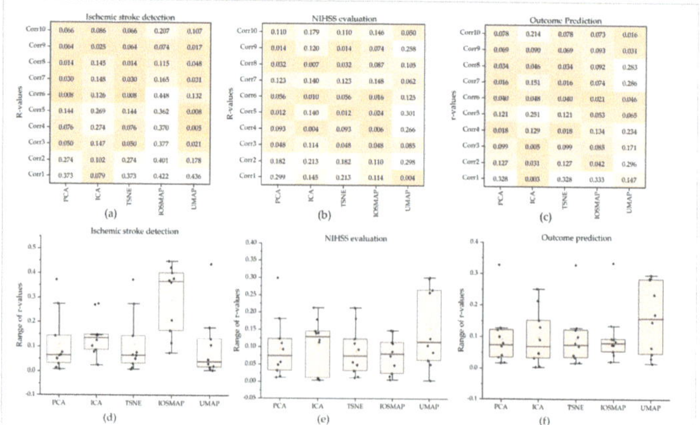

Figure 5. Dimension-reduction DRF for the three evaluation items. (**a–c**) are the Pearson correlation coefficients between dimension-reduction DRF and the ground truth for ischemic stroke detection, NIHSS evaluation, and outcome prediction. (**d–f**) are box plots of the Pearson correlation coefficients for the three evaluation items.

3.2.4. Selected Outstanding Dimension-Reduction DRF

This study used the Lasso algorithm to extract the outstanding dimension-reduction DRF from the dimension-reduction DRF in PCA, ICA, TSNE, ISMOP, and UMAP. As a result (seen in Figure 6), none of the dimension-reduction DRF in ICA were selected for the three evaluation items. Besides, for PCA, two outstanding dimension-reduction DRF (PCA0 and PCA1) for NIHSS evaluation and outcome prediction and four outstanding dimension-reduction DRF (PCA0, PCA1, PCA3, PCA4) for ischemic stroke detection were selected. Similar results were achieved for TSNE. TSNE0 and TSNE1 were selected for NIHSS evaluation and outcome prediction, and additional TSNE3 and TSNE4 were selected for ischemic stroke detection. For UMAP and ISOMAP, no outstanding dimension-reduction DRF was selected for outcome prediction. Furthermore, {UMAP0, UMAP3, UMAP4} and {ISOMAP0, ISOMAP1, ISOMAP5} were selected for ischemic stroke detection. UMAP1 and {ISOMAP0, ISOMAP1, ISOMAP7} were selected for NIHSS evaluation.

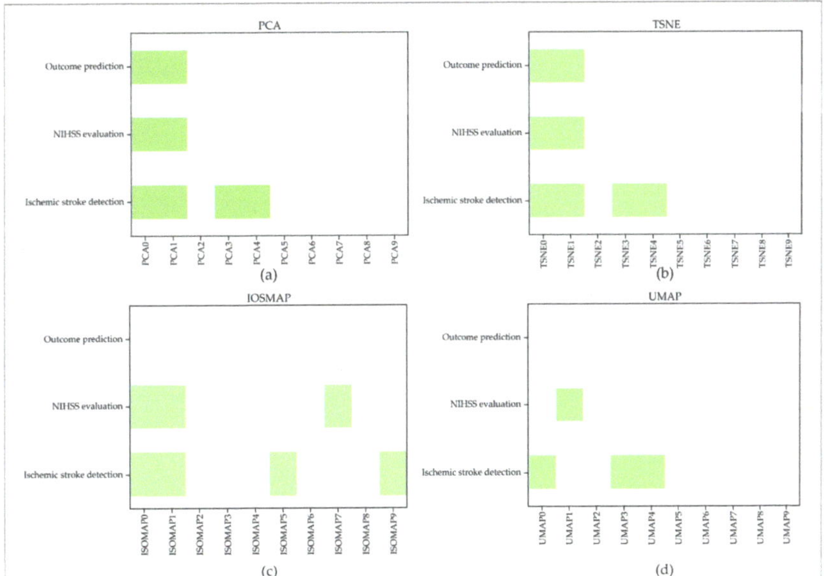

Figure 6. Outstanding dimension-reduction DRF for the three evaluation items. (**a–d**) are the selected outstanding dimension-reduction DRF by PCA, TSNE, ISOMAP, and UMAP for the three evaluation items; the dark green represents the selected outstanding dimension-reduction DRF.

3.3. Performance of Four Experimental Groups

This study evaluated the performance of the four constructed experimental groups in classifying and predicting neurological impairment by ischemic stroke detection, NIHSS evaluation, and outcome prediction. Among the four experiments, the combination DRF in experimental group_D achieved the best score in the three evaluation items. In particular, the combinations of Lasso + PCA_Lasso and Lasso + TSNE_Lasso achieved the same highest score, respectively, which proved the potentially significant value of DRF in the clinical treatment and prognosis analysis of stroke. The detailed results were introduced as follows.

For experimental group_A (seen in Table 6), the outstanding DRF from Lasso performed best, and the five dimension-reduction DRF achieved a similar score, with the significant DRF $F_{t\text{-test}}$ selected by *t*-test analysis in the three items. In detail, the significant DRF achieved the best AUC of 0.731 for ischemic stroke detection, 0.652 for NIHSS assessment, and 0.679 for outcome prediction. In terms of ischemic stroke detection, the AUCs of dimension-reduction DRF obtained by PCA, TSNE, UMAP, ICA, and ISOMAP ranged from

0.672 ± 0.033, 0.664 ± 0.036, 0.681 ± 0.040, 0.684 ± 0.034, and 0.687 ± 0.027, respectively. In contrast, all of the AUCs of outstanding DRF selected by Lasso were better than 0.731, ranging from 0.819 ± 0.064, and the best AUC was 0.837. In terms of NIHSS assessment, the AUCs of dimension-reduction DRF obtained by PCA, TSNE, UMAP, ICA, and ISOMAP ranged from 0.540 ± 0.038, 0.529 ± 0.025, 0.502 ± 0.050, 0.531 ± 0.045, and 0.551 ± 0.050, and the best one of them had an AUC of 0.649. The outstanding DRF from Lasso achieved the best AUC of 0.795, and the performance of significant DRF (best AUC = 0.652) was better than all of the dimension-reduction DRF (best AUC = 0.649). In terms of outcome prediction, the AUCs of dimension-reduction DRF obtained by PCA, TSNE, UMAP, ICA, and ISOMAP ranged from 0.544 ± 0.025, 0.544 ± 0.025, 0.560 ± 0.049, 0.562 ± 0.029, and 0.602 ± 0.036. Similarly, the best AUC of 0.818 was achieved by outstanding DRF from Lasso, and the performance of significant DRF (best AUC = 0.679) was better than all of the dimension-reduction DRF (best AUC = 0.646).

Table 6. The performance of DRF in experimental group_A.

Item	Classifier	Lasso	PCA	TSNE	UMAP	ICA	IOSMAP	*t*-Test
Stroke detection	SVM	0.861	0.691	0.691	0.710	0.686	0.731	0.716
	nn	0.861	0.680	0.634	0.680	0.666	0.646	0.713
	RF	0.783	0.711	0.666	0.660	0.688	0.704	0.698
	DT	0.662	0.615	0.596	0.613	0.632	0.647	0.601
	KNN	0.873	0.669	0.669	0.692	0.666	0.705	0.719
	Ada	0.797	0.642	0.642	0.639	0.739	0.669	0.679
	LR	0.854	0.704	0.704	0.723	0.729	0.699	0.723
	NB	0.840	0.639	0.639	0.722	0.646	0.685	0.653
	GBDT	0.791	0.659	0.685	0.648	0.677	0.683	0.687
	DA	0.867	0.710	0.710	0.722	0.710	0.692	0.731
NIHSS evaluation	SVM	0.727	0.482	0.482	0.500	0.528	0.492	0.500
	nn	0.743	0.619	0.562	0.464	0.610	0.533	0.652
	RF	0.692	0.548	0.521	0.486	0.504	0.533	0.587
	DT	0.663	0.587	0.568	0.521	0.494	0.557	0.603
	KNN	0.677	0.523	0.523	0.428	0.522	0.480	0.536
	Ada	0.731	0.526	0.526	0.491	0.476	0.597	0.574
	LR	0.795	0.532	0.532	0.496	0.512	0.568	0.629
	NB	0.755	0.527	0.527	0.618	0.607	0.649	0.667
	GBDT	0.687	0.513	0.509	0.531	0.518	0.527	0.606
	DA	0.783	0.541	0.541	0.489	0.541	0.574	0.607
Outcome prediction	SVM	0.818	0.546	0.546	0.500	0.549	0.573	0.576
	nn	0.766	0.551	0.554	0.635	0.605	0.646	0.647
	RF	0.684	0.553	0.526	0.572	0.540	0.588	0.578
	DT	0.595	0.515	0.525	0.596	0.550	0.592	0.521
	KNN	0.694	0.592	0.592	0.584	0.553	0.667	0.638
	Ada	0.681	0.504	0.504	0.525	0.562	0.622	0.553
	LR	0.797	0.546	0.546	0.503	0.510	0.559	0.679
	NB	0.818	0.526	0.526	0.616	0.606	0.597	0.664
	GBDT	0.681	0.544	0.561	0.562	0.580	0.618	0.593
	DA	0.676	0.563	0.563	0.508	0.563	0.556	0.571

For experimental group_B (Table 7), the performance can be improved when combining the outstanding DRF with each dimension-reduction DRF in ischemic stroke detection and NIHSS assessment. However, these combinations failed in outcome prediction. In detail, in terms of ischemic stroke detection, when combining Lasso with PCA, the best AUC increased from 0.711 (PCA) to 0.899 (Lasso + PCA); when combining Lasso with TNSE, the best AUC increased from 0.710 (TNSE) to 0.905 (Lasso + TNSE); when combining Lasso with UMAP, the best AUC increased from 0.723 (UMAP) to 0.873 (Lasso + UMAP); when combining Lasso with ICA, the best AUC increased from 0.739 (ICA) to 0.893 (Lasso + ICA); when combining Lasso with ISOMAP, the AUC increased from 0.731 (ISOMAP) to 0.874

(Lasso + ISOMAP). Besides, the above combination performed better than the best score of 0.873 (Lasso) in experimental group_A. In terms of NIHSS assessment, when combining Lasso with PCA and TSNE, the best AUC increased from 0.583 (PCA, TSNE) to 0.835 (Lasso + PCA, Lasso + TSNE); when combining Lasso with UMAP, the best AUC increased from 0.618 (UMAP) to 0.786 (Lasso + UMAP); when combining Lasso with ICA, the best AUC increased from 0.610 (ICA) to 0.835 (Lasso + ICA); when combining Lasso with ISOMAP, the AUC increased from 0.668 (ISOMAP) to 0.812 (Lasso + UMAP). Besides, the Lasso + UMAP obtained a lower score than the best score of 0.795 (Lasso) in experimental group_A, and the other four combinations achieved better scores. Regarding outcome prediction, when combining Lasso with PCA, TSNE, and ICA, the best AUCs were 0.806; when combining Lasso with UMAP and ISOMAP, the best AUCs were 0.795 and 0.814, respectively. In this item, all of the combinations of Lasso and each dimension-reduction DRF were inferior to the outstanding DRF (Lasso) but better than the single dimension-reduction DRF and significant DRF (*t*-test) in experimental group_A.

Table 7. The performance of DRF in experimental group_B.

	Classifier	Lasso + PCA	Lasso + TSNE	Lasso + UMAP	Lasso + ICA	Lasso + IOSMAP
	SVM	0.691	0.691	0.840	0.861	0.731
	nn	0.730	0.685	0.848	0.842	0.743
	RF	0.808	0.802	0.796	0.808	0.809
	DT	0.684	0.700	0.648	0.682	0.641
Stroke	KNN	0.669	0.669	0.853	0.873	0.705
detection	Ada	0.766	0.766	0.772	0.784	0.753
	LR	0.899	0.905	0.873	0.874	0.874
	NB	0.847	0.847	0.820	0.839	0.846
	GBDT	0.790	0.790	0.778	0.758	0.777
	DA	0.893	0.893	0.843	0.893	0.837
	SVM	0.482	0.482	0.658	0.727	0.492
	nn	0.586	0.636	0.782	0.780	0.573
	RF	0.674	0.684	0.665	0.662	0.667
	DT	0.633	0.650	0.607	0.657	0.660
NIHSS	KNN	0.536	0.536	0.622	0.677	0.480
evaluation	Ada	0.701	0.701	0.676	0.684	0.747
	LR	0.824	0.824	0.776	0.805	0.800
	NB	0.760	0.760	0.732	0.745	0.744
	GBDT	0.667	0.667	0.684	0.686	0.671
	DA	0.835	0.835	0.786	0.835	0.812
	SVM	0.555	0.555	0.732	0.818	0.573
	nn	0.616	0.616	0.793	0.770	0.697
	RF	0.662	0.684	0.694	0.657	0.669
	DT	0.634	0.618	0.623	0.673	0.622
Outcome	KNN	0.594	0.594	0.639	0.694	0.667
prediction	Ada	0.715	0.715	0.696	0.689	0.697
	LR	0.770	0.770	0.795	0.802	0.765
	NB	0.804	0.804	0.795	0.796	0.814
	GBDT	0.642	0.654	0.697	0.663	0.660
	DA	0.806	0.806	0.735	0.806	0.716

For experimental group_C (Table 8), the selected outstanding dimension-reduction DRF (PCA_Lasso, TSNE_Lasso, UMAP_Lasso, and IOSMAP_Lasso) achieved a better AUC than the dimension-reduction DRF in experimental group_A. However, it was still inferior to the Lasso in experimental group_A and the combinations in experimental group_B. In detail, in terms of ischemic stroke detection, the best AUCs of the outstanding dimension-reduction DRF in PCA, TSNE, UMAP, and ISOMAP were 0.742, 0.742, 0.748, and 0.742, which were better than the AUCs of the original dimension-reduction DRF in experimental group_A, 0.711, 0.710, 0.723, and 0.731, but lower than the best performance, 0.873, in experimental group_A and 0.905 in experimental group_B. In terms of NIHSS assessment, the best AUCs of the outstanding dimension-reduction DRF in PCA, TSNE, and UMAP (0.645, 0.660, and 0.536) were better than the AUCs of the original dimension-reduction DRF (0.619, 0.568, and 0.610) in experimental group_A. In contrast, the best AUC (0.581) of the outstanding dimension-reduction DRF in ISOMAP was inferior to the AUC (0.649) of ISOMAP in experimental group_A. The best AUC in experimental group_C was lower than in experimental group_A and experimental group_B. Regarding outcome prediction, the best AUC's 0.596 of outstanding dimension-reduction DRF in PCA and TSNE was better than that of the original dimension-reduction DRF in experimental group_A but inferior to the best performance 0.818 in experimental group_A and 0.814 in experimental group_B.

Table 8. The performance of DRF in experimental group_C.

	Classifier	PCA_Lasso	TSNE_Lasso	UMAP_Lasso	IOSMAP_Lasso
	SVM	0.717	0.717	0.722	0.724
	nn	0.634	0.633	0.704	0.707
	RF	0.670	0.677	0.705	0.742
	DT	0.618	0.598	0.652	0.672
Stroke	KNN	0.704	0.704	0.690	0.679
detection	Ada	0.647	0.647	0.588	0.704
	LR	0.742	0.742	0.748	0.712
	NB	0.658	0.658	0.737	0.717
	GBDT	0.665	0.633	0.665	0.716
	DA	0.730	0.730	0.730	0.718
	SVM	0.496	0.496	0.500	0.501
	nn	0.550	0.613	0.524	0.557
	RF	0.645	0.655	0.536	0.502
	DT	0.602	0.627	0.536	0.534
NIHSS	KNN	0.621	0.621	0.472	0.581
evaluation	Ada	0.618	0.618	0.433	0.532
	LR	0.550	0.550	0.482	0.574
	NB	0.570	0.570	0.492	0.580
	GBDT	0.641	0.660	0.507	0.514
	DA	0.541	0.541	0.508	0.554
	SVM	0.532	0.532		
	nn	0.573	0.575		
	RF	0.571	0.556		
	DT	0.524	0.506		
Outcome	KNN	0.555	0.555		
prediction	Ada	0.505	0.505		
	LR	0.596	0.596		
	NB	0.577	0.577		
	GBDT	0.531	0.531		
	DA	0.591	0.591		

For experimental group_D (Table 9), the combination of outstanding DRF and outstanding dimension-reduction DRF provided a chance to improve their performance further. In terms of ischemic stroke detection, the combination of outstanding DRF and outstanding dimension-reduction DRF in PCA, TSNE, UMAP, and ISOMAP obtained the best AUCs, 0.925, 0.925, 0.873, and 0.887, which were the highest scores among all of the experimental groups. In terms of NIHSS assessment, the combination of outstanding DRF and outstanding dimension-reduction DRF in PCA and TSNE achieved the best AUC of 0.853 in all four experimental groups. The combination of outstanding DRF and outstanding dimension-reduction DRF in UMAP and ISOMAP failed to surpass the best performance in experimental group_B, with the best AUC of 0.787 and 0.829, respectively. Finally, regarding outcome prediction, the combination of Lasso and outstanding dimension-reduction DRF in PCA and TSNE achieved the best AUC of 0.828, which also was the best score in the four experimental groups. Besides, among all of the ten classification models, the performance of DA and LR was better than the other models, and they achieved almost all of the best AUCs.

Table 9. The performance of DRF in experimental group_D.

	Classifier	Lasso + PCA_Lasso	Lasso + Tsne_Lasso	Lasso + UMAP_Lasso	Lasso + Iosmap_Lasso
Stroke detection	SVM	0.717	0.717	0.872	0.724
	nn	0.795	0.803	0.861	0.732
	RF	0.809	0.802	0.796	0.814
	DT	0.623	0.663	0.614	0.662
	KNN	0.704	0.704	0.834	0.679
	Ada	0.776	0.776	0.773	0.766
	LR	0.905	0.905	0.873	0.887
	NB	0.847	0.847	0.833	0.847
	GBDT	0.790	0.778	0.772	0.770
	DA	0.925	0.925	0.862	0.874
NIHSS evaluation	SVM	0.496	0.496	0.761	0.501
	nn	0.788	0.735	0.777	0.741
	RF	0.703	0.692	0.656	0.684
	DT	0.675	0.612	0.636	0.630
	KNN	0.616	0.616	0.662	0.577
	Ada	0.726	0.726	0.756	0.686
	LR	0.846	0.846	0.770	0.829
	NB	0.759	0.759	0.736	0.740
	GBDT	0.673	0.673	0.672	0.657
	DA	0.853	0.853	0.787	0.822
Outcome prediction	SVM	0.522	0.522		
	nn	0.808	0.808		
	RF	0.669	0.699		
	DT	0.592	0.615		
	KNN	0.541	0.541		
	Ada	0.705	0.705		
	LR	0.803	0.803		
	NB	0.828	0.828		
	GBDT	0.629	0.647		
	DA	0.756	0.756		

4. Discussion

The ability to accurately detect ischemic stroke and predict its neurological recovery is of great clinical value [50], which can help to prepare appropriate treatment plans and improve the prognosis and recovery of patients. Previous studies have proved the value of medical images (CT, DWI, SWI, PWI, CTP, and ASL) in ischemic stroke detection [51–55], outcome prediction [56–58], and the association between NIHSS and prognosis [59–61]. In addition, some scholars have introduced radiomics technology into the above studies [36,40,41]. However, these studies generally focused on the relationship between local features in lesions while ignoring global information in the whole brain. Furthermore, few studies directly used the dynamic perfusion information in DSC-PWI images to perform the above works. This study intended to explore the role of whole-brain DRF in evaluating the state of brain tissue and neurological function, especially in ischemic stroke diagnosis, the assessment of neurological impairment, and outcome prediction. As a result, the highest AUC was 0.925 for ischemic stroke diagnosis, 0.846 for the assessment of neurological impairment (NIHSS), and 0.835 for outcome prediction (90-day mRS). Therefore, the reproducibility and applicability of this study indicate the feasibility of whole-brain DRF-based radiomics in detecting and assessing ischemic stroke and predicting the neurological recovery of ischemic stroke patients.

This paper is an exploratory work based on the DRF of the whole brain. In DSC-PWI images, the intensity of voxels changes under the action of the contrast agent, resulting in changes in 3D image features. Therefore, the 3D image features can reflect the process of the propagation of the contrast agent in the brain tissue and then indicate the state of blood flow propagation. Furthermore, ischemic stroke is a vascular disease that causes tissue and neurological damage due to vascular blockage. Therefore, there is an inevitable relationship between the extracted time-related DRF and ischemic stroke. Some studies reported that the intensity drop of the ischemic tissue was less than normal tissue, and the decline rate was slower than in normal tissues [62]. In addition, although ischemic stroke is caused by local tissue ischemia, when the local tissue is in an ischemic state, the blood flow characteristics of the whole brain will also change accordingly, since blood flow propagation is carried out in the whole brain. Therefore, global cerebral blood flow features may provide a chance to diagnose and treat ischemic stroke. Based on the above, this study used radiomics technology to extract the radiomics features of each 3D image in the time series of DSC-PWI images. Then DRF that reflects the blood flow transmission changes of the whole brain were generated. From the results of this study, the original significant DRF $F_{t\text{-test}}$ can obtain an AUC of 0.731 for ischemic stroke, 0.652 for NIHSS assessment, and 0.679 for outcome prediction. It can be seen that DRF can reflect the changes in blood flow and assess the state of brain tissue and the degree of damage to brain nerve function.

Lasso and dimension-reduction algorithms have been used in various scenes of ischemic stroke analysis [48,49,63,64]. This study used Lasso and five dimension-reduction algorithms to generate four experimental groups, evaluating the role of DRF in the diagnosis, assessment, and outcome prediction of ischemic stroke. Based on the results in experimental group_A, the performance of outstanding DRF from Lasso was better than dimension-reduction DRF from PCA, ICA, TSNE, UMAP, and ISOMAP, and the best score of dimension-reduction DRF was similar to the original significant DRF from T-test. Therefore, it may mean that Lasso was the best among all of the feature selection methods. Then, this study combined the outstanding DRF from Lasso with each dimension-reduction DRF from PCA, ICA, TSNE, UMAP, and ISOMAP and evaluated them in experimental group_B. Based on the results in experimental group_B, the combination features achieved a better AUC than single features in experimental group_A for stroke diagnosis and NIHSS assessment and failed to surpass single features in experimental group_A for outcome prediction. Besides, this study used Lasso to select outstanding DRF from dimension-reduction DRF (PCA, ICA, TSNE, UMAP, and ISOMA), which were shown as experimental group_C. Based on the results in experimental group_C, the performance of outstanding DRF in experimental group_A performed better than outstanding dimension-reduction DRF in PCA, ICA, TSNE, UMAP, and ISOMAP, and the best score of dimension-reduction DRF was similar to the original significant DRF from t-test. This also means that Lasso was the best one among all of the feature selection methods. Finally, the outstanding DRF in experimental group_A and outstanding dimension-reduction DRF in experimental group_C were combined, achieving the best score for the three evaluation items. Thus, it can be concluded that different combinations of Lasso and dimension-reduction algorithms could achieve different results for stroke detection, NIHSS assessment, and outcome prediction.

Although the results showed that the combination of outstanding DRF and outstanding dimension-reduction DRF selected from Lasso could improve performance in stroke detection, NIHSS assessment, and outcome prediction, further optimization of the models can be regarded as one future work. We will validate our improved method with the more extensive and varied datasets before applying it to clinical trials in future work. The results in this study do not mean that the models can be used alone for stroke treatment decision-making. Instead, this study should be considered a support tool in stroke treatment guidance.

5. Conclusions

In conclusion, although different feature selection methods on whole-brain DRF achieved diverse performance, the critical role of whole-brain DRF in ischemic stroke detection, NIHSS assessment, and outcome prediction has been proven. From experimental group_A to experimental group_D, it can be concluded that the combination of outstanding DRF with outstanding dimension-reduction DRF in experimental group_D performed best in all of the three evaluation items. Comparing the best AUC of $F_{t\text{-test}}$ in experimental group_A and the best_AUC experimental group_D, the AUC in stroke detection increased by 19.4% (from 0.731 to 0.925), the AUC in NIHSS assessment increased by 20.1% (from 0.652 to 0.853), and the AUC in prognosis prediction increased by 14.9% (from 0.679 to 0.828). This study provided a potential clinical tool for detailed clinical diagnosis and outcome prediction before treatment.

Author Contributions: Conceptualization, Y.K. and Y.L. (Yu Luo); methodology, Y.G., Y.Y., Y.L. (Yu Luo), J.G. and Y.K.; software, Y.G., X.M., Y.L. (Yang Liu), J.G. and X.Z.; validation, F.C., X.Z. and Y.G.; formal analysis, Y.G., M.W. and X.M.; investigation, M.W., A.Z. and F.C.; data curation, M.W. and Y.L. (Yu Luo); writing—original draft preparation, Y.G., Y.Y. and J.L.; writing—review and editing, A.Z., Y.L. (Yang Liu), J.L., F.C. and Y.G.; funding acquisition, Y.K. All authors have read and agreed to the published version of the manuscript.

Funding: This research was funded by the National Natural Science Foundation of China, Grant Number: 62071311; the Stable Support Plan for Colleges and Universities in Shenzhen of China, Grant Number: SZWD2021010; the Scientific Research Fund of Liaoning Province of China, Grant Number: JL201919; the special program for key fields of colleges and universities in Guangdong Province (biomedicine and health) of China, Grant Number: 2021ZDZX2008.

Institutional Review Board Statement: The studies involving human participants were reviewed and approved by the Ethics Committee of Shanghai Fourth People's Hospital affiliated with the Tongji University School of Medicine, China (Approval Code: 20200066-01; Approval Date, 15 May 2020).

Informed Consent Statement: Patient consent was waived due to the nature of the retrospective study.

Data Availability Statement: The data supporting this study's findings are available from the corresponding author upon reasonable request.

Acknowledgments: We would like to thank the Department of Radiology, Shanghai Fourth People's Hospital, affiliated with the Tongji University School of Medicine, for providing the datasets and clinical opinions. We also appreciate the support of the College of Medicine and Biological Information Engineering, Northeastern University, Department of Psychiatry, Columbia University, School of Applied Technology, Shenzhen University, and Medical Device Innovation Center, Shenzhen Technology University.

Conflicts of Interest: All authors have no conflicts of interest to report.

References

1. Puzio, M.; Moreton, N.; O'Connor, J.J. Neuroprotective Strategies for Acute Ischemic Stroke: Targeting Oxidative Stress and Prolyl Hydroxylase Domain Inhibition in Synaptic Signalling. *Brain Disord.* **2022**, *5*, 100030. [CrossRef]
2. Rha, J.H.; Saver, J.L. The impact of recanalization on ischemic stroke outcome: A meta-analysis. *Stroke* **2007**, *38*, 967–973. [CrossRef] [PubMed]
3. Zaidat, O.O.; Suarez, J.I.; Sunshine, J.L.; Tarr, R.W.; Alexander, M.J.; Smith, T.P.; Enterline, D.S.; Selman, W.R.; Dennis, D.M. Thrombolytic therapy of acute ischemic stroke: Correlation of angiographic recanalization with clinical outcome. *Am. J. Neuroradiol.* **2005**, *26*, 880–884.
4. Kaur, M.; Sakhare, S.R.; Wanjale, K.; Akter, F. Early stroke prediction methods for prevention of strokes. *Behav. Neurol.* **2022**, *2022*, 7725597. [CrossRef]
5. Hinman, J.D.; Rost, N.S.; Leung, T.W.; Montaner, J.; Muir, K.W.; Brown, S.; Arenillas, J.F.; Feldmann, E.; Liebeskind, D.S. Principles of precision medicine in stroke. *J. Neurol. Neurosurg. Psychiatry* **2017**, *88*, 54–61. [CrossRef]
6. Baird, A.E.; Dambrosia, J.; Janket, S.J.; Eichbaum, Q.; Chaves, C.; Silver, B.; Barber, P.A.; Parsons, M.; Darby, D.; Davis, S.; et al. A three-item scale for the early prediction of stroke recovery. *Lancet* **2001**, *357*, 2095–2099. [CrossRef]
7. Ogoh, S.; Tarumi, T. Cerebral blood flow regulation and cognitive function: A role of arterial baroreflex function. *J. Physiol. Sci.* **2019**, *69*, 813–823. [CrossRef]
8. You, W.; Andescavage, N.N.; Kapse, K.; Donofrio, M.T.; Jacobs, M.; Limperopoulos, C. Hemodynamic responses of the placenta and brain to maternal hyperoxia in fetuses with congenital heart disease by using blood oxygen–level dependent MRI. *Radiology* **2020**, *294*, 141. [CrossRef]
9. Conti, E.; Piccardi, B.; Sodero, A.; Tudisco, L.; Lombardo, I.; Fainardi, E.; Nencini, P.; Sarti, C.; Allegra Mascaro, A.L.; Baldereschi, M. Translational Stroke Research Review: Using the Mouse to Model Human Futile Recanalization and Reperfusion Injury in Ischemic Brain Tissue. *Cells* **2021**, *10*, 3308. [CrossRef]
10. Lv, Y.; Zhang, Y.; Wu, J. Diffusion-Weighted Imaging Image Combined with Transcranial Doppler Ultrasound in the Diagnosis of Patients with Cerebral Infarction and Vertigo. *Contrast Media Mol. Imaging* **2022**, *2022*, 5313238. [CrossRef]

11. Guyon, J.; Chapouly, C.; Andrique, L.; Bikfalvi, A.; Daubon, T. The normal and brain tumor vasculature: Morphological and functional characteristics and therapeutic targeting. *Front. Physiol.* **2021**, *12*, 622615. [CrossRef] [PubMed]

12. Arvanitis, C.D.; Ferraro, G.B.; Jain, R.K. The blood–brain barrier and blood–tumour barrier in brain tumours and metastases. *Nat. Rev. Cancer* **2020**, *20*, 26–41. [CrossRef] [PubMed]

13. Gregori-Pla, C.; Blanco, I.; Camps-Renom, P.; Zirak, P.; Serra, I.; Cotta, G.; Maruccia, F.; Prats-Sánchez, L.; Martínez-Domeño, A.; Busch, D.R.; et al. Early microvascular cerebral blood flow response to head-of-bed elevation is related to outcome in acute ischemic stroke. *J. Neurol.* **2019**, *266*, 990–997. [CrossRef] [PubMed]

14. Demeestere, J.; Wouters, A.; Christensen, S.; Lemmens, R.; Lansberg, M.G. Review of perfusion imaging in acute ischemic stroke: From time to tissue. *Stroke* **2020**, *51*, 1017–1024. [CrossRef] [PubMed]

15. Kuriakose, D.; Xiao, Z. Pathophysiology and treatment of stroke: Present status and future perspectives. *Int. J. Mol. Sci.* **2020**, *21*, 7609. [CrossRef]

16. De Jong, D.L.K.; de Heus, R.A.A.; Rijpma, A.; Donders, R.; Olde Rikkert, M.G.; Günther, M.; Lawlor, B.A.; van Osch, M.J.; Claassen, J.A. Effects of nilvadipine on cerebral blood flow in patients with Alzheimer disease: A randomized trial. *Hypertension* **2019**, *74*, 413–420. [CrossRef]

17. De Heus RA, A.; de Jong DL, K.; Sanders, M.L.; van Spijker, G.J.; Oudegeest-Sander, M.H.; Hopman, M.T.; Lawlor, B.A.; Rikkert, M.G.M.O.; Claassen, J.A.H.R. Dynamic regulation of cerebral blood flow in patients with Alzheimer disease. *Hypertension* **2018**, *72*, 139–150. [CrossRef]

18. Olivot, J.M.; Mlynash, M.; Vincent, N. Relationships between cerebral perfusion and reversibility of acute diffusion lesions in DEFUSE: Insights from RADAR. *Stroke* **2009**, *40*, 1692–1697. [CrossRef]

19. Wang, Y.R.; Li, Z.S.; Huang, W.; Yang, H.Q.; Gao, B.; Chen, Y.T. The value of susceptibility-weighted imaging (SWI) in evaluating the ischemic penumbra of patients with acute cerebral ischemic stroke. *Neuropsychiatr. Dis. Treat.* **2021**, *17*, 1745. [CrossRef]

20. Ermine, C.M.; Bivard, A.; Parsons, M.W.; Baron, J. The ischemic penumbra: From concept to reality. *Int. J. Stroke* **2021**, *16*, 497–509. [CrossRef]

21. Fischer, U.; Arnold, M.; Nedeltchev, K.; Brekenfeld, C.; Ballinari, P.; Remonda, L.; Schroth, G.; Mattle, H.P. NIHSS score and arteriographic findings in acute ischemic stroke. *Stroke* **2005**, *36*, 2121–2125. [CrossRef]

22. Kwah, L.K.; Diong, J. National institutes of health stroke scale (NIHSS). *J. Physiother.* **2014**, *60*, 61. [CrossRef] [PubMed]

23. Martin-Schild, S.; Albright, K.C.; Tanksley, J.; Pandav, V.; Jones, E.B.; Grotta, J.C.; Savitz, S.I. Zero on the NIHSS does not equal the absence of stroke. *Ann. Emerg. Med.* **2011**, *57*, 42–45. [CrossRef] [PubMed]

24. Ahuja, A.S. The impact of artificial intelligence in medicine on the future role of the physician. *PeerJ* **2019**, *7*, e7702. [CrossRef] [PubMed]

25. Lea-Pereira, M.C.; Amaya-Pascasio, L.; Martínez-Sánchez, P.; Rodríguez Salvador, M.D.M.; Galván-Espinosa, J.; Téllez-Ramírez, L.; Reche-Lorite, F.; Sánchez, M.-J.; García-Torrecillas, J.M. Predictive Model and Mortality Risk Score during Admission for Ischaemic Stroke with Conservative Treatment. *Int. J. Environ. Res. Public Health* **2022**, *19*, 3182. [CrossRef]

26. Xie, Y.; Jiang, B.; Gong, E.; Li, Y.; Zhu, G.; Michel, P.; Wintermark, M.; Zaharchuk, G. Use of gradient boosting machine learning to predict patient outcome in acute ischemic stroke on the basis of imaging, demographic, and clinical information. *Am. J. Roentgenol.* **2019**, *212*, 44–51. [CrossRef]

27. Brugnara, G.; Neuberger, U.; Mahmutoglu, M.A.; Foltyn, M.; Herweh, C.; Nagel, S.; Schönenberger, S.; Heiland, S.; Ulfert, C.; Ringleb, P.A.; et al. Multimodal predictive modeling of endovascular treatment outcome for acute ischemic stroke using machine-learning. *Stroke* **2020**, *51*, 3541–3551. [CrossRef]

28. Brooks, S.D.; Spears, C.; Cummings, C.; VanGilder, R.L.; Stinehart, K.R.; Gutmann, L.; Domico, J.; Culp, S.; Carpenter, J.; Rai, A.; et al. Admission neutrophil–lymphocyte ratio predicts 90 day outcome after endovascular stroke therapy. *J. Neurointerv. Surg.* **2014**, *6*, 578–583. [CrossRef]

29. Barrett, K.M.; Ding, Y.H.; Wagner, D.P.; Kallmes, D.F.; Johnston, K.C. Change in diffusion-weighted imaging infarct volume predicts neurologic outcome at 90 days: Results of the Acute Stroke Accurate Prediction (ASAP) trial serial imaging substudy. *Stroke* **2009**, *40*, 2422–2427. [CrossRef]

30. Isensee, F.; Kickingereder, P.; Wick, W.; Bendszus, M.; Maier-Hein, K.H. Brain tumor segmentation and radiomics survival prediction: Contribution to the brats. In Proceedings of the International MICCAI Brainlesion Workshop, Quebec City, QC, Canada, 14 September 2017; Springer: Berlin, Germany, 2017; pp. 287–297.

31. Zhou, M.; Scott, J.; Chaudhury, B.; Hall, L.; Goldgof, D.; Yeom, K.W.; Iv, M.; Ou, Y.; Kalpathy-Cramer, J.; Napel, S.; et al. Radiomics in brain tumor: Image assessment, quantitative feature descriptors, and machine-learning approaches. *Am. J. Neuroradiol.* **2018**, *39*, 208–216. [CrossRef]

32. Zhu, H.; Jiang, L.; Zhang, H.; Luo, L.; Chen, Y.; Chen, Y. An automatic machine learning approach for ischemic stroke onset time identification based on DWI and FLAIR imaging. *NeuroImage Clin.* **2021**, *31*, 102744. [CrossRef] [PubMed]

33. Gillies, R.J.; Kinahan, P.E.; Hricak, H. Radiomics: Images are more than pictures, they are data. *Radiology* **2016**, *278*, 563–577. [CrossRef] [PubMed]

34. Gatenby, R.A.; Grove, O.; Gillies, R.J. Quantitative imaging in cancer evolution and ecology. *Radiology* **2013**, *269*, 8–14. [CrossRef] [PubMed]

35. Lambin, P.; Leijenaar, R.T.H.; Deist, T.M.; Peerlings, J.; de Jong, E.E.C.; van Timmeren, J.; Sanduleanu, S.; Larue, R.T.H.M.; Even, A.J.G.; Jochems, A.; et al. Radiomics: The bridge between medical imaging and personalized medicine. *Nat. Rev. Clin. Oncol.* **2017**, *14*, 749–762. [CrossRef]
36. Chen, Q.; Xia, T.; Zhang, M.; Xia, N.; Liu, J.; Yang, Y. Radiomics in Stroke Neuroimaging: Techniques, Applications, and Challenges. *Aging Dis.* **2021**, *12*, 143–154. [CrossRef] [PubMed]
37. Yang, Y.; Li, W.; Kang, Y.; Guo, Y.; Yang, K.; Li, Q.; Liu, Y.; Yang, C.; Chen, R.; Chen, H.; et al. A novel lung radiomics feature for characterizing resting heart rate and COPD stage evolution based on radiomics feature combination strategy. *Math. Biosci. Eng.* **2022**, *19*, 4145–4165. [CrossRef] [PubMed]
38. Ortiz-Ramón, R.; Hernández, M.D.C.V.; González-Castro, V.; Makin, S.; Armitage, P.A.; Aribisala, B.S.; Bastin, M.E.; Deary, I.J.; Wardlaw, J.M.; Moratal, D. Identification of the presence of ischaemic stroke lesions by means of texture analysis on brain magnetic resonance images. *Comput. Med. Imaging Graph.* **2019**, *74*, 12–24. [CrossRef]
39. Tang, T.-Y.; Jiao, Y.; Cui, Y.; Zhao, D.-L.; Zhang, Y.; Wang, Z.; Meng, X.-P.; Yin, X.-D.; Yang, Y.-J.; Teng, G.-J.; et al. Penumbra-based radiomics signature as prognostic biomarkers for thrombolysis of acute ischemic stroke patients: A multicenter cohort study. *J. Neurol.* **2020**, *267*, 1454–1463. [CrossRef]
40. Guo, Y.; Yang, Y.; Cao, F.; Li, W.; Wang, M.; Luo, Y.; Guo, J.; Zaman, A.; Zeng, X.; Miu, X.; et al. Novel Survival Features Generated by Clinical Text Information and Radiomics Features May Improve the Prediction of Ischemic Stroke Outcome. *Diagnostics* **2022**, *12*, 1664. [CrossRef]
41. Wang, H.; Sun, Y.; Ge, Y.; Wu, P.-Y.; Lin, J.; Zhao, J.; Song, B. A Clinical-Radiomics Nomogram for Functional Outcome Predictions in Ischemic Stroke. *Neurol. Ther.* **2021**, *10*, 819–832. [CrossRef]
42. Sohn, J.; Jung, I.Y.; Ku, Y.; Kim, Y. Machine-learning-based rehabilitation prognosis prediction in patients with ischemic stroke using brainstem auditory evoked potential. *Diagnostics* **2021**, *11*, 673. [CrossRef] [PubMed]
43. Studholme, C.; Hill, D.; Hawkes, D. An overlap invariant entropy measure of 3D medical image alignment. *Pattern Recognit.* **1999**, *32*, 71–86. [CrossRef]
44. Smith, S.M.; Jenkinson, M.; Woolrich, M.W.; Beckmann, C.F.; Behrens, T.E.; Johansen-Berg, H.; Bannister, P.R.; De Luca, M.; Drobnjak, I.; Flitney, D.E.; et al. Advances in functional and structural MR image analysis and implementation as FSL. *Neuroimage* **2004**, *23*, S208–S219. [CrossRef] [PubMed]
45. Fan, S.; Bian, Y.; Wang, E.; Kang, Y.; Wang, D.; Yang, Q.; Ji, X. An Automatic Estimation of Arterial Input Function Based on Multi-Stream 3D CNN. *Front. Neuroinform.* **2019**, *13*, 49. [CrossRef]
46. Van Griethuysen, J.J.M.; Fedorov, A.; Parmar, C.; Hosny, A.; Aucoin, N.; Narayan, V.; Beets-Tan, R.G.H.; Fillion-Robin, J.-C.; Pieper, S.; Aerts, H.J.W.L. Computational Radiomics System to Decode the Radiographic Phenotype. *Cancer Res.* **2017**, *77*, e104–e107. [CrossRef]
47. Zhang, Y.; Zhang, B.; Liang, F.; Liang, S.; Zhang, Y.; Yan, P.; Ma, C.; Liu, A.; Guo, F.; Jiang, C. Radiomics features on non-contrast-enhanced CT scan can precisely classify AVM-related hematomas from other spontaneous intraparenchymal hematoma types. *Eur. Radiol.* **2019**, *29*, 2157–2165. [CrossRef]
48. Ghosh, P.; Azam, S.; Jonkman, M.; Karim, A.; Shamrat, F.J.M.; Ignatious, E.; Shultana, S.; Beeravolu, A.B.; De Boer, F. Efficient prediction of cardiovascular disease using machine learning algorithms with relief and LASSO feature selection techniques. *IEEE Access* **2021**, *9*, 19304–19326. [CrossRef]
49. Muthukrishnan, R.; Rohini, R. LASSO: A feature selection technique in predictive modeling for machine learning. In Proceedings of the IEEE International Conference on Advances in Computer Applications (ICACA), Coimbatore, India, 24 October 2016. [CrossRef]
50. Ezzeddine, M.A.; Lev, M.H.; McDonald, C.T.; Rordorf, G.; Oliveira-Filho, J.; Aksoy, F.G.; Farkas, J.; Segal, A.Z.; Segal, L.H.; Schwamm, R.; et al. CT angiography with whole brain perfused blood volume imaging: Added clinical value in the assessment of acute stroke. *Stroke* **2002**, *33*, 959–966. [CrossRef]
51. Chawla, M.; Sharma, S.; Sivaswamy, J.; Kishore, L. A method for automatic detection and classification of stroke from brain CT images. In Proceedings of the Annual International Conference of the IEEE Engineering in Medicine and Biology Society, Minneapolis, MN, USA, 3–6 September 2009. [CrossRef]
52. Edlow, B.L.; Hurwitz, S.; Edlow, J.A. Diagnosis of DWI-negative acute ischemic stroke: A meta-analysis. *Neurology* **2017**, *89*, 256–262. [CrossRef]
53. Lansberg, M.G.; Albers, G.W.; Beaulieu, C.; Marks, M.P. Comparison of diffusion-weighted MRI and CT in acute stroke. *Neurology* **2000**, *54*, 1557–1561. [CrossRef]
54. Abdelgawad, E.A.; Amin, M.F.; Abdellatif, A.; Mourad, M.A.; Abusamra, M.F. Value of susceptibility weighted imaging (SWI) in assessment of intra-arterial thrombus in patients with acute ischemic stroke. *Egypt. J. Radiol. Nucl. Med.* **2021**, *52*, 270. [CrossRef]
55. Zhang, S.; Yao, Y.; Zhang, S.; Zhu, W.; Tang, X.; Yu, Q.; Zhao, L.; Liu, C.; Zhu, W. Comparative study of DSC-PWI and 3D-ASL in ischemic stroke patients. *J. Huazhong Univ. Sci. Technol. Med. Sci.* **2015**, *35*, 923–927. [CrossRef] [PubMed]
56. Barber, P.A.; Darby, D.G.; Desmond, P.M.; Yang, Q.; Gerraty, R.P.; Jolley, D.; Donnan, G.A.; Tress, B.M.; Davis, S.M. Prediction of stroke outcome with echoplanar perfusion-and diffusion-weighted MRI. *Neurology* **1998**, *51*, 418–426. [CrossRef] [PubMed]
57. Dejobert, M.; Cazals, X.; Annan, M.; Debiais, S.; Lauvin, M.; Cottier, J. Susceptibility–diffusion mismatch in hyperacute stroke: Correlation with perfusion–diffusion mismatch and clinical outcome. *J. Stroke Cerebrovasc. Dis.* **2016**, *25*, 1760–1766. [CrossRef] [PubMed]

58. Beaulieu, C.; De Crespigny, A.; Tong, D.C.; Moseley, M.E.; Albers, G.W.; Marks, M.P. Longitudinal magnetic resonance imaging study of perfusion and diffusion in stroke: Evolution of lesion volume and correlation with clinical outcome. *Ann. Neurol. Off. J. Am. Neurol. Assoc. Child Neurol. Society* **1999**, *46*, 568–578. [CrossRef]

59. Bruno, A.; Saha, C.; Williams, L.S. Percent change on the National Institutes of Health Stroke Scale: A useful acute stroke outcome measure. *J. Stroke Cerebrovasc. Dis.* **2009**, *18*, 56–59. [CrossRef]

60. Kasner, S.E. Clinical interpretation and use of stroke scales. *Lancet Neurol.* **2006**, *5*, 603–612. [CrossRef]

61. Fernandez-Lozano, C.; Hervella, P.; Mato-Abad, V.; Rodríguez-Yáñez, M.; Suárez-Garaboa, S.; López-Dequidt, I.; Estany-Gestal, A.; Sobrino, T.; Campos, F.; Castillo, J.; et al. Random forest-based prediction of stroke outcome. *Sci. Rep.* **2021**, *11*, 10071. [CrossRef]

62. Wu, X.; Wang, H.; Chen, F.; Jin, L.; Li, J.; Feng, Y.; DeKeyzer, F.; Yu, J.; Marchal, G.; Ni, Y. Rat model of reperfused partial liver infarction: Characterization with multiparametric magnetic resonance imaging, microangiography, and histomorphology. *Acta Radiol.* **2009**, *50*, 276–287. [CrossRef]

63. Bonkhoff, A.K.; Xu, T.; Nelson, A.; Gray, R.; Jha, A.; Cardoso, J.; Ourselin, S.; Hans, G.R.; Jäger, R.; Nachev, P. Reclassifying stroke lesion anatomy. *Cortex* **2021**, *145*, 1–12. [CrossRef]

64. Nanga, S.; Bawah, A.T.; Acquaye, B.A.; Billa, M.; Baeta, F.D.; Odai, N.A.; Obeng, S.K.; Nsiah, A.D. Review of Dimension Reduction Methods. *J. Data Anal. Inf. Process.* **2021**, *9*, 189–231. [CrossRef]

Journal of
Clinical Medicine

MDPI

Article

An Intermodal Correlation Study among Imaging, Histology, Procedural and Clinical Parameters in Cerebral Thrombi Retrieved from Anterior Circulation Ischemic Stroke Patients

Rebeka Viltužnik [1,2], Franci Bajd [3], Zoran Milošević [4], Igor Kocijančič [4], Miran Jeromel [5], Andrej Fabjan [2], Eduard Kralj [6], Jernej Vidmar [1,2] and Igor Serša [1,*]

[1] Jožef Stefan Institute, 1000 Ljubljana, Slovenia
[2] Institute of Physiology, Faculty of Medicine, University of Ljubljana, 1000 Ljubljana, Slovenia
[3] Faculty of Mathematics and Physics, University of Ljubljana, 1000 Ljubljana, Slovenia
[4] Department of Diagnostic and Interventional Neuroradiology, Clinical Institute of Radiology, University Medical Centre, 1000 Ljubljana, Slovenia
[5] Department of Diagnostic and Interventional Radiology, General Hospital Slovenj Gradec, 2380 Slovenj Gradec, Slovenia
[6] Institute of Forensic Medicine, Faculty of Medicine, University of Ljubljana, 1000 Ljubljana, Slovenia
* Correspondence: igor.sersa@ijs.si; Tel.: +386-1-477-3696

Citation: Viltužnik, R.; Bajd, F.; Milošević, Z.; Kocijančič, I.; Jeromel, M.; Fabjan, A.; Kralj, E.; Vidmar, J.; Serša, I. An Intermodal Correlation Study among Imaging, Histology, Procedural and Clinical Parameters in Cerebral Thrombi Retrieved from Anterior Circulation Ischemic Stroke Patients. *J. Clin. Med.* **2022**, *11*, 5976. https://doi.org/10.3390/jcm11195976

Academic Editors: Gabriel Broocks and Lukas Meyer

Received: 25 August 2022
Accepted: 8 October 2022
Published: 10 October 2022

Publisher's Note: MDPI stays neutral with regard to jurisdictional claims in published maps and institutional affiliations.

Abstract: The precise characterization of cerebral thrombi prior to an interventional procedure can ease the procedure and increase its success. This study investigates how well cerebral thrombi can be characterized by computed tomography (CT), magnetic resonance (MR) and histology, and how parameters obtained by these methods correlate with each other as well as with the interventional procedure and clinical parameters. Cerebral thrombi of 25 patients diagnosed by CT with acute ischemic stroke were acquired by mechanical thrombectomy and, subsequently, scanned by a high spatial-resolution 3D MRI including T_1-weighted imaging, apparent diffusion coefficient (ADC), T_2 mapping and then finally analyzed by histology. Parameter pairs with Pearson correlation coefficient more than 0.5 were further considered by explaining a possible cause for the correlation and its impact on the difficulty of the interventional procedure and the treatment outcome. Significant correlations were found between the variability of ADC and the duration of the mechanical recanalization, the deviation in average Hounsfield units (HU) and the number of passes with the thrombectomy device, length of the thrombus, its RBC content and many others. This study also demonstrates the clinical potentials of high spatial resolution multiparametric MRI in characterization of thrombi and its use for interventional procedure planning.

Keywords: ischemic stroke; cerebral thrombi; thrombectomy; computed tomography; multiparametric MRI; correlation analysis

1. Introduction

Treatment of the cardiovascular diseases has been improved considerably over the last decades and the death rate due to these diseases has also declined; however, the burden to society remains high [1]. A significant contribution to this decline is in much better treatment of the large vessel occlusions, especially in acute ischemic stroke. Modern interventional treatment of ischemic stroke is based on the improved, more accurate, as well as faster, diagnostic imaging, either by computed tomography (CT) or magnetic resonance imaging (MRI) to confirm the diagnosis, which is immediately followed by vessel occlusion recanalization through biochemical thrombus degradation in thrombolysis [2] and/or by its mechanical removal in thrombectomy [3]. The origin of thrombi causing the occlusions may be in the heart, most often due to atrial fibrillation, or in atherosclerotic lesions within the affected vessel or proximal to it. However, it is not easy to distinguish between these two different thrombi etiologies, namely cardioembolic or arteriopathic, since they have practically identical histological characteristics [4]. In this analysis, MRI could be advantageous because it can produce images of various contrasts, which depend on different chemical and mechanical characteristics of thrombi. Thus, MRI is accurate in thrombi localization and can precisely determine thrombus structure and composition, which makes it an interesting in vivo alternative to conventional histology [5].

Platelets and red blood cells (RBCs) are two main cell components of thrombi. However, there is also a fibrin meshwork that acts as a bond with the cells and makes the thrombus structure rigid and resilient to recanalization procedures. The presence of blood cells in the thrombus is responsible for its porous structure and also defines its porosity [6]. A higher density of blood cells reduces the porosity and, therefore, makes the thrombus less permeable [7]. In addition, the fibrin meshwork can be densified, reducing the clot permeability in regions with an increased platelet concentration and also, in the case of establishing a fibrin meshwork cross-linking [8,9]. Regions with the denser fibrin meshwork have also an increased rigidity and are mechanically more stable [10]. Obviously, there is a link between the microscopic properties of thrombi and their susceptibility to recanalization, i.e., porosity of the thrombus structure defines its susceptibility to thrombolysis [11,12], while the rigidity of the fibrin meshwork determines the thrombus resistance to mechanical thrombectomy [13]. As the success of recanalization of the occluded artery in ischemic stroke critically depends on thrombus structure and composition, accurate characterization of these two properties prior to the recanalization procedure could significantly ease this interventional procedure and increase its success [14,15].

The most-often used imaging modality in the diagnosis of stroke is CT [16]. While native CT provides relatively poor contrasted images of patients suspected of stroke, its contrasted version used for CT angiography can clearly depict the site of the occluded artery and the extent of penumbra. Its rival, MRI, cannot compete with CT in terms of speed; however, it provides much better contrasted images. In MRI, the origin of image contrast is quite different than in CT. It is primarily based on NMR relaxation times that depend on the molecular mobility, chemical environment, local magnetic centers, etc., [17]. In addition, the image contrast can be made sensitive to diffusion, perfusion and also to the environmental factors, i.e., temperature, pH, etc., [18]. In CT, contrast is based on X-ray attenuation of tissues in the body. This increases with higher atomic number tissues and, therefore, denser tissues [19]. In the case of MRI, a very efficient method in diagnosis of ischemic stroke is the diffusion-weighted imaging (DWI) or its quantitative version, the apparent diffusion coefficient (ADC) mapping [20]. This detection of ischemic stroke is based on slow water diffusion in the ischemic regions of the brain due to an increased intra- to extra-cellular water ratio [21]. The affected regions of the brain appear brighter in DWI and exhibit lower ADC values than normal tissue. The DWI imaging sequence utilizes a pair of strong bipolar magnetic field gradients that result in additional signal attenuation by a factor $\exp(-bD)$ where D is the apparent diffusion coefficient and b is the instrumental diffusion-weighting factor dependent on the bipolar gradient pulse. Ischemic stroke was also investigated by $T_2{}^*$ and T_2 mapping, where it was found that $T_2{}^*$ is sensitive on the oxygenation state of the affected brain region [22], while T_2 correlates with the time from the stroke onset [23]. Both MRI methods, namely T_2 and ADC mapping, have been used previously for the characterization of blood clots, e.g., to study changes in the structure and composition of artificial blood clots during their dissolution [24] and extracted pulmonary thromboemboli [25].

This study is a continuation of our previous studies on cerebral thrombi retrieved by mechanical thrombectomy from patients diagnosed with acute ischemic stroke. In these studies, the thrombi were evaluated in vitro after their thrombectomy by multiparametric MRI [26] or assessed in vivo by CT [27] and correlation was made with either of these techniques, and the results of histology, clinical and the interventional procedure data was tested. Herein, all the available data on the thrombi (MRI, CT, histology, interventional procedure and clinical data) are evaluated as a whole, providing opportunity for a comprehensive correlation study of these parameters characterizing thrombi.

2. Materials and Methods

2.1. Patient Selection and Study Design

The study was designed in a way to minimally interfere with the normal clinical interventional procedure for the management of patients with the suspicion of acute stroke. Upon admission to the Neurology Clinic of University Medical Center Ljubljana, due to neurological symptoms suggesting brain stroke, these patients underwent standard diagnostic procedure that included urgent clinical examination followed by an urgent multimodal CT scan. Patients diagnosed with acute ischemic stroke received a standard full dose of the rt-PA (0.9 mg/kg, maximum 90 mg, 10% in bolus first and then the remaining 90% intravenously in 1 h) systemic thrombolytic treatment within 4.5 h of onset. If the clinical stroke signs persisted after the thrombolytic treatment, the patients received further therapy by mechanical thrombectomy at an average of 130 min after initiation of thrombolysis. This was undertaken by the standard mechanical recanalization procedure performed by the skilled interventional neuroradiologist using the thrombectomy device (Trevo®stent retriever, 4 × 20 mm, Stryker Neurovascular, Kalamazoo, MI, USA).

All the relevant clinical and procedure parameters, e.g., age, time to thrombolysis, duration of mechanical recanalization, number of passes with the thrombectomy device, were recorded for each patient. The thrombi, retrieved from the M1 segment of the middle cerebral artery (MCA), of $n = 25$ patients (mean age $= 73 \pm 11$ years, 16 males and 9 females) were preserved and additionally examined by high spatial resolution multiparametric MRI and immunohistochemistry. Prior to the MRI, the samples were rinsed with isotonic saline of 0.9% w/v of NaCl, pH 7.4 and closed in Teflon tubes to prevent tissue desiccation during the MR scanning, which was performed within 24 h after the retrieval of the thrombus. Thereafter, samples were processed for histological analysis. All the three imaging modalities (CT, MRI and histology), as well as the interventional procedure and treatment outcome, were the source of several different parameters that were then analyzed pairwise for possible correlations. The design of this study is shown in Figure 1.

This study was approved by the Ethical Committee of the National Ministry of Health of the Republic of Slovenia, approval No. 0120-99/2021/7 from 21 May 2021. The study was performed in agreement with the informed-consent policy.

2.2. CT Imaging

The CT examination of patients suspected with stroke included non-contrast enhanced (NCE) sequential CT scans and contrast CT angiography (CTA) scans. The NCE CT scan was a two-part scan: scull-base and cerebral. Both scans were sequential and had the following parameters for the scull base region: 120 kV, 265 mAs, matrix 1024 × 1024, slice thickness 3 mm, collimation 20 × 0.6, rotation time 1 s, window width 90–190, window center 38, and for the cerebral region: 120 kV, 310 mAs, matrix 1024 × 1024, slice thickness 4.8 mm, collimation 24 × 1.2, rotation time 1 s, window width 80, window center 38. CT scanning was performed on a Siemens Sensation Open 40 CT scanner (Siemens, Erlangen, Germany) at the Neurology Clinic of University Medical Centre Ljubljana.

2.3. Multi-Parametric MRI

The MR imaging was performed on an experimental MRI scanner consisting of a 2.35 T (100 MHz proton frequency) horizontal bore superconducting magnet (Oxford Instruments, Abingdon, UK) equipped with a Bruker micro-imaging system (Bruker, Ettlingen, Germany) for MR microscopy with a maximum imaging gradient of 300 mT/m and an Apollo spectrometer (Tecmag, Houston, TX, USA). Each thrombus sample was inserted into a 10 mm micro-imaging probe and then analyzed by a multi-parametric MRI protocol, which consisted of the following 3D imaging sequences: T_1-weighted, ADC mapping and T_2 mapping. The corresponding imaging sequences were based on the spin-echo, pulsed gradient spin-echo (PGSE) [28] and Carr–Purcell–Meiboom–Gill (CPMG) multi-echo [29] sequences, respectively. MR imaging was performed with the field of view of $20 \times 10 \times 10$ mm^3 and the imaging matrix equal to $128 \times 64 \times 64$ (T_1-weighted) or to $128 \times 64 \times 16$ (mapping) so that the resolution was equal to 156 μm isotropic (T_1-weighted) or 156 μm in-plane (mapping). Other sequence specific parameters were equal to: TE/TR = 5/100 ms, signal averages 10 for the 3D spin-echo sequence, TE/TR = 34/1035 ms, b = 0, 260, 620, 1250 s/mm^2, signal averages 2 for the 3D PGSE sequence and iTE/TR = 16/1930 ms, number of echoes 8, signal averages 2 for the 3D multi-echo sequence. MRI scanning of thrombi was performed at a constant room temperature of 22 °C.

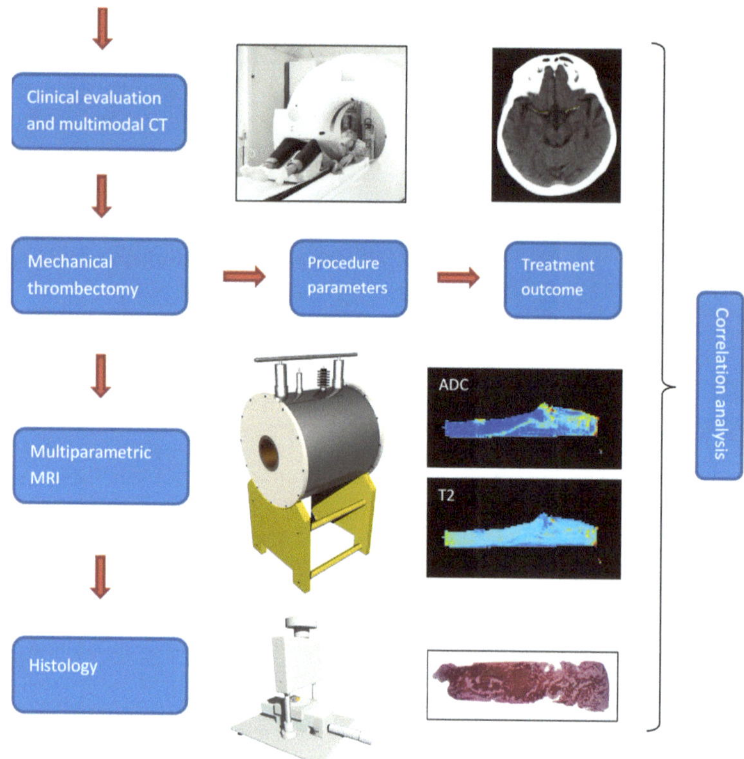

Figure 1. Study design. Patients suspected with an acute stroke underwent clinical examination and a CT scan. If ischemic stroke was confirmed and the thrombolysis was unsuccessful, the patients received further therapy by mechanical thrombectomy. The thrombi retrieved from the MCA vessel segment were further analyzed by high spatial resolution multiparametric MRI and histology. Different parameters obtained from imaging (CT, MRI and histology) and the interventional procedure and treatment were then analyzed for possible correlations.

2.4. Histology

For histological analysis, which followed the MRI examination, the thrombi were fixed in 4% buffered formaldehyde for 48 h, cut longitudinally and, if necessary, transversely, processed on the Tissue processor Thermo Scientific Excelsior ES and embedded in paraffin. Finally, serial 5-μm-thick sections were cut from each sample and stained by hematoxylin-eosin (HE) for the determination of a general thrombi composition, by Monoclonal Mouse Anti-Human CD235a Glycophorin A, DakoCytomation, Denmark (GPA) for RBC content determination and by Monoclonal Mouse Anti-Human CD61, DakoCytomation, Denmark (CD61) for platelet content determination. The immunohistochemical (IHC) stainings were performed using a Ventana BenchMark automatic stainer with a streptavidin-biotin peroxidase complex and diaminobenzidine tetrahydrochloride as a chromogen. Stained sections of the thrombi were examined by a Nikon Eclipse E600 optical microscope (Nikon, Düsseldorf, Germany) equipped with Nikon Plan Fluor objectives and with a high-resolution CCD camera. The system was controlled by the Nikon NIS Elements software allowing precise microphotography of the samples with optimal image contrast.

2.5. Processing of Images

CT images were processed by the ImageJ program (NIH, Bethesda, MD, USA) in order to extract information on the thrombus location, its length (L) and especially its radio-density measured in the units of Hounsfield (HU). The first two parameters, i.e., location and length were determined from CTA images, while the HU values were measured from NCE CT images. In order to obtain more accurate HU values, images of three consecutive slices central to the thrombus were stacked and the HU value profile along the line of the central part of the thrombus was measured. Special care was taken for the line not to include the HU values from the vessel wall or possible calcifications next to the thrombus. The same operation was repeated on the non-occluded symmetrically located MCA artery to obtain a reference HU value profile. Both profiles, from the occluded and non-occluded site, were then analyzed further to determine the average HU value (HU_avg) of the profile and its standard deviation (HU_var). Finally, differences of these two parameters between the occluded and non-occluded MCA site were calculated, thus obtaining ΔHU_avg and ΔHU_var parameters.

MR and histological images were processed by using in-house written image-analysis software that was developed within the Matlab programming environment (MathWorks, Inc., Natick, MA, USA). In the MR image processing software, ADC and T_2 values were calculated pixelwise from the masked DWI and multi-spin-echo images of all slices through the entire thrombi volumes by fitting the data to the exponentially decaying function. Calculated ADC and T_2 maps were analyzed for average and within-sample variation of their values in the thrombus region (in all slices) thus obtaining four parameters: ADC_avg, ADC_var, T2_avg and T2_var. The within-sample coefficient of variation can be considered as a measure for heterogeneity of the thrombus.

Red blood cell (RBC) proportions of each thrombus was determined from its standard (hematoxylin-eosin stained) histological image of the central slice across the thrombus. This analysis included the following steps: correction of uneven illumination [30], calculation of the image histogram to determine optimal threshold for the discrimination between RBC-rich and platelet-rich part of the thrombus and, finally, the calculation of the RBC proportion as the ratio between the area of the thresholded RBC region and the total thrombus area.

2.6. Clinical and Intervention Procedure Parameters

On admission to the urgent medical-care center and at discharge from the hospital, the neurological status of each patient was graded according to the modified Ranking Scale (mRS) for the stroke and also to the NIH stroke scale (NIHSS). This enabled following of four clinical parameters: imRS, ΔmRS, iNIHSS and ΔNIHSS, where character i denotes initial (grade on admission) and Δ difference between final grades at discharge and initial grades on admission.

In addition to the clinical parameters of each patient, the corresponding interventional procedure parameters were also recorded. These parameters include the time to thrombolysis (tt_Lysis), duration of mechanical recanalization (t_MeR) and number of passes with the thrombectomy device (# Passes). t_MeR implies successful recanalization of the occluded vessel and it was determined as the time from groin puncture to recanalization through the occluded artery by complete mechanical removal of the occluding thrombus.

2.7. Statistical Analysis

Parameters obtained in the study: MRI (ADC_avg, ADC_var, T2_avg, T2_var), CT (ΔHU_avg, ΔHU_var, L), histology (RBC %), procedure (tt_Lysis, t_MeR, # Passes) and clinical (imRS, ΔmRS, iNIHSS, ΔNIHSS) were tested for the possible correlations. Specifically, the Pearson and Spearman correlation coefficients were calculated for each possible pair of the parameters using Excel statistical tools (Microsoft, Seattle, WA, USA). The correlations were considered important when the absolute value of the correlation coefficient was higher than 0.5 and indicative when it was higher than 0.3.

3. Results

CT brain imaging is the most common imaging modality in stroke diagnosis. In Figure 2 are shown two such images, one contrasted (CTA, a) and one non-contrasted (NCT, b). Both images are of the same slice positioned along both middle cerebral arteries (MCAs). In the CTA image, all the major blood vessels as well as the location of the occluded MCA to where points a red arrow are clearly seen. The NCT image is displayed with a different HU brightness window that is adjusted for displaying of soft tissues. Along the occluded MCA vessel, a yellow line is drawn and then another such line is drawn in the symmetric nonoccluded MCA site. Along these lines, HU values were measured and then processed to obtain ΔHU_avg and ΔHU_var parameters.

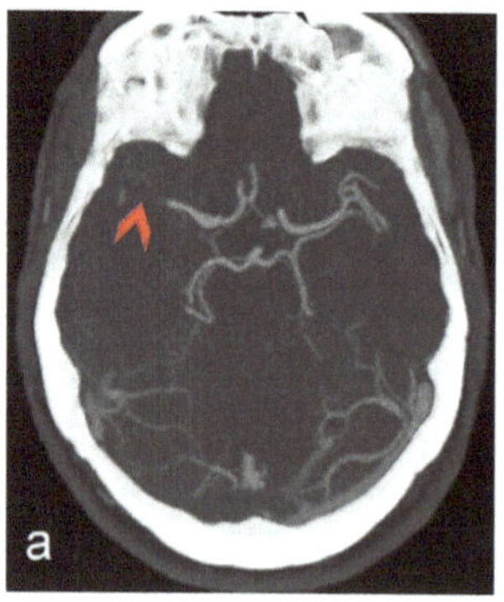
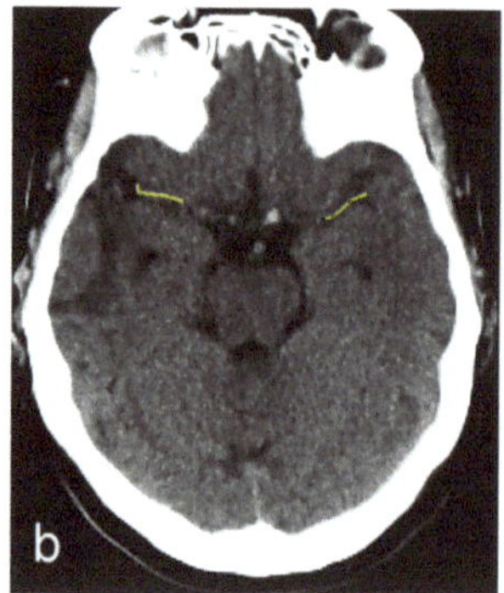

Figure 2. Representative brain CT images of an ischemic stroke patient in a slice with the middle cerebral (MCA) artery: (**a**) CT angiograph (CTA) image and (**b**) non-contrasted CT (NCT) image of the same slice. Red arrow in the CTA image points to the occlusion (thrombus), while yellow lines in the NCT image correspond to the lines along which HU values were measured in the occluded site (left) and in the symmetrical positioned non-occluded site (right).

Figure 3 depicts T_1-weighted images and ADC and T_2 maps along with the corresponding histological background-corrected HE/GPA/CD61 images of the two representative cerebral thrombi of a different structure and composition. It can be seen that both samples show quite significant variability of ADC values across the thrombi, while their T_2 values are somewhat more uniform. In general, discrimination among different regions of low water mobility is better in ADC maps than in the corresponding T_2 maps. From the comparison between ADC/T_2 maps and IHC images, it can be seen in Sample 1 that regions with low ADC/T_2 values correspond to the regions abundant with RBCs. This reduction of ADC/T_2 values can be explained by the reduced mobility of water molecules due to a high RBC compaction resulting in a significant reduction of the extracellular space. Another mechanism of the ADC/T_2 reduction can be seen in Sample 2, where high platelet concentration locally affects fibrin fiber architecture, resulting in a fibrin meshwork contraction and, therefore, a pore size reduction, and also in an extracellular serum expulsion [9]. The regions with intermediate ADC/T_2 values are mainly composed of a mixture of platelet-rich and RBC-scarce regions. In addition, the central part of Sample 1 also consists of a few smaller inclusions with higher ADC/T_2 values. The inclusions correspond to regions of entrapped serum that can be clearly identified in the underlying.

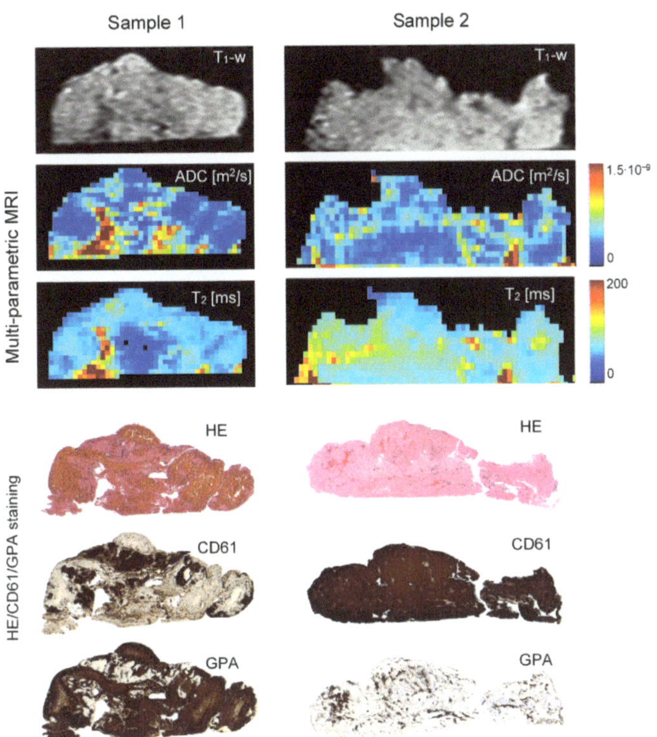

Figure 3. Two representative cerebral thrombi of a different structure are presented by central-slice T_1-weighetd images, ADC and T_2 maps along with the corresponding histological (HE/CD61/GPA) images. Lower ADC and T_2 values, shown in dark blue, in maps of Sample 1 correspond to compact RBC-rich regions, while lower ADC and T_2 values in Sample 2 correspond to compact platelet-fibrin-rich regions. CD61 staining is specific to platelets (dark brown), while GPA staining is specific to RBC components (dark brown).

Histological images as tissue voids. Differences between both samples are also well observed in histological and IHC images. Specifically, in CD61 images a positive reaction to the platelet content, and in GPA images a positive reaction to RBCs are shown in brown. These two thrombi samples were also significantly different in RBC proportions, $63 \pm 7\%$ (Sample 1) vs. $5 \pm 3\%$ (Sample 2), as determined from the corresponding histological/IHC images.

A relation between the duration of mechanical thrombectomy procedure (t_MeR) and the thrombus length (L) as measured in vivo from CT images prior to the thrombus retrieval is shown by a graph in Figure 4. The figure also shows histological images of the thrombi ordered lengthwise. The result supports existing findings of other groups, namely that the duration of mechanical thrombectomy procedure increases with the thrombus length [31]. However, it can also be seen that this general trend has also some isolated exceptions where shorter thrombi have been hard to retrieve or longer thrombi were easy to retrieve. This indicates that, in addition to the thrombus length, its composition also plays a very important role in the mechanical thrombectomy procedure. Large diversity of thrombi composition can be seen well in the presented histological images. The histological analysis of the thrombi also shows that the longer thrombi have in average more complex composition, namely, they often contain several inclusions of compact RBC-rich areas surrounded by thin platelet-fibrin layers.

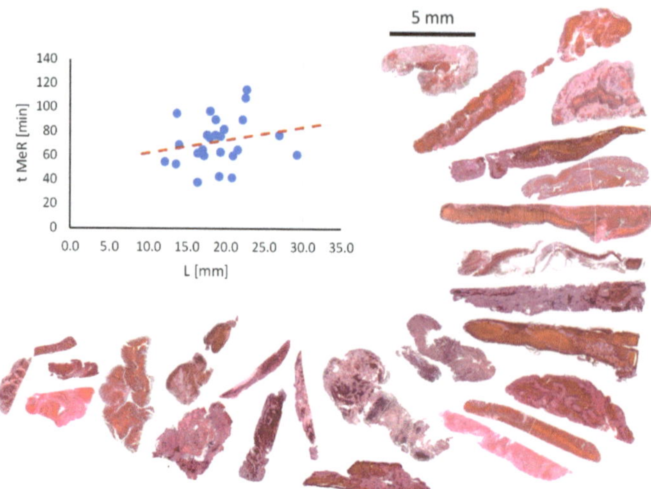

Figure 4. A correlation plot between the mechanical thrombectomy procedure time and the thrombus length. From the plot it can be seen that the procedure time increases with the thrombus length. The figure also shows histological images of all the thrombi in the study ordered by their lengths.

All data of the thrombi qualified for the study are presented in Table 1. The data are grouped according to the categories into MRI, CT, histology, interventional procedure and clinical parameters; all together there are 16 parameters on each thrombus. This data then enabled the calculation of correlations between different pairs of parameters. The calculated correlation coefficients are displayed in Table 2 with Pearson correlation coefficients above the diagonal cells and Spearman correlation coefficient below the diagonal cells. The correlation pair of a cell is identified by the parameters corresponding to the intersection of the row and column of the cell. Cells with correlation coefficients considered significant are highlighted in green ($0.5 \leq |\rho| < 1$), the semi significant correlation coefficients are highlighted in yellow ($0.3 \leq |\rho| < 0.5$), while those that are considered insignificant are left unmarked ($|\rho| < 0.3$).

For easier interpretation of the results in Table 2, all significant correlations according to the Pearson correlation coefficient are presented by a diagram in Figure 5. In the diagram, all the parameters with significant correlations are organized by their category and interconnected with a green line, for a significant positive correlation, or a red line, for a significant negative correlation. From this diagram, can also be seen that some of the parameters have quite substantial number of significant correlations with other parameters, e.g., ΔHU_avg, ΔHU_var and # Passes, while others are more isolated, e.g., L and RBC %. It can also be seen that the negative correlations dominate over positive correlations.

Some of the more important and characteristic correlations are shown in Figure 6. The variabilities of both MRI parameters i.e., T_2 and ADC, correlate positively and so do the number of passes with the thrombectomy device and the difference in HU between the occluded and symmetric normal site, as well as the proportion of RBCs and the thrombus length. The remaining of the presented correlations in Figure 6 are negative. Such correlations are found between the duration of mechanical recanalization and the variability of ADC, the number of passes with the thrombectomy device and the difference between the final and initial modified Rankin score, the initial modified Rankin score and the variability of T_2. Red dashed lines in the plots indicate the trend lines.

Table 1. MRI, CT, histological, interventional procedure and clinical parameters of patients diagnosed with ischemic stroke due to occlusion in the MCA artery that were qualified for the study.

Pt. #	MR Parameters				CT Parameters			Histol.	Procedure Parameters			Clinical Parameters					
	ADC Avg $[10^{-9}\,\mathrm{m^2/s}]$	ADC Var $[10^{-9}\,\mathrm{m^2/s}]$	T_2 Avg [ms]	T_2 Var [ms]	ΔHU Avg	ΔHU Var	L [mm]	RBC % [%]	Tt Lysis [min]	t MeR [min]	# Passes	Age [yrs]	Tx before Stroke	iNIHSS	ΔNIHSS	imRS	ΔmRS
1	0.50	0.26	89	21	−3.3	−0.22	20.8	11.5	110	42	1	77	/	21	20	4	3
2	0.48	0.24	90	17	12.8	−4.40	16.3	45.1	240	62	2	77	/	17	9	5	1
3	0.72	0.30	79	21	3.4	0.80	17.0	12.0	120	65	1	63	/	23	16	4	1
4	0.64	0.25	59	9	2.5	7.08	17.5	41.4	95	77	1	83	AA	18	15	5	3
5	0.38	0.15	78	12	−0.5	−0.31	17.9	65.9	145	97	3	43	/	26	20	5	2
6	0.61	0.28	85	23	5.9	1.95	16.3	61.0	89	38	3	78	/	7	7	3	3
7	0.57	0.24	73	19	−4.2	−0.49	22.2	57.3	50	90	1	58	/	13	10	4	3
8	0.73	0.23	94	15	−1.1	−1.63	26.9	50.3	107	77	1	77	/	14	11	4	1
9	0.48	0.20	85	18	8.1	4.60	22.7	56.3	110	115	2	85	ACAA	5	3	3	1
10	0.61	0.23	64	20	44.1	−24.60	19.7	52.4	165	82	4	81	/	26	−16	5	−1
11	0.55	0.23	83	21	−1.1	1.83	17.1	19.1	148	60	1	66	/	12	12	3	3
12	0.52	0.21	76	18	1.7	−1.83	13.9	31.8	185	69	1	72	AA	6	3	3	2
13	0.44	0.20	73	16	12.3	−0.46	22.5	74.7	125	108	2	62	/	19	7	5	1
14	0.69	0.28	101	24	76.9	−53.00	19.3	38.9	81	63	5	79	/	18	−2	4	−1
15	0.50	0.24	85	20	64.7	−44.40	17.9	48.8	120	75	2	91	/	42	0	5	−1
16	0.50	0.17	73	11	1.7	−1.30	18.6	21.3	140	77	1	73	/	15	11	5	1
17	0.71	0.25	88	21	1.8	−0.63	13.5	14.8	67	53	1	72	AA	14	12	5	2
18	0.71	0.27	73	18	15.4	0.87	19.1	42.4	90	43	1	85	AC	22	19	5	1
19	0.56	0.26	109	29	3.8	1.18	20.9	37.7	60	60	1	73	/	3	2	1	1
20	0.62	0.26	77	16	8.8	0.56	19.3	56.0	120	76	3	86	AC	11	−29	4	−2
21	0.36	0.19	93	15	8.7	−1.99	29.2	79.1	90	61	1	79	/	19	9	5	0
22	0.37	0.16	71	19	2.6	0.24	13.6	9.3	110	95	3	70	/	16	7	5	0
23	0.53	0.27	84	19	6.9	0.67	21.5	80.4	25	65	2	65	/	16	13	4	2
24	0.65	0.29	89	27	5.6	−1.20	12.1	57.1	75	55	1	53	/	13	10	4	3
25	0.63	0.26	80	19	61.3	−42.60	18.7	46.8	105	90	4	69	AA	42	0	5	−1

#—index/number; ADC—Apparent diffusion coefficient, T_2—Transversal NMR relaxation time, HU—Hounsfield units, avg—Sample mean value, var—Within sample variability (standard deviation), RBC %—Percentage of red blood cells in the thrombus, /—None, AA—Anti-aggregation, AC—Anticoagulant, NIHSS—NIH stroke scale, mRS—Modified Rankin scale, i—Initial, Δ—Difference between final and initial.

Table 2. Correlation coefficients between different pairs of thrombi parameters. A correlation pair is defined by parameters of row and column of the corresponding cell. Above the table diagonal cells are shown Pearson correlation coefficients, while below them are shown Spearman correlation coefficients. Cells with the absolute value of the correlation coefficient higher than 0.5 are highlighted in green, while those between 0.3 and 0.5 are highlighted in yellow. To better distinguish between Pearson and Spearman correlation coefficients, the diagonal elements are written in red.

Pearson / Spearman	ADC Avg	ADC Var	T_2 Avg	T_2 Var	ΔHU Avg	ΔHU Var	L	RBC %	Tt Lysis	t MeR	# Passes	Age	iNIHSS	ΔNIHSS	imRS	ΔmRS
ADC avg	1.00	0.77	0.05	0.29	0.18	−0.16	−0.16	−0.22	−0.28	−0.40	−0.02	0.18	−0.01	−0.09	−0.12	0.02
ADC var	0.71	1.00	0.24	0.57	0.26	−0.21	−0.18	−0.06	−0.35	−0.59	0.01	0.19	0.04	−0.10	−0.26	0.06
T_2 avg	0.06	0.24	1.00	0.60	0.12	−0.18	0.23	0.03	−0.23	−0.37	−0.04	0.04	−0.22	0.04	−0.52	−0.05
T_2 var	0.33	0.62	0.43	1.00	0.21	−0.22	−0.24	−0.16	−0.33	−0.43	0.08	−0.03	−0.20	−0.14	−0.61	0.03
ΔHU avg	0.01	0.30	0.08	0.10	1.00	−0.97	0.00	0.12	0.04	0.09	0.70	0.35	0.64	−0.44	0.25	−0.68
ΔHU var	0.05	0.14	−0.24	0.05	−0.35	1.00	0.02	−0.03	−0.04	−0.07	−0.66	−0.24	−0.66	0.38	−0.23	0.62
L	−0.15	−0.19	0.14	−0.25	0.08	−0.01	1.00	0.50	−0.23	0.21	−0.08	0.21	0.01	−0.03	−0.03	−0.21
RBC %	−0.22	−0.06	0.04	−0.25	0.28	−0.06	0.47	1.00	−0.20	0.25	0.20	−0.10	0.04	−0.19	0.08	−0.10
tt Lysis	−0.38	−0.50	−0.30	−0.38	0.01	−0.19	−0.19	−0.21	1.00	0.15	0.10	0.07	0.12	−0.15	0.21	−0.17
t MeR	−0.32	−0.58	−0.48	−0.54	0.01	−0.13	0.31	0.24	0.34	1.00	0.27	−0.15	0.15	−0.23	0.21	−0.30
# Passes	−0.18	0.01	−0.09	0.04	0.60	−0.26	0.02	0.29	0.17	0.38	1.00	0.16	0.34	−0.55	0.19	−0.60
Age	0.08	0.06	0.07	−0.13	0.46	−0.02	0.21	−0.05	0.10	−0.19	0.09	1.00	0.04	−0.42	0.01	−0.45
iNIHSS	−0.04	0.00	−0.19	−0.17	0.37	−0.42	0.05	0.02	0.22	0.17	0.28	0.03	1.00	0.01	0.65	−0.43
ΔNIHSS	0.08	0.01	−0.09	−0.16	−0.60	0.40	−0.12	−0.15	−0.10	−0.26	−0.57	−0.37	0.17	1.00	0.11	0.72
imRS	−0.19	−0.31	−0.36	−0.52	0.30	−0.41	−0.09	0.04	0.20	0.28	0.19	0.07	0.73	0.12	1.00	−0.27
ΔmRS	0.06	0.10	−0.03	0.15	−0.70	0.49	−0.26	−0.01	−0.22	−0.34	−0.54	−0.45	−0.35	0.66	−0.35	1.00

\#—index/number; ADC—Apparent diffusion coefficient, T_2—Transversal NMR relaxation time, HU—Hounsfield units, avg—Sample mean value, var—Within sample variability (standard deviation), RBC %—Percentage of red blood cells in the thrombus, /—None, AA—Anti-aggregation, AC—Anticoagulant, NIHSS—NIH stroke scale, mRS—Modified Rankin scale, i—Initial, Δ—Difference between final and initial.

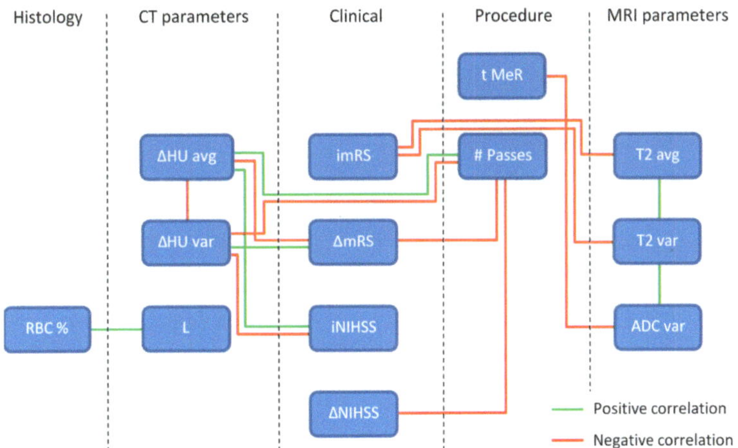

Figure 5. A diagram of significant correlations among different thrombi parameters. The parameters are organized according to their category in histology, CT, clinical, procedure and MRI parameter groups and existing correlations between parameter pairs are plotted with a green line for positive correlations and a red line for negative correlations.

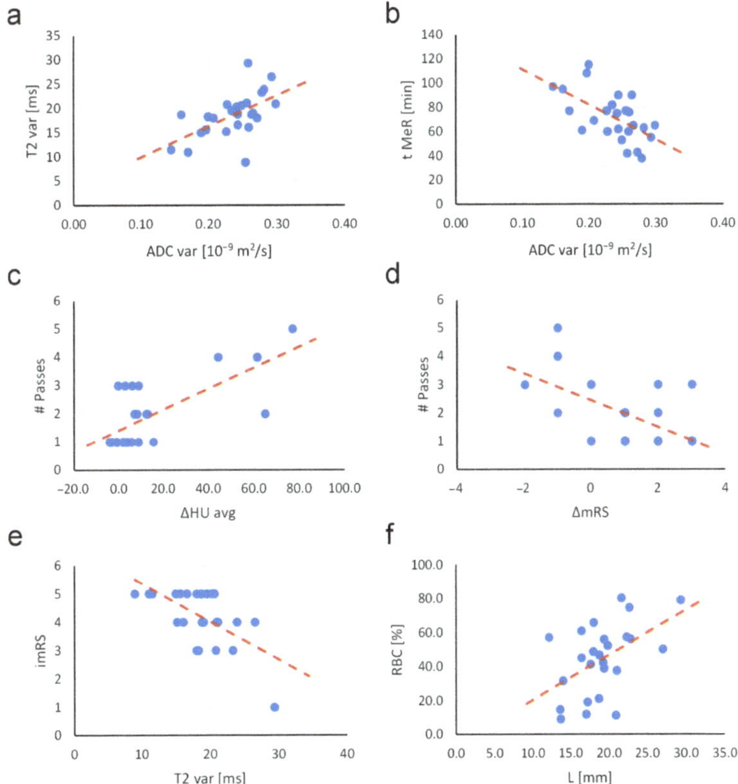

Figure 6. Plots of correlations between selected pairs of thrombi parameters: (**a**) T2_var vs. ADC_var, (**b**) t_MeR vs. ADC_var, (**c**) # Passes vs. ΔHU_avg, (**d**) # Passes vs. ΔmRS, (**e**) imRS vs. T2_var, (**f**) RBC % vs. L. Red dashed lines indicate trend lines of the data.

4. Discussion

In this study, the retrieved cerebral thrombi were characterized by means of native (non-contrasted) CT, multi-parametric MRI and histology. In addition, the characterization by these imaging modalities was complemented with clinical and interventional procedure data and, thus, enabled a vast selection of parameters on each thrombus. Specifically, 16 parameters were measured: four MRI, three CT, one histological, three interventional procedure and five clinical. This enabled a comprehensive investigation of relations among these parameters by two standard correlation tests, Pearson and Spearman. Correlation analyses were possible since the thrombi were characterized quantitatively by the imaging methods of which contrast depends exclusively on the intrinsic tissue properties. Two such methods, that were used in this study are ADC and T_2 mapping that produce images of water mobility and T_2 NMR relaxation time. In several existing studies involving MRI, cerebral thrombi were scanned by the non-quantitative methods, e.g., T_2-weighetd, FLAIR and GRE [32] or T_2*-weighted GRE and SWI [5]. These methods produce images in which contrast, in addition to the tissue properties, also depends on the scanning parameters, and so would the obtained correlation coefficients. The relations among some of the parameters analyzed in this study were also known from some previous studies; however, these studies were focused to a more limited set of parameters or were performed on other type of blood clots, e.g., relation between thrombus length and the number of stent retrievals [31], relation between RBC content and T_1 and T_2 NMR relaxation times in artificial blood clots [33].

Table 2 clearly shows which parameter correlations among these are significant (highlighted in green) or indicative (highlighted in yellow). For the most part, these results meet expectations, e.g., more compact thrombi have lower T_2, ADC and higher ΔHU average values and also higher variability of these parameters due to their often-heterogeneous structure. Such values of ADC and T_2 of the compact thrombi can be explained by an increased tortuosity and decreased porosity that is responsible for ADC decrease [34], while an increase in surface-to-volume ratio in these thrombi causes an increased surface-induced relaxation; therefore, T_2 decreases [35]. The origin of surface-induced relaxation can also be calcifications in thrombi [36]. However, as can be seen from the histological sections in Figure 4, thrombi included in this study did not contain any noticeable (macroscopic) calcifications. Compacted thrombi also have higher radio-density (HU values) mainly due to the higher iron concentration in closely packed RBCs [37]. Since compact thrombi lyse slower and pose greater challenge for mechanical recanalization, all these parameters are also associated with longer duration of mechanical recanalization (t_MeR), higher number of passes with thrombectomy device (# Passes) and longer times to thrombolysis (tt_Lysis), which is also supported by the results in Tables 1 and 2. In addition, from these results, it can be seen that the thrombi with such MRI and CT parameters have higher initial NIH stroke scale and modified Rankin scale values (iNIHSS and imRS) and poorer decrease in these values after the treatment (ΔNIHSS and ΔmRS). Among the parameters that can already be obtained in vivo is also the thrombus length (L). This parameter correlates positively with t_MeR (Figure 4) and interestingly also with RBC % (Figure 6f). The impact of thrombus histological characteristics (structure, composition, RBC proportion and length) on the performance of the thrombectomy device was discussed already in [14] where the link between the histological characteristics, the mechanical properties and therefore mechanical recanalization was exposed. Thrombus length is not only important for the success of mechanical thrombectomy, but also has a great impact on thrombolysis [38]. In our study, this link cannot be seen because thrombolysis was unsuccessful and was in all cases continued with mechanical thrombectomy as soon as it was assessed that the progress of thrombolysis is poor.

The correlations discussed above, and in Section 3, refer to the Pearson correlation coefficients, i.e., above diagonal elements in Table 2. Below the diagonal elements of this table are Spearman (rank) correlations coefficients. From the results in this table, it can be seen that in most cases the Pearson and Spearman correlation tests yield the same pairs of parameters with significant correlations, and only the significance of the correlations differs between the two.

An interesting addition to our ADC and T_2 mapping MRI techniques for characterization of thrombus composition would be magnetization transfer (MT) imaging [39]. In MT, selective off-resonance radio-frequency irradiation of the protein-bound protons enables the detection of the protein-rich fibrin-to-platelet regions. MT enables the quantitative determination of the individual protein components in thrombi. This is an entirely different characterization method than ADC or T_2 mapping, which are sensitive to changes in water diffusion in thrombus porous structure or to microstructure-induced changes in T_2 NMR relaxation time. Recently, the MT technique was efficiently employed for the characterization of the human arterial thrombi ex vivo [40].

This study was performed on the M1 segment of the MCA for two reasons. The first is that this segment can be optimally visible on CT images and the second is that thrombi can be best retrieved from this site, i.e., mostly without their disintegration. This was important for the success of the study, namely, the second part the study was designed for in vitro investigation of thrombi structure and composition by histology and by MR microscopy, which could not be undertaken on the disintegrated thrombi.

Perhaps, the biggest limitation of this study is the relatively low number of patients involved in the study ($n = 25$). The small sample group could result in a reduced accuracy of the calculated correlation coefficients [41]. Correlation bias is higher with parameter pairs with lower correlation coefficients. From this perspective, the results of this study can be considered as preliminary results. An interesting expansion of the sample group would be the inclusion of patients who were treated only by mechanical thrombectomy. This would allow comparison of the measured parameters between our existing group (thrombolysis followed by thrombectomy) with this group (thrombectomy only). However, the narrow selection criteria would have made this group even smaller, so no such comparison can be made with the available data. Another limitation of this study is its two-way design that includes in the first stage in vivo experiments (CT) and in the second stage in vitro experiments (MRI, histology). In case of clinical application of the results of this study, all data can only be acquired in vivo.

As can be seen from Table 1, the selection of patients for the study was such that all range of stroke severity was covered (see initial NIHSS and mRS grades). Thus, the study group includes also two patients with most severe symptoms (iNIHSS = 42) and two with very mild symptoms (iNIHSS < 6) where mechanical thrombectomy is usually not performed due to lack of validation; however, it is not contraindicated. In our opinion, such selection of the study group enables reaching extremes of the measured parameters and, thus, contributes to accuracy of the studied correlations among the parameters.

The prediction of the recanalization procedure duration and optimization of the means for achieving the recanalization could help to improve the planning of interventions. This would include optimal selection among thrombolysis alone, in combination with thrombectomy or thrombectomy alone, finer thrombolytic treatment adjustment and optimization in selection of the thrombectomy device. Based on the methods used and the results of this study, predictive models for the optimal intervention procedure can be designed. However, the use of such a model in the clinical environment would mainly be hampered by limitations in clinical diagnostic imaging. Current technology of MRI does not enable clinical imaging with microscopic resolution, especially not in acceptable time. Therefore, there is currently no efficient substitute for the standard histology. However, there is constant improvement in diagnostic imaging technology, not only in MRI, but also in CT, especially in soft tissue contrast. Therefore, the future prospects for possible optimization of the interventional procedure based on the accurate characterization of the thrombus before the intervention with diagnostic imaging are good.

5. Conclusions

Multi-modality imaging of the cerebral thrombi, by CT, MRI and histology, enables the precise assessment of their intrinsic properties, such as composition, structure and compactness. Furthermore, it was shown that various parameters obtained by imaging, e.g., variability of T_2 and ADC, difference in HU between occluded and normal site, correlate well with some of the interventional procedure parameters, e.g., number of passes with the thrombectomy device, duration of the mechanical recanalization. The study was designed in vitro; however, its results have also a clinical relevance as they can help better planning of the interventional procedures in the patients diagnosed with ischemic stroke.

Author Contributions: Conceptualization, I.S. and J.V.; methodology, F.B. and I.S.; software, F.B.; formal analysis, R.V., F.B. and E.K.; investigation, R.V.; resources, Z.M., I.K., M.J. and A.F.; writing—original draft preparation, I.S. and R.V.; visualization, F.B.; supervision, I.S. and J.V. All authors have read and agreed to the published version of the manuscript.

Funding: This research was funded by the Slovenian Research Agency grant J3-9288 "Optimization of magnetic resonance imaging techniques to predict outcome of thrombolysis".

Institutional Review Board Statement: The study was conducted in accordance with the Declaration of Helsinki and approved by the Institutional Review Board (or Ethics Committee) of Slovenian Medical Ethics Committee (approval no. 0120-99/2021/7 from 21 May 2021) at Ministry of Health.

Informed Consent Statement: Informed consent was obtained from all subjects involved in the study.

Data Availability Statement: The data presented in this study are available on request from the corresponding author.

Acknowledgments: The author thank Kanza Awais for proofreading the manuscript.

Conflicts of Interest: The authors declare no conflict of interest.

References

1. Roger, V.L.; Go, A.S.; Lloyd-Jones, D.M.; Benjamin, E.J.; Berry, J.D.; Borden, W.B.; Bravata, D.M.; Dai, S.F.; Ford, E.S.; Fox, C.S.; et al. Executive Summary: Heart Disease and Stroke Statistics-2012 Update A Report from the American Heart Association. *Circulation* **2012**, *125*, 188–197. [CrossRef]
2. Berkhemer, O.A.; Fransen, P.S.S.; Beumer, D.; van den Berg, L.A.; Lingsma, H.F.; Yoo, A.J.; Schonewille, W.J.; Vos, J.A.; Nederkoorn, P.J.; Wermer, M.J.H.; et al. A Randomized Trial of Intraarterial Treatment for Acute Ischemic Stroke. *N. Engl. J. Med.* **2015**, *372*, 11–20. [CrossRef] [PubMed]
3. Xavier, A.R.; Farkas, J. Catheter-based recanalization techniques for acute ischemic stroke. *Neuroimag. Clin. N. Am.* **2005**, *15*, 441–453. [CrossRef] [PubMed]
4. Marder, V.J.; Chute, D.J.; Starkman, S.; Abolian, A.M.; Kidwell, C.; Liebeskind, D.; Ovbiagele, B.; Vinuela, F.; Duckwiler, G.; Jahan, R.; et al. Analysis of thrombi retrieved from cerebral arteries of patients with acute ischemic stroke. *Stroke* **2006**, *37*, 2086–2093. [CrossRef] [PubMed]
5. Gasparian, G.G.; Sanossian, N.; Shiroishi, M.S.; Liebeskind, D.S. Imaging of occlusive thrombi in acute ischemic stroke. *Int. J. Stroke* **2015**, *10*, 298–305. [CrossRef]
6. Lipinski, B.; Pretorius, E.; Oberholzer, H.M.; van der Spuy, W.J. Interaction of Fibrin with Red Blood Cells: The Role of Iron. *Ultrastruct. Pathol.* **2012**, *36*, 79–84. [CrossRef]
7. Fang, J.; Tsui, P.H. Evaluation of thrombolysis by using ultrasonic imaging: An in vitro study. *Sci. Rep.* **2015**, *5*, 11669. [CrossRef]
8. Bajd, F.; Vidmar, J.; Fabjan, A.; Blinc, A.; Kralj, E.; Bizjak, N.; Sersa, I. Impact of altered venous hemodynamic conditions on the formation of platelet layers in thromboemboli. *Thromb. Res.* **2012**, *129*, 158–163. [CrossRef]
9. Weisel, J.W. Structure of fibrin: Impact on clot stability. *J. Thromb. Haemost.* **2007**, *5*, 116–124. [CrossRef]
10. Gennisson, J.L.; Lerouge, S.; Cloutier, G. Assessment by transient elastography of the viscoelastic properties of blood during clotting. *Ultrasound Med. Biol.* **2006**, *32*, 1529–1537. [CrossRef]
11. Varin, R.; Mirshahi, S.; Mirshahi, P.; Klein, C.; Jamshedov, J.; Chidiac, J.; Perzborn, E.; Mirshahi, M.; Soria, C.; Soria, J. Whole blood clots are more resistant to lysis than plasma clots—Greater efficacy of rivaroxaban. *Thromb. Res.* **2013**, *131*, E100–E109. [CrossRef] [PubMed]
12. Wohner, N.; Sotonyi, P.; Machovich, R.; Szabo, L.; Tenekedjiev, K.; Silva, M.M.C.G.; Longstaff, C.; Kolev, K. Lytic Resistance of Fibrin Containing Red Blood Cells. *Arterioscl. Throm. Vas.* **2011**, *31*, 2306–2313. [CrossRef] [PubMed]
13. Hashimoto, T.; Hayakawa, M.; Funatsu, N.; Yamagami, H.; Satow, T.; Takahashi, J.C.; Nagatsuka, K.; Ishibashi-Ueda, H.; Kira, J.; Toyoda, K. Histopathologic Analysis of Retrieved Thrombi Associated With Successful Reperfusion after Acute Stroke Thrombectomy. *Stroke* **2016**, *47*, 3035–3037. [CrossRef] [PubMed]

14. Yuki, I.; Kan, I.; Vinters, H.V.; Kim, R.H.; Golshan, A.; Vinuela, F.A.; Sayre, J.W.; Murayanna, Y.; Vinuela, F. The Impact of Thromboemboli Histology on the Performance of a Mechanical Thrombectomy Device. *Am. J. Neuroradiol.* **2012**, *33*, 643–648. [CrossRef] [PubMed]

15. Mokin, M.; Morr, S.; Natarajan, S.K.; Lin, N.; Snyder, K.V.; Hopkins, L.N.; Siddiqui, A.H.; Levy, E.I. Thrombus density predicts successful recanalization with Solitaire stent retriever thrombectomy in acute ischemic stroke. *J. Neurointerv. Surg.* **2015**, *7*, 104–107. [CrossRef]

16. White, P.; Nanapragasam, A. What is new in stroke imaging and intervention? *Clin. Med.* **2018**, *18*, s13–s16. [CrossRef]

17. Nitz, W.R.; Reimer, P. Contrast mechanisms in MR imaging. *Eur. Radiol.* **1999**, *9*, 1032–1046. [CrossRef]

18. Minati, L.; Weglarz, W.P. Physical foundations, models, and methods of diffusion magnetic resonance imaging of the brain: A review. *Concept. Magn. Reson. A* **2007**, *30*, 278–307. [CrossRef]

19. Hawkes, D.J.; Jackson, D.F. An accurate parametrisation of the x-ray attenuation coefficient. *Phys. Med. Biol.* **1980**, *25*, 1167–1171. [CrossRef]

20. van Everdingen, K.J.; van der Grond, J.; Kappelle, L.J.; Ramos, L.M.; Mali, W.P. Diffusion-weighted magnetic resonance imaging in acute stroke. *Stroke* **1998**, *29*, 1783–1790. [CrossRef]

21. Le Bihan, D.; Breton, E.; Lallemand, D.; Grenier, P.; Cabanis, E.; Laval-Jeantet, M. MR imaging of intravoxel incoherent motions: Application to diffusion and perfusion in neurologic disorders. *Radiology* **1986**, *161*, 401–407. [CrossRef] [PubMed]

22. Bauer, S.; Wagner, M.; Seiler, A.; Hattingen, E.; Deichmann, R.; Noth, U.; Singer, O.C. Quantitative T2'-mapping in acute ischemic stroke. *Stroke* **2014**, *45*, 3280–3286. [CrossRef]

23. Duchaussoy, T.; Budzik, J.F.; Norberciak, L.; Colas, L.; Pasquini, M.; Verclytte, S. Synthetic T_2 mapping is correlated with time from stroke onset: A future tool in wake-up stroke management? *Eur. Radiol.* **2019**, *29*, 7019–7026. [CrossRef]

24. Vidmar, J.; Blinc, A.; Sersa, I. A comparison of the ADC and T_2 mapping in an assessment of blood-clot lysability. *NMR Biomed.* **2010**, *23*, 34–40. [CrossRef] [PubMed]

25. Vidmar, J.; Kralj, E.; Bajd, F.; Sersa, I. Multiparametric MRI in characterizing venous thrombi and pulmonary thromboemboli acquired from patients with pulmonary embolism. *J. Magn. Reson. Imaging* **2015**, *42*, 354–361. [CrossRef]

26. Vidmar, J.; Bajd, F.; Milosevic, Z.V.; Kocijancic, I.J.; Jeromel, M.; Sersa, I. Retrieved cerebral thrombi studied by T_2 and ADC mapping: Preliminary results. *Radiol. Oncol.* **2019**, *53*, 427–433. [CrossRef] [PubMed]

27. Viltuznik, R.; Vidmar, J.; Fabjan, A.; Jeromel, M.; Milosevic, Z.V.; Kocijancic, I.J.; Sersa, I. Study of correlations between CT properties of retrieved cerebral thrombi with treatment outcome of stroke patients. *Radiol. Oncol.* **2021**, *55*, 409–417. [CrossRef]

28. Stejskal, E.O.; Tanner, J.E. Spin Diffusion Measurements: Spin Echoes in the Presence of a Time-Dependent Field Gradient. *J. Chem. Phys.* **1965**, *42*, 288–292. [CrossRef]

29. Carr, H.Y.; Purcell, E.M. Effects of Diffusion on Free Precession in Nuclear Magnetic Resonance Experiments. *Phys. Rev. Lett.* **1954**, *94*, 630–638. [CrossRef]

30. Leong, F.J.W.M.; Brady, M.; McGee, J.O. Correction of uneven illumination (vignetting) in digital microscopy images. *J. Clin. Pathol.* **2003**, *56*, 619–621. [CrossRef]

31. Jindal, G.; Miller, T.; Shivashankar, R.; Mitchell, J.; Stern, B.J.; Yarbrough, K.; Gandhi, D. Relationship of thrombus length to number of stent retrievals, revascularization, and outcomes in acute ischemic stroke. *J. Vasc. Interv. Radiol.* **2014**, *25*, 1549–1557. [CrossRef] [PubMed]

32. Fujimoto, M.; Salamon, N.; Mayor, F.; Yuki, I.; Takemoto, K.; Vinters, H.V.; Vinuela, F. Characterization of Arterial Thrombus Composition by Magnetic Resonance Imaging in a Swine Stroke Model. *Stroke* **2013**, *44*, 1463–1465. [CrossRef] [PubMed]

33. Vidmar, J.; Sersa, I.; Kralj, E.; Tratar, G.; Blinc, A. Discrimination between red blood cell and platelet components of blood clots by MR microscopy. *Eur. Biophys. J.* **2008**, *37*, 1235–1240. [CrossRef] [PubMed]

34. Pisani, L. Simple Expression for the Tortuosity of Porous Media. *Transport. Porous. Med.* **2011**, *88*, 193–203. [CrossRef]

35. Brownstein, K.R.; Tarr, C.E. Spin-Lattice Relaxation in a System Governed by Diffusion. *J. Magn. Reson.* **1977**, *26*, 17–24. [CrossRef]

36. Wu, Z.; Mittal, S.; Kish, K.; Yu, Y.; Hu, J.; Haacke, E.M. Identification of calcification with MRI using susceptibility-weighted imaging: A case study. *J. Magn. Reson. Imaging* **2009**, *29*, 177–182. [CrossRef]

37. Velasco Gonzalez, A.; Buerke, B.; Gorlich, D.; Fobker, M.; Rusche, T.; Sauerland, C.; Meier, N.; Jeibmann, A.; McCarthy, R.; Kugel, H.; et al. Clot Analog Attenuation in Non-contrast CT Predicts Histology: An Experimental Study Using Machine Learning. *Transl. Stroke Res.* **2020**, *11*, 940–949. [CrossRef]

38. Riedel, C.H.; Zimmermann, P.; Jensen-Kondering, U.; Stingele, R.; Deuschl, G.; Jansen, O. The importance of size: Successful recanalization by intravenous thrombolysis in acute anterior stroke depends on thrombus length. *Stroke* **2011**, *42*, 1775–1777. [CrossRef]

39. Phinikaridou, A.; Andia, M.E.; Saha, P.; Modarai, B.; Smith, A.; Botnar, R.M. In Vivo Magnetization Transfer and Diffusion-Weighted Magnetic Resonance Imaging Detects Thrombus Composition in a Mouse Model of Deep Vein Thrombosis. *Circ-Cardiovasc. Imaging* **2013**, *6*, 433–440. [CrossRef]

40. Qiao, Y.; Hallock, K.J.; Hamilton, J.A. Magnetization transfer magnetic resonance of human atherosclerotic plaques ex vivo detects areas of high protein density. *J. Cardiovasc. Magn. R* **2011**, *13*, 73. [CrossRef]

41. Hulley, S.B.; Cummings, S.R.; Browner, W.S.; Grady, D.; Newman, T.B. *Designing Clinical Research*, 4th ed.; Lippincott Williams & Wilkins (LWW): Philadelphia, PA, USA, 2013; pp. 55–83.

Journal of
Clinical Medicine

Article

Prospective Assessment of Cerebral Microbleeds with Low-Field Magnetic Resonance Imaging (0.55 Tesla MRI)

Thilo Rusche [1,*], Hanns-Christian Breit [1], Michael Bach [1], Jakob Wasserthal [1], Julian Gehweiler [2], Sebastian Manneck [3], Johanna M. Lieb [1], Gian Marco De Marchis [4], Marios Psychogios [1,†] and Peter B. Sporns [1,5,6,†]

[1] Department of Radiology, Clinic of Radiology & Nuclear Medicine, University Hospital of Basel, University of Basel, 4001 Basel, Switzerland
[2] Imamed Radiologie Nordwest AG, 4051 Basel, Switzerland
[3] Department of Radiology, Gesundheitszentrum Fricktal, 4310 Rheinfelden, Switzerland
[4] Department of Neurology, University Hospital of Basel, University of Basel, 4001 Basel, Switzerland
[5] Department of Diagnostic and Interventional Neuroradiology, University Medical Center Hamburg-Eppendorf, 20251 Hamburg, Germany
[6] Department of Radiology and Neuroradiology, Stadtspital Zürich, 8063 Zürich, Switzerland
[*] Correspondence: rusche.thilo@googlemail.com; Tel.: +41-(0)-762-957-963; Fax: +41-(0)-612-654-354
[†] These authors contributed equally to this work.

Abstract: Purpose: Accurate detection of cerebral microbleeds (CMBs) on susceptibility-weighted (SWI) magnetic resonance imaging (MRI) is crucial for the characterization of many neurological diseases. Low-field MRI offers greater access at lower costs and lower infrastructural requirements, but also reduced susceptibility artifacts. We therefore evaluated the diagnostic performance for the detection of CMBs of a whole-body low-field MRI in a prospective cohort of suspected stroke patients compared to an established 1.5 T MRI. Methods: A prospective scanner comparison was performed including 27 patients, of whom 3 patients were excluded because the time interval was >1 h between acquisition of the 1.5 T and 0.55 T MRI. All SWI sequences were assessed for the presence, number, and localization of CMBs by two neuroradiologists and additionally underwent a Likert rating with respect to image impression, resolution, noise, contrast, and diagnostic quality. Results: A total of 24 patients with a mean age of 74 years were included (11 female). Both readers detected the same number and localization of microbleeds in all 24 datasets (sensitivity and specificity 100%; interreader reliability $\varkappa = 1$), with CMBs only being observed in 12 patients. Likert ratings of the sequences at both field strengths regarding overall image quality and diagnostic quality did not reveal significant differences between the 0.55 T and 1.5 T sequences ($p = 0.942$; $p = 0.672$). For resolution and contrast, the 0.55 T sequences were even significantly superior ($p < 0.0001$; $p < 0.0003$), whereas the 1.5 T sequences were significantly superior ($p < 0.0001$) regarding noise. Conclusion: Low-field MRI at 0.55 T may have similar accuracy as 1.5 T scanners for the detection of microbleeds and thus may have great potential as a resource-efficient alternative in the near future.

Keywords: low-field MRI; MRI; reading study; scanner comparison; cerebral microbleeds; 0.55 T

Citation: Rusche, T.; Breit, H.-C.; Bach, M.; Wasserthal, J.; Gehweiler, J.; Manneck, S.; Lieb, J.M.; De Marchis, G.M.; Psychogios, M.; Sporns, P.B. Prospective Assessment of Cerebral Microbleeds with Low-Field Magnetic Resonance Imaging (0.55 Tesla MRI). *J. Clin. Med.* **2023**, *12*, 1179. https://doi.org/10.3390/jcm12031179

Academic Editor: Giovanni Rizzo

Received: 19 December 2022
Revised: 22 January 2023
Accepted: 31 January 2023
Published: 2 February 2023

1. Introduction

Cerebral microbleeds (CMBs) are small (2–10 mm diameter), round, or ovoid hypointense foci with associated blooming with enhanced visibility on MRI sequences sensitive to susceptibility artifacts [1,2]. They can be observed in patients with cognitive complaints and stroke but also in healthy individuals. Technically, local magnetic field inhomogeneities caused by paramagnetic iron in CMBs result in signal loss on MRI sequences, such as T2*-weighted gradient echo sequences or susceptibility-weighted imaging (SWI). SWI is derived from gradient echo sequences with additional post-processing to improve contrast resolution and is usually acquired in three dimensions to increase spatial resolution with flow compensation in all three planes to reduce artifacts. SWI has increased sensitivity and reliability for CMBs compared with T2*-weighted gradient echo but requires a longer acquisition time [3]. Further, technical aspects such as a low flip angle, long echo time, and long repetition time increase the sensitivity to susceptibility effects [2]. Moreover, the susceptibility effect and signal-to-noise ratio have been described to increase with higher magnetic field strength [4].

On the other hand, MRI accessibility is low and extremely inhomogeneous around the world, because MRI installations require expensive infrastructure (e.g., site preparation to host the large magnets, magnetic/radiofrequency shielding, and emergency helium exhaust conduit), high maintenance costs (i.e., for helium refill), have high operational costs for specialized radiographic technicians, and require huge amounts of electricity and water leaving a large ecological footprint [5,6]. Thus, the distribution of MRI scanners is concentrated mainly within high-income countries, and ~70% of the world's population has little to no access to MRI (OECD (2022), magnetic resonance imaging (MRI) units (indicator); DOI: 10.1787/1a72e7d1-en). Moreover, even in high-income countries, clinical MRI scanners are mostly located in highly specialized radiology departments, large and centralized imaging centers, and housed on the ground floors of hospitals and clinics, excluding easy access to neurology clinics, trauma centers, surgical suites, neonatal/pediatric centers, and community clinics [5]. With this in mind, low-field MRI is increasingly coming into focus, offering MR imaging at a much lower cost, offering a considerable energy-saving potential but also reducing possible complications with metallic implants. Therefore, the aim of our study was to evaluate the performance of a 0.55 T low-field MRI in a prospective cohort of suspected stroke patients and to directly compare the diagnostic value for the detection of CMBs to a standard 1.5 T MRI.

2. Materials and Methods

This prospective study was reviewed and approved by the cantonal (Basel-Stadt, Switzerland) ethics committee (BASEC2021-00166). All included patients signed an informed consent form.

2.1. Data Acquisition

Data acquisition was performed from 1 May 2021 to 30 June 2021 at the Division of Neuroradiology, Clinic of Radiology and Nuclear Medicine, University Hospital Basel, Switzerland. All patients who presented to the emergency room with suspected ischemic stroke or transient ischemic attack (TIA) and who underwent MRI using a 1.5 T scanner (Siemens MAGNETOM Avanto FIT 1.5 T; Siemens Healthineers; Erlangen, Germany) as part of the diagnostic stroke workup were included. Immediately afterwards, patients were examined using a 0.55 T scanner (Siemens MAGNETOM FreeMax 0.55 T; Siemens Healthineers; Erlangen, Germany). The 1.5 T scanning protocol was in accordance with the hospital's internal standard protocol for emergency stroked diagnostics including SWI sequences (Table 1). The 0.55 T SWI protocol was adapted to the 1.5 T SWI protocol as far as technically possible (same slice thickness (ST) and slice spacing (SP); comparable in-plane resolution) to ensure the most objective scanner comparison. After subsequent verification with respect to data completeness (scan protocols with complete image acquisition) and image quality (artifacts, image contrast), the datasets were transferred to the Picture Archiving and Communication System (PACS; General Electric (GE); Waukesha, WI, USA) for further analysis.

2.2. Data Analysis

Data analysis was performed in a two-step procedure: First, the 0.55 T and 1.5 T SWI datasets were evaluated using Likert rating. Second, a reading study was performed regarding identification, localization, and number of CMBs.

Table 1. Scan protocols Siemens MAGNETOM Avanto FIT 1.5 T and Siemens MAGNETOM FreeMax 0.55 T.

	Siemens MAGNETOM FreeMax 0.55 T	Siemens MAGNETOM Avanto Fit 1.5 T
FLAIR tra		
Field strength in T	0.55	1.5
Field of view (FOV) in mm^2	209 × 230	187 × 230
Slice thickness (ST) in mm	3	3
Slice spacing (SS)	3.6	3.6
Number of slices	40	40
Pixel spacing (PS) in mm^2	1.28 × 1.03	0.9 × 0.9
Repetition time (TR) in msec	7780	8510
Echo time (TE) in msec	96	112
Inversion delay (TI) in msec	2368.8	2460
Turbo factor	15	19
Time of acquisition (TA) in min	05:28	03:26
BW ((BW))	150	130
3D SWI tra		
Field strength in T	0.55	1.5
Sequence type	Multi-shot 3D EPI	3D FLASH
Field of view (FOV) in mm^2	201 × 230	194 × 230
Slice thickness (ST) in mm	3	3
Number of slices	40	48
Pixel spacing (PS) in mm^2	0.94 × 0.8	1.12 × 0.9
Repetition time (TR) in msec	172	48
Echo time (TE) in msec	100	40
Parallel imaging	-	GRAPPA factor 2
Time of acquisition (TA) in min	02:23	02:17
BW ((BW))	276	80
Single-shot diffusion EPI tra		
Field strength in T	0.55	1.5
Field of view (FOV) in mm^2	220 × 220	230 × 230
Slice thickness (ST) in mm	3	3
Slice spacing (SS)	3.6	3.6
Number of slices	40	40
Pixel spacing (PS) in mm^2	1.67 × 1.67	1.44 × 1.44
b-values in s/mm^2	0, 1000	0, 1000
Repetition time (TR) in msec	7400	6200
Echo time (TE) in msec	102	103
Parallel imaging	GRAPPA factor 2	GRAPPA factor 2
Time of acquisition (TA) in min	04:35	02:04
BW ((BW))	842	1490

2.3. Likert Rating

Likert rating was performed by a neuroradiologist and a neuroradiologist in training with an experience of 9 and 5 years. Each acquired 0.55 T or 1.5 T SWI dataset was rated with respect to the following criteria with a numerical value between 1 and 10 (1 worst, 10 best):

(a) Overall image quality;
(b) Resolution;
(c) Noise;
(d) Contrast;
(e) Diagnostic quality.

Sample SWI sequences from a 3 T scanner (Siemens MAGNETOM Skyra 3 T; Siemens Healthineers; Erlangen, Germany) were set as the gold standard (numerical value = 10). Dataset assessment was PACS-based using a standardized bookmark. Both readers were blinded to the results of the other reader.

2.4. Reading Study

Reading of 0.55 T and 1.5 T datasets was performed PACS-based and blinded (no clinical information, no image information) by two neuroradiologists with 8 and 13 years of professional experience. PACS-based post-processing as part of image analysis was allowed. For each dataset, number of SWI lesions (0, 1, 2–10, >10) and SWI lesion localization were analyzed.

The final neuroradiological report was defined as the underlying gold standard for the accuracy of the reading study.

3. Statistical Analysis

For statistical evaluation of Likert rating, a mean of the ratings of Readers 1 and 2 was first calculated for each 0.55 T and 1.5 T patient dataset and evaluation points (a)–(e). Subsequently, a Wilcoxon signed-rank test was used to evaluate significant or non-significant differences in Likert rating between the 0.55 T and 1.5 T sequences. Then interreader comparisons were performed to determine intraclass correlation coefficients (ICC). Calculation of sensitivity and specificity of Readers 1 and 2 in the reading study was performed in relation to the gold standard.

4. Results

A total of 27 patients with complete and artifact-free datasets were prospectively included in this study, of whom 3 patients were excluded because the time interval between the 1.5 T scan and 0.55 T scan was >1 h. The mean age of the remaining 24 patients was 74 years (standard deviation 14 years), and 11 patients were female (46%). The mean time interval between the 1.5 T scans and 0.55 T scans was 36 ± 14 min. Baseline characteristics of the patient cohort are presented in Table 2.

Table 2. Detailed patient data.

Patient	Patient Age	CMB Yes/No	Number of CMB	(Main)-Localization of CMB	Time Gap between Scans in min
Patient 1	87	Yes	1	Left occipital	46
Patient 2	73	Yes	3	Right frontal/periventricular	25
Patient 3	88	Yes	4	Right temporal/parietal	37
Patient 4	29	No	0	-	33
Patient 5	82	No	0	-	93
Patient 6	70	No	0	-	44
Patient 7	87	Yes	2	Left occipital	32
Patient 8	74	No	0	-	25
Patient 9	60	No	0	-	21
Patient 10	44	No	0	-	49
Patient 11	84	Yes	1	Left occipital	33
Patient 12	58	No	0	-	40
Patient 13	80	No	0	-	35
Patient 14	65	Yes	1	Left Putamen	20
Patient 15	65	Yes	2	Left frontal	24
Patient 16	75	No	0	-	22
Patient 17	84	No	0	-	48
Patient 18	82	Yes	4	Left frontal	32
Patient 19	79	Yes	1	Left occipital	42
Patient 20	84	Yes	1	Left periventricular	32
Patient 21	86	Yes	>10	Bilateral Thalamus	25
Patient 22	83	No	0	-	31
Patient 23	89	No	0	-	38
Patient 24	69	Yes	3	Left periventricular	42
Patient 25	53	Excluded	-	-	916
Patient 26	59	Excluded	-	-	2936
Patient 27	46	Excluded	-	-	2812

Both readers detected the same number and localization of microbleeds in all 24 0.55 T and 1.5 T datasets (sensitivity and specificity 100%; interreader reliability $\varkappa = 1$). Ten 1.5 T datasets did not contain any microbleeds by analysis of the more experienced neuroradiologist and were therefore defined as the control group. No false-positive findings were observed by assessment of these images by Reader 2 (positive predictive value and negative predictive value 100%).

Likert ratings of the sequences with both field strengths regarding overall image quality (a) and diagnostic quality (e) did not reveal significant differences between the 0.55 T and 1.5 T sequences ($p = 0.942$ and $p = 0.672$, respectively; see Figure 1). Regarding the subjective evaluation of the spatial resolution (b) and contrast (d), the 0.55 T sequences were rated to be significantly superior: (b) $p < 0.0001$; (d) $p < 0.0003$. In contrast, the 1.5 T sequences were superior ($p < 0.0001$) regarding noise (c). Interreader comparisons showed moderate to high levels of agreement between Reader 1 and Reader 2 for the Likert ratings (ICC: (a) 0.91; (b) 0.93; (c) 0.60; (d) 0.87; (e) 0.88).

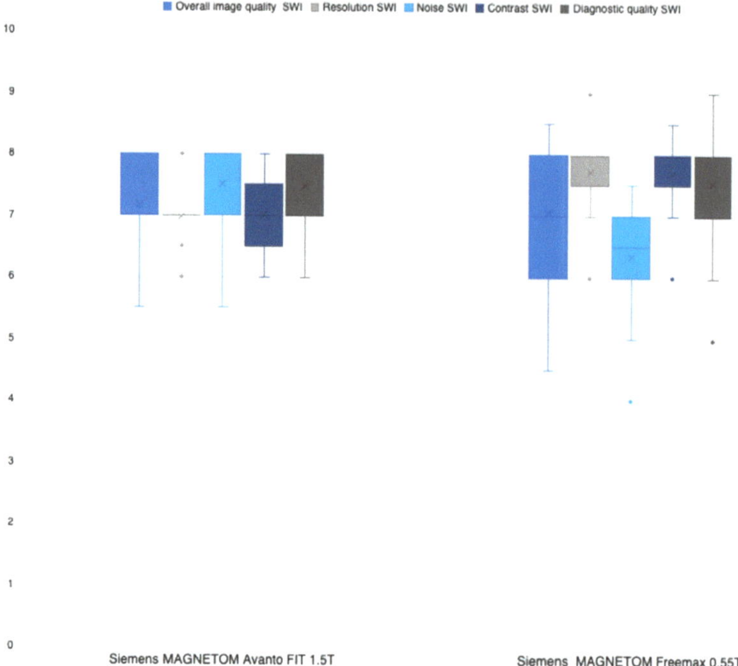

Figure 1. Average Likert scoring of Readers 1 and 2 SWI sequences.

5. Discussion

Our study shows that detection of microbleeds in SWI MRI sequences at a field strength of 0.55 T is possible with the same specificity and sensitivity compared to conventional 1.5 T MRI (for sample sequences, see Figure 2). Moreover, Likert ratings for the subjective evaluation of spatial resolution and contrast resolution were not significantly different for both 0.55 T and 1.5 T sequences, whereas for SWI noise, 0.55 T sequences were even superior.

To the best of our knowledge, no comparable study has performed a 0.55 T versus 1.5 T scanner comparison regarding the detection of microbleeds before. In a previous study, we showed that low-field MRI is not inferior to scanners with higher field strength for the detection of small infarcts in DWI and FLAIR sequences; however, in this study, there were minor limitations in the detection of very small infarcts [7]. Another group evaluated the performance of a modern 0.55 T MRI in the diagnosis of intracranial aneurysms in comparison to the gold standard digital subtraction angiography (DSA) [8]. This study included a total of 19 aneurysms in 16 patients, which were identified in both 0.55 T magnetic resonance angiography and DSA. Moreover, measurements of the two readers showed no significant differences between 0.55 T TOF MRA and DSA in the overall aneurysm size (calculated as the mean from height/width/neck), as well as in the mean width and neck values. The mean height was significantly larger in 0.55 T TOF MRA in comparison to DSA, whereas intermodality (1.5 T and 3 T TOF MRA) and interrater agreement were excellent (ICC > 0.94). Thus, the authors concluded that TOF MRA acquired with a modern 0.55 T MRI is a reliable tool for the detection and initial assessment of intracranial aneurysms. Moreover, in another study, we showed that patients perceived 0.55 T new-generation low-field MRI to be more comfortable than conventional 1.5 T MRI, given its larger bore opening and reduced noise levels during image acquisition, and concluded that new concepts regarding bore design and noise level reduction of MR scanner systems may help to reduce patient anxiety and improve well-being when undergoing MR imaging [9]. For microbleeds, the diagnostic accuracy of SWI sequences of the 0.55 T Magnetom FreeMax in our study seemed to be even higher, offering great potential for the characterization of associated diseases such as diffuse axonal injury or cerebral amyloid angiopathies [8].

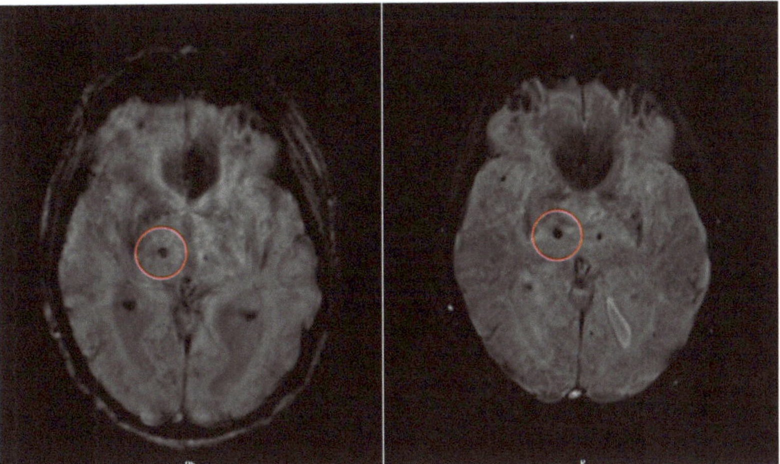

Figure 2. SWI lesion of the right thalamus. In both images (left, axial 1.5 T SWI sequence; right, axial 0.55 T SWI sequence), this lesion is clearly detectable, although it is even better delineated in the 0.55 T dataset.

In principle, the use of low-field MRI in clinical routine has several advantages. First, accessibility is low and extremely inhomogeneous around the world because MRI installations require expensive infrastructure, so the distribution of MRI scanners is concentrated mainly within high-income countries, and ~70% of the world's population has little to no access to MRI [5]. Moreover, even in high-income countries, clinical MRI scanners are mostly located in highly specialized radiology departments, large and centralized imaging centers, and in those are housed on the ground floors of hospitals [5]. In a recent analysis by our group, we could show that in terms of purchase price, the savings potential of a 0.55 T MRI compared to a 1.5 T MRI system is about 40–50% [10]. The 25% lower weight of the system additionally reduces the transportation costs incurred, and the smaller size of the unit allows for installation by a remotely controlled mobile robotic system without opening the exterior façade, if the operating site is at ground level. Together with the lack of need to install a quench pipe, this reduces the total cost of installation by up to 70%. The maintenance cost of a 0.55 T MRI is approximately 45% less than that of a 1.5 T unit with a comparable service contract. Further cost reductions result from the smaller room size and potentially lower energy consumption for examinations and cooling. In conclusion, the use of lower-field-strength MRI systems offers enormous economic and environmental potential for both hospitals and practice operators, as well as for the healthcare system as a whole. In this context, offering MR imaging at a lower cost and with fewer infrastructural requirements will be key to increasing access to MRI for many patients. In developed countries, the energy-saving potential [6], and the possibility to reduce complications and artifacts caused by metallic implants will be the major arguments for implementing low-field MR imaging. The increasing exploration of the potential and limitations of low-field MRI is crucial to guide this implementation at a larger scale without harming patients.

Limitations

There are several limitations that need to be addressed. First, a 1.5 T device of routine clinical use is in fact not the "gold standard" for the detection of CMBs. Various studies have shown that higher field strengths (3 T and 7 T) have the highest sensitivity for the detection of CMBs [2,4,11–17]. For example, Conijn et al. [4] showed that the detection of CMBs is more reliable at 7 T compared to 1.5 T. Stehling et al. [18] had similar results when comparing 1.5 T versus 3 T. Greenberg et al. [2] also described better CMB detection at higher field strengths (3 T or higher), as CMBs are better visible in this case due to stronger susceptibility artifacts and thus stronger blooming artifacts. These results were most recently supported by data from Bian et al. [14], who also confirmed a higher sensitivity in the diagnosis of radiotherapy-induced CMBs at 7 T compared to 3 T sequences. For this reason, the detection of CBMs with lower field strengths (0.55 T and 1.5 T) is already a priori suboptimal, and thus the definition of a 1.5 T scanner as a comparison scanner and "gold standard" is afflicted with deficiencies. Thus, future studies should compare 0.55 T versus 3 T or 7 T. Second, both scanners differ with respect to their gradient and coil system as well as the field strength. Third, the study cohort (only 12 patients with CMBs), nonetheless prospective, is still relatively small. Larger-scale studies to further define indications for the detection of microbleeds at 0.55 T are needed and should assess whether scanner choice has an impact on patient outcomes. Fourth, results may have been different if the study cohort would have been selected from a population where a high CMB burden is already likely due to expected or diagnosed underlying diseases (cerebral amyloid angiopathy (CAA) or Alzheimer's disease). Thus, more CMBs would be detectable and comparable in the collective as a whole.

6. Conclusions

Low-field MRI at 0.55 Tesla may have the same accuracy as 1.5 T MRI for the detection of microbleeds and thus may have great potential as a low-cost alternative in the near future.

Author Contributions: Conceptualization, T.R. and P.B.S.; Data curation, T.R., H.-C.B., M.B. and J.W.; Formal analysis, T.R., P.B.S., H.-C.B., J.G., S.M. and J.M.L.; Funding acquisition, T.R. and P.B.S.; Investigation, T.R.; Methodology, T.R., M.B. and P.B.S.; Project administration, T.R. and P.B.S.; Resources, T.R. and P.B.S.; Software, T.R. and J.W.; Supervision, T.R. and P.B.S.; Validation, T.R. and P.B.S.; Visualization, T.R. and P.B.S.; Writing—original draft, T.R.; Writing—review and editing, T.R., H.-C.B., M.B., G.M.D.M., M.P. and P.B.S. All authors have read and agreed to the published version of the manuscript.

Funding: This research received no external funding.

Institutional Review Board Statement: The study was approved under the provisions of the local ethics committee (BASEC2021-00166) in accordance with the Declaration of Helsinki.

Informed Consent Statement: Informed consent was obtained from all subjects involved in the study.

Data Availability Statement: The data presented in this study are available on request from the corresponding author.

Conflicts of Interest: The authors declare no conflict of interest.

Abbreviations

ADC	apparent diffusion coefficient
bSSFP	balanced steady-state free precession
BW	bandwidth
CAA	cerebral amyloid angiopathy
CMB	cerebral microbleed
CNR	contrast-to-noise ratio
CT	computed tomography
DWI	diffusion-weighted imaging
FLAIR	fluid-attenuated inversion recovery
ICC	intraclass correlation coefficient
MRI	magnetic resonance imaging
NIHSS	National Institutes of Health Stroke Scale
PACS	Picture Archiving and Communication System
QALY	quality-adjusted life-years
SNR	signal-to-noise ratio
SP	slice spacing
ST	slice thickness
SWI	susceptibility-weighted imaging
TIA	transient ischemic attack

References

1. Fazekas, F.; Kleinert, R.; Roob, G.; Kapeller, P.; Schmidt, R.; Hartung, H.P. Histopathologic analysis of foci of signal loss on gradient-echo T2*-weighted MR images in patients with spontaneous intracerebral hemorrhage: Evidence of microangiopathy-related microbleeds. *Am. J. Neuroradiol.* **1999**, *20*, 637–642. [PubMed]
2. Greenberg, S.M.; Vernooij, M.W.; Cordonnier, C.; Viswanathan, A.; Al-Shahi Salman, R.; Warach, S.; Launer, L.J.; Van Buchem, M.A.; Breteler, M.M.; Microbleed Study Group. Cerebral microbleeds: A guide to detection and interpretation. *Lancet Neurol.* **2009**, *8*, 165–174. [CrossRef] [PubMed]
3. Shams, S.; Martola, J.; Cavallin, L.; Granberg, T.; Shams, M.; Aspelin, P.; Wahlund, L.O.; Kristoffersen-Wiberg, M. SWI or T2*: Which MRI sequence to use in the detection of cerebral microbleeds? The Karolinska Imaging Dementia Study. *Am. J. Neuroradiol.* **2015**, *36*, 1089–1095. [CrossRef] [PubMed]
4. Conijn, M.M.A.; Geerlings, M.I.; Biessels, G.J.; Takahara, T.; Witkamp, T.D.; Zwanenburg, J.J.M.; Luijten, P.R.; Hendrikse, J. Cerebral microbleeds on MR imaging: Comparison between 1.5 and 7T. *Am. J. Neuroradiol.* **2011**, *32*, 1043–1049. [CrossRef]
5. Liu, Y.; Leong, A.T.L.; Zhao, Y.; Xiao, L.; Mak, H.K.F.; Tsang, A.C.O.; Lau, G.K.K.; Leung, G.K.K.; Wu, E.X. A low-cost and shielding-free ultra-low-field brain MRI scanner. *Nat. Commun.* **2021**, *12*, 7238. [CrossRef] [PubMed]
6. Heye, T.; Knoerl, R.; Wehrle, T.; Mangold, D.; Cerminara, A.; Loser, M.; Plumeyer, M.; Degen, M.; Lüthy, R.; Brodbeck, D.; et al. The energy consumption of Radiology: Energy- And cost-saving opportunities for CT and MRI operation. *Radiology* **2020**, *295*, 593–605. [CrossRef] [PubMed]
7. Rusche, T.; Breit, H.C.; Bach, M.; Wasserthal, J.; Gehweiler, J.; Manneck, S.; Lieb, J.M.; De Marchis, G.M.; Psychogios, M.N.; Sporns, P.B. Potential of Stroke Imaging Using a New Prototype of Low-Field MRI: A Prospective Direct 0.55 T/1.5 T Scanner Comparison. *J. Clin. Med.* **2022**, *11*, 2798. [CrossRef] [PubMed]
8. Osmanodja, F.; Rösch, J.; Knott, M.; Doerfler, A.; Grodzki, D.; Uder, M.; Heiss, R. Diagnostic Performance of 0.55 T MRI for Intracranial Aneurysm Detection. *Investig. Radiol.* **2023**, *58*, 121–125. [CrossRef] [PubMed]
9. Rusche, T.; Vosshenrich, J.; Winkel, D.J.; Donners, R.; Segeroth, M.; Bach, M.; Merkle, E.M.; Breit, H.-C. More Space, Less Noise-New-generation Low-Field Magnetic Resonance Imaging Systems Can Improve Patient Comfort: A Prospective 0.55 T–1.5 T-Scanner Comparison. *J. Clin. Med.* **2022**, *11*, 6705. [CrossRef] [PubMed]
10. Vosshenrich, J.; Breit, H.C.; Bach, M.; Merkle, E.M. Economic aspects of low-field magnetic resonance imaging: Acquisition, installation, and maintenance costs of 0.55 T systems. *Radiologe* **2022**, *62*, 400–404. [CrossRef] [PubMed]
11. Charidimou, A.; Shams, S.; Romero, J.R.; Ding, J.; Veltkamp, R.; Horstmann, S.; Eiriksdottir, G.; van Buchem, M.A.; Gudnason, V.; Himali, J.J.; et al. Clinical significance of cerebral microbleeds on MRI: A comprehensive meta-analysis of risk of intracerebral hemorrhage, ischemic stroke, mortality, and dementia in cohort studies (v1). *Int. J. Stroke* **2018**, *13*, 454–468. [CrossRef] [PubMed]
12. Charidimou, A.; Imaizumi, T.; Moulinm, S.; Biffi, A.; Samarasekera, N.; Yakushiji, Y.; Peeters, A.; Vandermeeren, Y.; Laloux, P.; Baron, J.C.; et al. Brain hemorrhage recurrence, small vessel disease type, and cerebral microbleeds: A meta-analysis. *Neurology* **2017**, *89*, 820–829. [CrossRef] [PubMed]
13. Charidimou, A.; Werring, D.J. Cerebral microbleeds and cognition in cerebrovascular disease: An update. *J. Neurol. Sci.* **2012**, *322*, 50–55. [CrossRef] [PubMed]
14. Bian, W.; Hess, C.P.; Chang, S.M.; Nelson, S.J.; Lupo, J.M. Susceptibility-weighted MR imaging of radiation therapy-induced cerebral microbleeds in patients with glioma: A comparison between 3 T and 7 T. *Neuroradiology* **2014**, *56*, 91–96. [CrossRef]

15. Hütter, B.O.; Altmeppen, J.; Kraff, O.; Maderwa, S.; Theysohn, J.M.; Ringelstein, A.; Wrede, K.H.; Dammann, P.; Quick, H.H.; Schlamann, M.; et al. Higher sensitivity for traumatic cerebral microbleeds at 7 T ultra-high field MRI: Is it clinically significant for the acute state of the patients and later quality of life? *Ther. Adv. Neurol. Disord.* **2020**, *13*, 1756286420911295. [CrossRef] [PubMed]
16. Conijn, M.M.; Hoogduin, J.M.; van der Graaf, Y.; Hendrikse, J.; Luijten, P.R.; Geerlings, M.I. Microbleeds, lacunar infarcts, white matter lesions and cerebrovascular reactivity—A 7 T study. *Neuroimage* **2012**, *59*, 950–956. [CrossRef]
17. Theysohn, J.M.; Kraff, O.; Maderwald, S.; Barth, M.; Ladd, S.C.; Forsting, M.; Ladd, M.E.; Gizewski, E.R. 7 tesla MRI of microbleeds and white matter lesions as seen in vascular dementia. *J. Magn. Reson. Imaging* **2011**, *33*, 782–791. [CrossRef] [PubMed]
18. Stehling, C.; Wersching, H.; Kloska, S.P.; Kirchhof, P.; Ring, J.; Nassenstein, I.; Allkemper, T.; Knecht, S.; Bachmann, R.; Heindel, W. Detection of asymptomatic cerebral microbleeds: A comparative study at 1.5 and 3.0 T. *Acad. Radiol.* **2008**, *15*, 895–900. [CrossRef] [PubMed]

Journal of
Clinical Medicine

Article

Location of Hyperintense Vessels on FLAIR Associated with the Location of Perfusion Deficits in PWI

Lisa D. Bunker [1,*] and Argye E. Hillis [2]

[1] Department of Neurology, Johns Hopkins University School of Medicine, Baltimore, MD 21287, USA
[2] Departments of Cognitive Science, Physical Medicine & Rehabilitation, and Neurology, Johns Hopkins University School of Medicine, Baltimore, MD 21287, USA
* Correspondence: lbunker3@jhmi.edu; Tel.: +1-443-287-6124

Abstract: Perfusion imaging is preferred for identifying hypoperfusion in the management of acute ischemic stroke, but it is not always feasible/available. An alternative method for quantifying hypoperfusion, using FLAIR-hyperintense vessels (FHVs) in various vascular regions, has been proposed, with evidence of a statistical relationship with perfusion-weighted imaging (PWI) deficits and behavior. However, additional validation is needed to confirm that areas of suspected hypoperfusion (per the location of FHVs) correspond to the location of perfusion deficits in PWI. We examined the association between the location of FHVs and perfusion deficits in PWI in 101 individuals with acute ischemic stroke, prior to the receipt of reperfusion therapies. FHVs and PWI lesions were scored as present/absent in six vascular regions (i.e., the ACA, PCA, and (four sub-regions of) the MCA territories). Chi-square analyses showed a significant relationship between the two imaging techniques for five vascular regions (the relationship in the ACA territory was underpowered). These results suggest that for most areas of the brain, the general location of FHVs corresponds to hypoperfusion in those same vascular territories in PWI. In conjunction with prior work, results support the use of estimating the amount and location of hypoperfusion using FLAIR imaging when perfusion imaging is not available.

Keywords: acute stroke; hypoperfusion; perfusion-weighted imaging (PWI); fluid-attenuated inversion recovery (FLAIR); hyperintense vessels

Citation: Bunker, L.D.; Hillis, A.E. Location of Hyperintense Vessels on FLAIR Associated with the Location of Perfusion Deficits in PWI. *J. Clin. Med.* **2023**, *12*, 1554. https://doi.org/10.3390/jcm12041554

Academic Editors: Georgios Tsivgoulis, Gabriel Broocks and Lukas Meyer

Received: 15 November 2022
Revised: 8 February 2023
Accepted: 12 February 2023
Published: 16 February 2023

1. Introduction

In acute ischemic stroke, current clinical standards in diagnosis and intervention planning include perfusion-weighted MRI or CT. These diagnostic tools can be used in various ways (e.g., to identify the core volume or a mismatch ratio across various imaging sequences) to rapidly identify if there is hypoperfused tissue that may benefit from reperfusion interventions, such as intravenous tissue-type plasminogen activator (IV tPA) or mechanical thrombectomy (MT). Furthermore, they can be useful to corroborate a neurological exam, or for research, to investigate brain–behavior relationships, and identify the areas of dysfunction (hypoperfusion and/or infarct) associated with specific deficits in acute stroke, before the opportunity for reorganization or recovery [1–3]. However, perfusion-weighted imaging may not be feasible because of contraindication for contrast agents [4–6] or other patient factors, or may not be useful due to technical difficulties. Arterial spin labelling (ASL) sequences are another option for identifying perfusion deficits, but ASL is technically complicated and sensitive to motion artifact [7–9], and thus may be difficult to acquire and interpret.

Reyes and colleagues proposed an alternative method for identifying hypoperfusion using fluid-attenuated inversion recovery (FLAIR) MRI sequences [10], which are routinely collected in stroke MRI protocols. Perfusion abnormalities on FLAIR present as a hyperintense signal in the arterial vessels, which appears as serpentine structures and/or isolated bright spots in the sulci (see Figure 1). FLAIR-hyperintense vessels (FHVs) are indicative of reduced blood flow [11,12], and thus may serve as an alternative avenue for the identification and quantification of hypoperfusion. Indeed, Reyes et al. developed the National Institutes of Health FHV (NIH-FHV) score to quantify the number and location of FHVs (in terms of the anterior cerebral artery (ACA), posterior cerebral artery (PCA), and four regions of the middle cerebral artery (MCA) territory). They demonstrated a strong significant association between the NIH-FHV scores and PWI lesion volume (i.e., one point on the NIH-FHV scale equated to about 12 mL of hypoperfusion on PWI). When controlling for lesion volume on diffusion-weighted imaging, the NIH-FVH was highly sensitive at detecting a mismatch ratio ≥ 1.8. However, Reyes and colleagues did not examine whether the location of FHVs corresponded to the location of the hypoperfusion in PWI.

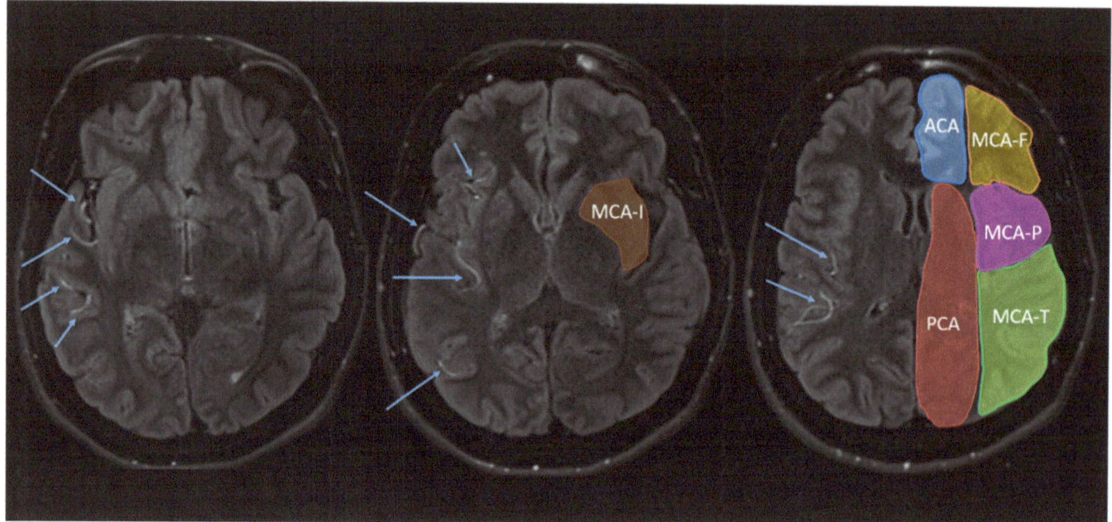

Figure 1. Examples of FHVs (indicated by the arrows) and vascular regions used for the NIH-FHV score: the anterior cerebral artery (ACA, blue), posterior cerebral artery (PCA, red), and four regions of the middle cerebral artery (MCA)—the MCA-frontal (MCA-F, orange), MCA-temporal (MCA-T, green), MCA-parietal (MCA-P, purple), and MCA-insular (MCA-I, brown). Note that no FLAIR imaging for the current study was shared by the NIH. Thus, this example of FHVs is from a participant in Bunker et al. [13].

To aid in the clinical interpretation of FHVs, it is necessary to establish a relationship between the location(s) of FHVs compared to the regions of the hypoperfusion identified using dynamic contrast PWI. That is, do FHVs in a specific vascular area actually correspond to a perfusion deficit in that same area, as seen in PWI? Without such validation, the NIH-FHV scale could not be reliably used in clinical practice or research. In a previous smaller study ($n = 73$), we found statistically significant relationships for FHV/PWI lesions in each vascular area, except the ACA territory [13], presumably due to low power (only three participants demonstrated FHVs in the ACA territory). The purpose of the current investigation was to examine the association between the location of FHVs on FLAIR and hypoperfusion in PWI in a larger cohort (i.e., the same cohort studied by Reyes et al.), to hopefully replicate our previous findings, and potentially identify an association for the ACA territory. The identification of significant relationships in the location of the hypoperfusion identified on the different MRI sequences will improve the clinical and research utility of the NIH-FHV score when PWI is not available.

2. Materials and Methods

PWI and FHV rating data are from the National Institutes of Health (NIH) Natural History of Stroke study (for protocol, see NCT00009243, https://www.clinicaltrials.gov). Enrollment and data collection (described in Reyes et al. [10]) was approved by the NIH institutional review board (IRB) and all participants or legally authorized representatives provided informed consent as per the Declaration of Helsinki. The current analysis, using de-identified data provided by the NIH, was approved by the Johns Hopkins Medicine IRB.

2.1. Participants

A total of 101 participants (53 male sex, 48 female sex; median (range) age = 73 (58–83)) with hyperacute unilateral (right or left) ischemic stroke were included in this analysis. They all completed an MRI with diffusion-weighted imaging (DWI), FLAIR and PWI, prior to receiving IV tPA or MT interventions. See Table 1 for additional sample characteristics (such as stroke severity, lesion volume, premorbid hypertension, hyperlipidemia, smoking history, etc.).

Table 1. Participant characteristics.

Summary Statistics (n = 101)	
Age (Median (Range))	73 (58–83)
Sex (female, male)	48, 53
PWI volume (mL; *M(SD)*)	37 (±56)
DWI volume (mL; *M(SD)*) †	18 (±24)
NIH-FHV (median (range))	4 (1–6)
NIHSS (median (range))	8 (4–17)
Stroke Risk Factors	
HTN	n = 73 (72.3%)
HLD	n = 34 (33.7%)
DM	n = 20 (19.8%)
CAD	n = 15 (14.9%)
A-Fib	n = 31 (30.7%)
Smoking	n = 12 (11.9%)

Notes: PWI volume = volume of hypoperfusion on perfusion-weighted imaging; DWI = infarct volume on diffusion-weighted imaging; NIH-FHV = National Institutes of Health FLAIR-Hyperintense Vessel score; NIHSS = National Institutes of Health Stroke Scale; HTN = hypertension; HLD = hyperlipidemia; DM = diabetes; CAD = coronary artery disease; A-Fib = arterial fibrillation. † DWI lesion calculated from the number of voxels within the PWI lesion (after co-registration) below 620 μm^2/s [10].

2.2. Imaging Acquisition

MRIs were acquired using 1.5T (GE Signa Scanner, General Electric Medical Systems) or 3T (Philips Achieva, Philips Healthcare; Siemens Skyra, Siemens AG) scanners. Parameters for FLAIR acquisition were: repetition time (TR), 9000 ms; echo time (TE), 120–145 ms; 3.5 mm slice thickness (40 slices). Dynamic susceptibility contrast PWI parameters were: TR, 1–1.5 s; TE, 25–45 ms; 7 mm slice thickness (20 slices with a full brain coverage); 40–80 dynamics. For PWI, participants received 0.1 mmol/kg of gadolinium–diethylenetriamine penta-acetic acid (Magnevist, Bayer Schering Pharma) or gadobenic acid MultiHance (Bracco Diagnostics) contrast at 5 mL/s flow rate.

2.3. Perfusion Deficit Identification

2.3.1. FLAIR-Hyperintense Vessels

For the initial NIH-FHV scoring, FLAIR images were reviewed slice by slice and the presence of FHVs was scored from 0 to 2 in each of the six vascular regions in the lesioned hemisphere. A score of '0' indicated no FHVs, a score of '1' indicated 1–2 FHVs on 1–2 slices, and a score of '2' indicated 3+ FHVs on one slice or 3+ slices with FHVs. Scores for each region were summed for a total score out of 12 for the hemisphere, with higher scores suggestive of a greater volumes of hypoperfusion. The six vascular regions (see Figure 1) were the ACA territory, the PCA territory, and four sub-regions of the MCA territory (frontal (i.e., MCA-F), temporal (i.e., MCA-T), parietal (i.e., MCA-P), and insular (i.e., MCA-I)). Since the purpose of this study was simply to identify the association between the location of hypoperfusion across FLAIR and PWI MR techniques, these scores were recoded as a binary variable, indicating that FHVs—or hypoperfusion—were/was present or absent in each region. Hence, a score of 1–2 in any of the six regions was recoded as 1, for 'present', or 0, for 'absent.' The relationship between total FHV scores (i.e., severity of hypoperfusion) and PWI volumes are reported elsewhere [10].

2.3.2. PWI Volume

Time-to-peak (TTP) maps were created from PWI sequences by subtracting the TTP (i.e., time until maximal contrast concentration) for each voxel of normal tissue in the contralesional hemisphere from corresponding voxels in the ipsilesional hemisphere. A perfusion deficit was defined as a greater than 4-s delay in the TTP. TTP maps were created by an NIH investigator, and then de-identified before being shared for this study. An example is provided in Figure 2. For this analysis, while blinded to the NIH-FHV score, perfusion deficit was coded as present or absent in each of the six vascular regions as described previously (viz., ACA, PCA, MCA-F, MCA-T, MCA-P, and MCA-I).

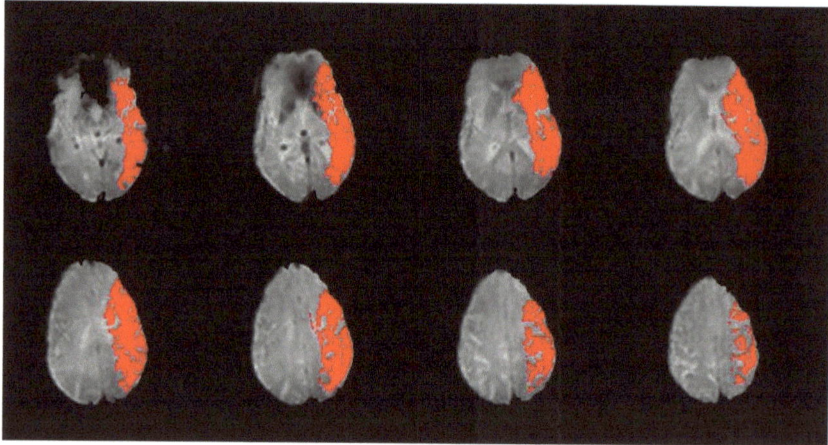

Figure 2. Sample slices from a participant's PWI scan with time-to-peak (TTP) map overlay. A TTP delay of ≥4 s is shown in red. Per the slices included here, this participant would be coded as having a perfusion deficit present in all four regions of the MCA territory (frontal (MCA-F), temporal (MCA-T), parietal (MCA-P), and insular (MCA-I)).

An association between the presence/absence of FHVs and hypoperfusion in PWI was calculated for each vascular region using a Pearson's chi-square (X^2), with the strength of association (comparable to Pearson's r) calculated using Cramér's V.

3. Results

Lesion metrics, such as DWI lesion volume and volume of hypoperfusion in PWI, and a median/range of NIH-FHV scores for the sample are reported in Table 1. With regards to the binary FHV score, 78% ($n = 78$) of the participants had FHVs in at least one vascular region, with 3% ($n = 3$) with FHVs in the ACA territory, 8% ($n = 8$) in the PCA territory, 54% ($n = 55$) in the MCA-F, 58% ($n = 59$) in the MCA-T, 33% ($n = 33$) in the MCA-P, and 60% ($n = 61$) in the MCA-I. Seventy-nine percent ($n = 80$) of the sample demonstrated perfusion deficits in PWI, with 5% ($n = 5$) of the sample showing hypoperfusion in the ACA territory, 12% (12) in the PCA territory, 40% ($n = 40$) in the MCA-F, 48% ($n = 48$) in the MCA-T, 60% ($n = 61$) in the MCA-P, and 37% in the MCA-I ($n = 37$). As NIH-FHV scores for this sample were completed by another lab, we also examined the agreement between binary FHV and PWI variables as an indication of an agreement across labs. There was a 92% point-to-point agreement on the presence/absence of hypoperfusion on both MRI sequences (i.e., scored as having FHVs and perfusion deficit in PWI, or the opposite: no FHVs and no PWI deficit) with 71% of the sample demonstrating both FHVs and perfusion deficit in PWI.

Results of the chi-square analysis (see Table 2) show a statistically significant association between the presence of FHVs and hypoperfusion in PWI for each vascular region, although the relationship in the ACA territory did not survive Bonferroni's correction (i.e., $p < 0.008$). That is, for ACA, $X^2 = 5.29$, $p = 0.02$; for PCA, $X^2 = 12.06$, $p = 0.001$, for MCA-F, $X^2 = 33.74$, $p < 0.000$; for MCA-T, $X^2 = 36.58$, $p < 0.000$; for MCA-P, $X^2 = 15.48$, $p < 0.000$; and for MCA-I, $X^2 = 28.55$, $p < 0.000$. For all analyses, df = 1 and $n = 101$. Per Cramer's V, the relationship between the presence of FHVs and perfusion deficit in PWI in the ACA territory is 'strong' (i.e., $V > 0.15$), with all other vascular regions having 'very strong' associations (i.e., $V > 0.25$) [14] between the two approaches. Because there were so few participants demonstrating FHVs or perfusion deficits in PWI in the ACA territory, we subsequently conducted a Fisher's exact test for this region, but it was also non-significant (Fisher's exact = 0.143).

Table 2. Chi-Square results (presence of FHVs and hypoperfusion in PWI).

Vascular Region	Chi-Square (X^2)	p-Value	Cramér's V
ACA	5.29	0.02	0.23
PCA	12.06	0.001 *	0.35
MCA-F	33.74	0.000 *	0.58
MCA-T	36.58	0.000 *	0.60
MCA-P	15.48	0.000 *	0.39
MCA-I	28.55	0.000 *	0.53

Notes: df = 1, $n = 101$ for all X^2 calculations; FHVs = FLAIR-hyperintense vessels; PWI = perfusion-weighted imaging; ACA = anterior cerebral artery; PCA = posterior cerebral artery; MCA = middle cerebral artery; F = frontal region of the MCA; T = temporal region of the MCA; P = parietal region of the MCA; I = insular region of the MCA. * statistically significant at $p < 0.008$.

4. Discussion

4.1. Clinical Importance of Hypoperfusion

4.1.1. Stroke Diagnosis and Localization

The pathophysiology of cerebrovascular ischemia varies for tissue where the blood flow is completely restricted compared to tissue where the blood flow is reduced (i.e., hypoperfused, or the ischemic penumbra), but both can cause deficits in neurological function [15]. A key difference, however, is the potential to restore perfusion and subsequent neurologic functions to the penumbra. In fact, the ischemic penumbra is the target for all acute stroke interventions [16]. Often, endovascular treatment is based on the volume of the penumbra, but the location of the penumbra should be considered in weighing the potential risks and benefits of intervention. To illustrate, reperfusion of the non-infarcted left temporal gyrus is likely to restore the ability to understand language [17] and to produce meaningful speech, which may be considered a potential benefit that would outweigh substantial risks. On the other hand, reperfusion of the right superior temporal gyrus would be likely to restore more subtle cortical functions, such as understanding the emotional tone of voice [18], recognizing facial expressions [19,20], and empathy [21]. While these functions are important for social interactions, the potential benefit of restoring these functions might not be weighed favorably against the risks of intervention in cases where the intervention carries the risk of hemorrhagic conversion of a relatively large infarct (outside the penumbra). Likewise, reperfusion of the right occipital cortex would be likely to restore the left visual field, which would be especially important for individuals who depend on full vision, such as pilots and truck drivers. Identification of the location of the penumbra is therefore essential for clinicians to provide valuable information to individuals with stroke and their caregivers, to allow a fully informed consent for treatment on the basis of predicted risks and benefits. Our findings will assist clinicians in providing this information based on the location of hypoperfusion, even when perfusion imaging is not available (because of allergy to contrast, inadequate IV access, or lack of necessary technology).

In fact, the window for reperfusion treatment has been expanded well beyond the initial 4–6 h in cases where a penumbra can be demonstrated. The benefit of endovascular treatment has been demonstrated up to 24 h (e.g., DAWN trial, [22]); the benefit of reperfusion with temporary blood pressure elevation to reperfuse the penumbra has been demonstrated up to 1 week after the onset of symptoms in cases of large vessel stenosis and large penumbra [23]; and the benefit of encephaloduroarteriosynangiosis, where there is penumbra in the presence of the moya-moya syndrome weeks after the onset of stroke [24]. However, in these cases, it is also essential for the individual to fully understand the potential benefits in terms of functions that can be restored, in order to make informed decisions about treatment.

Thus, the need to identify and differentiate penumbra from infarct has become a critical need in acute stroke management, both to determine appropriate interventions, and to assess the response to those interventions [15,25]. Identification of hypoperfused tissue on imaging is also important in differential diagnosis through the confirmation of localization of symptoms seen on neurological exam. While perfusion imaging is a core component of the current guidelines for the management of acute ischemic stroke [26], here we have shown, in conjunction with other research [27–29], that FLAIR MRI is another useful tool for the diagnosis and management of (hyper) acute ischemic stroke, and is also much more routinely collected. With the discovery of hyperintense vessels, and their association with abnormal hemodynamic function [11,12,30], FLAIR imaging has become an alternative tool for identifying hypoperfusion when PWI is not an appropriate or successful option. Reyes and colleagues [10] demonstrated a novel use for FLAIR imaging to identify and quantify perfusion deficits as a proxy for PWI. However, additional investigation for this new tool was needed to support its clinical application during stroke diagnosis and care management. That is, validation of the location of FHVs relative to the perfusion deficits in PWI was needed.

In our prior work, we showed a significant relationship between the location of FHVs and perfusion deficits in PWI [13], but that study was in a retrospective analysis of imaging data from a subset of participants in a prospective, longitudinal study examining stroke recovery in right and left hemisphere strokes. Additionally, because the study was retrospective, some participants' FLAIR imaging was acquired under different specifications (i.e., with different slice thicknesses), so some participants had more opportunities to demonstrate FHVs than others. Although we reported significant associations with regards to localization in five of six vascular regions, the limitations warranted additional investigation to validate the NIH-FHV tool. We subsequently aimed to replicate our findings by analyzing this larger sample of prospectively collected data.

4.1.2. Investigation of Neuroanatomical Function and Organization

Many investigations of neurological (dys) function, utilizing brain lesions, have been conducted in individuals in chronic stages of recovery, after reorganization has likely occurred. Thus, examinations of lesion–deficit relationships would ideally be conducted in premorbidly healthy individuals as soon as possible following injury onset, such as in an acute stroke. Since behavioral presentation associated with neural dysfunction corresponds not only to regions of frank lesion, but regions of hypoperfused tissue as well, it is helpful and appropriate for researchers to incorporate measures of hypoperfusion when undertaking investigations in acute/early subacute populations, before reperfusion therapy [3,31–33]. Furthermore, the strongest evidence that an area of the brain is critical to a function is obtained when the function is impaired when the neural region is hypoperfused, and recovers when the neural region is reperfused [32]. Although better methods of measuring hypoperfusion may be available (i.e., PWI or ASL), there are various reasons why an investigator may want to adopt the NIH-FHV scale utilizing FLAIR imaging, such as cost, safety (with regards to administering contrast), and accessibility. While Reyes et al. [10] showed a strong association between the estimated amount of hypoperfusion on FLAIR and the actual volume of hypoperfusion identified in PWI, they did not investigate any associations in location. Without clear evidence that the location on one corresponded to the location on the other, the findings of brain–behavior relationships, in terms of localization of function, using the NIH-FHV might be, at worst, erroneous, or, at best, speculative.

This analysis demonstrates that for five of the six vascular regions—as defined by the NIH-FHV—the general location of FHVs does in fact significantly correspond with general areas of PWI deficits of ≥ 4 s TTP delays. Unfortunately, we failed to identify a significant association between FHV and PWI lesion in the ACA territory (after correcting for multiple comparisons). As with our prior study, we suspect this was due to having too few participants with FHVs or PWI lesions in the ACA territory for statistical comparison (i.e., only $n = 3$ with FHVs and $n = 5$ with PWI deficits). This finding was unsurprising given the fact that ischemic stroke due to ACA vessel occlusion is relatively uncommon (i.e., less than 2% of ischemic strokes per some studies) [34,35]. Thus, additional investigation is still needed to determine if there is a relationship between the two MRI methods in this specific vascular territory.

With validation of the location of hypoperfusion in these regions, the current study offers some validation of our previous work examining associations between NIH-FHV and behavioral outcomes [13,36]. For example, we reported a functional relationship between FHV scores in the MCA frontal region and performance on the NIH Stroke Scale (NIHSS) [37], which has been shown to be biased towards anterior lesions [38] and/or motor and language modalities [39,40]. NIHSS scores were collected for this sample, but per our data-sharing agreement with the NIH, we did not have access to this data in order to examine any association between NIH-FHV scores and NIHSS scores, so we are unable to replicate this finding with the current sample.

We have also demonstrated a relationship between NIH-FHV scores in the MCA parietal region, and content production during a discourse task and picture naming, independent of the lesion volume and age/education [13]. In another study, we found that various tests of hemispatial neglect were associated with NIH-FHV scores in different regions (e.g., NIH-FHV score in the MCA-P region was independently associated with line bisection, while viewer- and stimulus-centered errors on a gap detection task were associated with scores in the MCA-T and MCA-F areas, respectively) [36]. Although these various findings were statistically significant, the full impact of the results depended on more clear evidence that the location of FHVs does indeed correspond with hypoperfusion in those same regions, which this study provides. Again, we were not provided with any individual behavioral data for this sample, if it was collected, so we could not examine any relationships between NIH-FHV and behavior as part of this analysis. Regardless, our prior studies do show the utility of using the NIH-FHV score to examine brain–behavior relationships beyond the clinical examination.

4.2. Conclusions and Future Directions

These results, in conjunction with our previous study [13], indicate that FHVs in the MCA and PCA territories may be reliably associated with hypoperfusion in those same areas. Thus, not only can the NIH-FHV score be used to quantify the volume of hypoperfusion (see [10]), it can be used to identify the general location of the hypoperfusion in the MCA and PCA vascular territories as well (although additional investigation in the ACA territory is needed). Information about the location of the hypoperfusion allows the clinician to predict what functions can be restored with intervention to restore blood flow, which is information that is critical for informed decision making. These findings are important for confirming the localization of deficits seen on neurological exam. These results also validate the associations with behavioral performance we reported previously, such as the significant independent relationship between FHVs in the MCA-F region and

NIHSS scores [13,36]. Since individual NIHSS scores—or other behavioral outcomes—for this data set were not available to us, we were unable to replicate our findings with regards to the behavioral performance in this cohort. Nonetheless, collectively, these studies support use of the NIH-FHV scale in both clinical and research settings. Although perfusion imaging would be the ideal method for identifying the location and volume of perfusion deficits, and likely available for many patients with concern for stroke, care providers and investigators may be able to identify the location and estimate the volume of hypoperfusion using FLAIR imaging. This tool may also allow investigators to examine brain–behavior relationships in acute/early subacute patients, before reorganization has occurred.

Author Contributions: Conceptualization, L.D.B. and A.E.H.; Methodology, L.D.B. and A.E.H.; Formal Analysis, L.D.B.; Writing—Original Draft Preparation, L.D.B.; Writing—Review & Editing, L.D.B. and A.E.H. All authors have read and agreed to the published version of the manuscript.

Funding: There was no funding support for this specific study, but the authors are supported by the NIDCD grants R01 DC05375, R01 DC015466 and P50 DC014664.

Institutional Review Board Statement: The initial ethical review and study approval was completed by the Institutional Review Board (IRB) of the National Institutes of Health (NIH), as part of the National Institute of Neurological Disorders and Stroke's Natural History of Stroke study (01N0007), in accordance with the Declaration of Helsinki. Although all shared data were de-identified, our study was also approved by the Johns Hopkins Medicine IRB (IRB00339817; 20 July 2022).

Informed Consent Statement: Informed consent was obtained from all subjects involved in the study (see [10]).

Data Availability Statement: Quantitative data are available upon request, but restrictions apply to the data obtained from the NIH. It may be available from the authors with the permission of the NIH.

Acknowledgments: We would like to acknowledge the generosity of the NIH and Richard Leigh for collecting and sharing data for this analysis. We are also grateful to the participants for their invaluable contributions to science.

Conflicts of Interest: The authors report no conflict of interest.

References

1. Hillis, A.E.; Barker, P.B.; Beauchamp, N.J.; Gordon, B.; Wityk, R.J. MR Perfusion Imaging Reveals Regions of Hypoperfusion Associated with Aphasia and Neglect. *Neurology* **2000**, *55*, 782–788. [CrossRef]
2. Hillis, A.E.; Wityk, R.J.; Barker, P.B.; Beauchamp, N.J.; Gailloud, P.; Murphy, K.; Cooper, O.; Metter, E.J. Subcortical Aphasia and Neglect in Acute Stroke: The Role of Cortical Hypoperfusion. *Brain* **2002**, *125*, 1094–1104. [CrossRef]
3. Hillis, A.E.; Wityk, R.J.; Barker, P.B.; Caramazza, A. Neural Regions Essential for Writing Verbs. *Nat. Neurosci.* **2003**, *6*, 19–20. [CrossRef]
4. Ramalho, J.; Semelka, R.C.; Ramalho, M.; Nunes, R.H.; AlObaidy, M.; Castillo, M. Gadolinium-Based Contrast Agent Accumulation and Toxicity: An Update. *Am. J. Neuroradiol.* **2016**, *37*, 1192–1198. [CrossRef]
5. Hasebroock, K.M.; Serkova, N.J. Toxicity of MRI and CT Contrast Agents. *Expert Opin. Drug Metab. Toxicol.* **2009**, *5*, 403–416. [CrossRef]
6. Kanda, T.; Ishii, K.; Kawaguchi, H.; Kitajima, K.; Takenaka, D. High Signal Intensity in the Dentate Nucleus and Globus Pallidus on Unenhanced T1-Weighted MR Images: Relationship with Increasing Cumulative Dose of a Gadolinium-Based Contrast Material. *Radiology* **2014**, *270*, 834–841. [CrossRef]
7. Wang, Z. Improving Cerebral Blood Flow Quantification for Arterial Spin Labeled Perfusion MRI by Removing Residual Motion Artifacts and Global Signal Fluctuations. *Magn. Reson. Imaging* **2012**, *30*, 1409–1415. [CrossRef]
8. Xie, D.; Li, Y.; Yang, H.; Bai, L.; Wang, T.; Zhou, F.; Zhang, L.; Wang, Z. Denoising Arterial Spin Labeling Perfusion MRI with Deep Machine Learning. *Magn. Reson. Imaging* **2020**, *68*, 95–105. [CrossRef]
9. Jezzard, P.; Chappell, M.A.; Okell, T.W. Arterial Spin Labeling for the Measurement of Cerebral Perfusion and Angiography. *J. Cereb. Blood. Flow Metab.* **2018**, *38*, 603–626. [CrossRef]
10. Reyes, D.; Simpkins, A.N.; Hitomi, E.; Lynch, J.K.; Hsia, A.W.; Nadareishvili, Z.; Luby, M.; Latour, L.L.; Leigh, R. Estimating Perfusion Deficits in Acute Stroke Patients without Perfusion Imaging. *Stroke* **2022**, *53*, 3439–3445. [CrossRef]
11. Toyoda, K.; Ida, M.; Fukuda, K. Fluid-Attenuated Inversion Recovery Intraarterial Signal: An Early Sign of Hyperacute Cerebral Ischemia. *Am. J. Neuroradiol.* **2001**, *22*, 1021–1029.
12. Kamran, S.; Bates, V.; Bakshi, R.; Wright, P.; Kinkel, W.; Miletich, R. Significance of Hyperintense Vessels on FLAIR MRI in Acute Stroke. *Neurology* **2000**, *55*, 265–269. [CrossRef]

13. Bunker, L.D.; Walker, A.; Meier, E.; Goldberg, E.; Leigh, R.; Hillis, A.E. Hyperintense Vessels on Imaging Account for Neurological Function Independent of Lesion Volume in Acute Ischemic Stroke. *NeuroImage Clin.* **2022**, *34*, 102991. [CrossRef]
14. Akoglu, H. User's Guide to Correlation Coefficients. *Turk. J. Emerg. Med.* **2018**, *18*, 91–93. [CrossRef]
15. Astrup, J.; Siesjo, B.K.; Symon, L. Thresholds in Cerebral Ischemia—The Ischemic Penumbra. *Stroke* **1981**, *12*, 723–725. [CrossRef]
16. Hillis, A.E.; Baron, J.-C. Editorial: The Ischemic Penumbra: Still the Target for Stroke Therapies? *Front. Neurol.* **2015**, *6*, 85. [CrossRef]
17. Hillis, A.E.; Rorden, C.; Fridriksson, J. Brain Regions Essential for Word Comprehension: Drawing Inferences from Patients. *Ann. Neurol.* **2017**, *81*, 759–768. [CrossRef]
18. Ross, E.D.; Monnot, M. Neurology of Affective Prosody and Its Functional-Anatomic Organization in Right Hemisphere. *Brain Lang.* **2008**, *104*, 51–74. [CrossRef]
19. Tippett, D.C.; Godin, B.R.; Oishi, K.; Oishi, K.; Davis, C.; Gomez, Y.; Trupe, L.A.; Kim, E.H.; Hillis, A.E. Impaired Recognition of Emotional Faces after Stroke Involving Right Amygdala or Insula. *Semin. Speech Lang.* **2018**, *39*, 87–100. [CrossRef]
20. Sheppard, S.M.; Meier, E.L.; Zezinka Durfee, A.; Walker, A.; Shea, J.; Hillis, A.E. Characterizing Subtypes and Neural Correlates of Receptive Aprosodia in Acute Right Hemisphere Stroke. *Cortex* **2021**, *141*, 36–54. [CrossRef]
21. Leigh, R.; Oishi, K.; Hsu, J.; Lindquist, M.; Gottesman, R.F.; Jarso, S.; Crainiceanu, C.; Mori, S.; Hillis, A.E. Acute Lesions That Impair Affective Empathy. *Brain* **2013**, *136*, 2539–2549. [CrossRef] [PubMed]
22. Nogueira, R.G.; Jadhav, A.P.; Haussen, D.C.; Bonafe, A.; Budzik, R.F.; Bhuva, P.; Yavagal, D.R.; Ribo, M.; Cognard, C.; Hanel, R.A.; et al. Thrombectomy 6 to 24 Hours after Stroke with a Mismatch between Deficit and Infarct. *N. Engl. J. Med.* **2018**, *378*, 11–21. [CrossRef] [PubMed]
23. Hillis, A.E.; Ulatowski, J.A.; Barker, P.B.; Torbey, M.; Ziai, W.; Beauchamp, N.J.; Oh, S.; Wityk, R.J. A Pilot Randomized Trial of Induced Blood Pressure Elevation: Effects on Function and Focal Perfusion in Acute and Subacute Stroke. *Cereb. Dis.* **2003**, *16*, 236–246. [CrossRef] [PubMed]
24. Wityk, R.J.; Hillis, A.; Beauchamp, N.; Barker, P.B.; Rigamonti, D. Perfusion-Weighted Magnetic Resonance Imaging in Adult Moyamoya Syndrome: Characteristic Patterns and Change after Surgical Intervention: Case Report. *Neurosurgery* **2002**, *51*, 1499–1505. [CrossRef] [PubMed]
25. Donnan, G.A.; Davis, S.M. Neuroimaging, the Ischaemic Penumbra, and Selection of Patients for Acute Stroke Therapy. *Lancet Neurol.* **2002**, *1*, 417–425. [CrossRef]
26. Powers, W.J.; Rabinstein, A.A.; Ackerson, T.; Adeoye, O.M.; Bambakidis, N.C.; Becker, K.; Biller, J.; Brown, M.; Demaerschalk, B.M.; Hoh, B.; et al. Guidelines for the Early Management of Patients with Acute Ischemic Stroke: 2019 Update to the 2018 Guidelines for the Early Management of Acute Ischemic Stroke: A Guideline for Healthcare Professionals From the American Heart Association/American Stroke Association. *Stroke* **2019**, *50*, e344–e418. [CrossRef]
27. Noguchi, K.; Ogawa, T.; Inugami, A.; Fujita, H.; Hatazawa, J.; Shimosegawa, E.; Okudera, T.; Uemura, K.; Seto, H. MRI of Acute Cerebral Infarction: A Comparison of FLAIR and T2-Weighted Fast Spin-Echo Imaging. *Neuroradiology* **1997**, *39*, 406–410. [CrossRef]
28. Thomalla, G.; Rossbach, P.; Rosenkranz, M.; Siemonsen, S.; Krützelmann, A.; Fiehler, J.; Gerloff, C. Negative Fluid-Attenuated Inversion Recovery Imaging Identifies Acute Ischemic Stroke at 3 Hours or Less. *Ann. Neurol.* **2009**, *65*, 724–732. [CrossRef]
29. Lee, K.Y.; Latour, L.L.; Luby, M.; Hsia, A.W.; Merino, J.G.; Warach, S. Distal Hyperintense Vessels on FLAIR. *Neurology* **2009**, *72*, 1134–1139. [CrossRef]
30. Legrand, L.; Tisserand, M.; Turc, G.; Naggara, O.; Edjlali, M.; Mellerio, C.; Mas, J.-L.; Méder, J.-F.; Baron, J.-C.; Oppenheim, C. Do FLAIR Vascular Hyperintensities beyond the DWI Lesion Represent the Ischemic Penumbra? *AJNR Am. J. Neuroradiol.* **2015**, *36*, 269–274. [CrossRef]
31. Hillis, A.E.; Barker, P.B.; Beauchamp, N.J.; Winters, B.D.; Mirski, M.; Wityk, R.J. Restoring Blood Pressure Reperfused Wernicke's Area and Improved Language. *Neurology* **2001**, *56*, 670–672. [CrossRef] [PubMed]
32. Hillis, A.E.; Kane, A.; Tuffiash, E.; Ulatowski, J.A.; Barker, P.B.; Beauchamp, N.J.; Wityk, R.J. Reperfusion of Specific Brain Regions by Raising Blood Pressure Restores Selective Language Functions in Subacute Stroke. *Brain Lang.* **2001**, *79*, 495–510. [CrossRef] [PubMed]
33. Hillis, A.E.; Barker, P.B.; Wityk, R.J.; Aldrich, E.M.; Restrepo, L.; Breese, E.L.; Work, M. Variability in Subcortical Aphasia Is Due to Variable Sites of Cortical Hypoperfusion. *Brain Lang.* **2004**, *89*, 524–530. [CrossRef] [PubMed]
34. Kazui, S.; Sawada, T.; Naritomi, H.; Kuriyama, Y.; Yamaguchi, T. Angiographic Evaluation of Brain Infarction Limited to the Anterior Cerebral Artery Territory. *Stroke* **1993**, *24*, 549–553. [CrossRef]
35. Bogousslavsky, J. Anterior Cerebral Artery Territory Infarction in the Lausanne Stroke Registry: Clinical and Etiologic Patterns. *Arch. Neurol.* **1990**, *47*, 144–150. [CrossRef]
36. Stein, C.; Bunker, L.; Chu, B.; Leigh, R.; Faria, A.; Hillis, A.E. Various Tests of Left Neglect Are Associated with Distinct Territories of Hypoperfusion in Acute Stroke. *Brain Commun.* **2022**, *4*, fcac064. [CrossRef]
37. National Institute of Neurological Disorders and Stroke. *NIH Stroke Scale*; National Institute of Neurological Disorders and Stroke, Department of Health and Human Services: Washington, DC, USA, 2011.
38. Sato, S.; Toyoda, K.; Uehara, T.; Toratani, N.; Yokota, C.; Moriwaki, H.; Naritomi, H.; Minematsu, K. Baseline NIH Stroke Scale Score Predicting Outcome in Anterior and Posterior Circulation Strokes. *Neurology* **2008**, *70*, 2371–2377. [CrossRef]

39. Woo, D.; Broderick, J.P.; Kothari, R.U.; Lu, M.; Brott, T.; Lyden, P.D.; Marler, J.R.; Grotta, J.C. Does the National Institutes of Health Stroke Scale Favor Left Hemisphere Strokes? *Stroke* **1999**, *30*, 2355–2359. [CrossRef]
40. Fink, J.N.; Selim, M.H.; Kumar, S.; Silver, B.; Linfante, I.; Caplan, L.R.; Schlaug, G. Is the Association of National Institutes of Health Stroke Scale Scores and Acute Magnetic Resonance Imaging Stroke Volume Equal for Patients with Right- and Left-Hemisphere Ischemic Stroke? *Stroke* **2002**, *33*, 954–958. [CrossRef]

Journal of
Clinical Medicine

Article

External Validation and Retraining of DeepBleed: The First Open-Source 3D Deep Learning Network for the Segmentation of Spontaneous Intracerebral and Intraventricular Hemorrhage

Haoyin Cao [1,*], Andrea Morotti [2], Federico Mazzacane [3,4], Dmitriy Desser [5], Frieder Schlunk [5], Christopher Güttler [5], Helge Kniep [6], Tobias Penzkofer [1,7], Jens Fiehler [6], Uta Hanning [6], Andrea Dell'Orco [5,†] and Jawed Nawabi [1,5,7,†]

1   Department of Radiology, Charité—Universitätsmedizin Berlin, Freie Universität Berlin, Humboldt-Universität zu Berlin, Charitéplatz 1, 10117 Berlin, Germany
2   Neurology Unit, Department of Neurological Sciences and Vision, ASST-Spedali Civili, 25123 Brescia, Italy
3   Department of Brain and Behavioral Sciences, University of Pavia, 27100 Pavia, Italy
4   U.C. Malattie Cerebrovascolari e Stroke Unit, IRCCS Fondazione Mondino, 27100 Pavia, Italy
5   Department of Neuroradiology, Charité School of Medicine and University Hospital Berlin, 10117 Berlin, Germany
6   Department of Diagnostic and Interventional Neuroradiology, University Medical Center Hamburg Eppendorf, 20246 Hamburg, Germany
7   Berlin Institute of Health (BIH), BIH Biomedical Innovation Academy, 10178 Berlin, Germany
*   Correspondence: haoyin.cao@charite.de
†   These authors contributed equally to this work.

Citation: Cao, H.; Morotti, A.; Mazzacane, F.; Desser, D.; Schlunk, F.; Güttler, C.; Kniep, H.; Penzkofer, T.; Fiehler, J.; Hanning, U.; et al. External Validation and Retraining of DeepBleed: The First Open-Source 3D Deep Learning Network for the Segmentation of Spontaneous Intracerebral and Intraventricular Hemorrhage. *J. Clin. Med.* **2023**, *12*, 4005. https://doi.org/10.3390/jcm12124005

Academic Editor: Hugues Chabriat

Received: 10 May 2023
Revised: 3 June 2023
Accepted: 7 June 2023
Published: 12 June 2023

Abstract: Background: The objective of this study was to assess the performance of the first publicly available automated 3D segmentation for spontaneous intracerebral hemorrhage (ICH) based on a 3D neural network before and after retraining. Methods: We performed an independent validation of this model using a multicenter retrospective cohort. Performance metrics were evaluated using the dice score (DSC), sensitivity, and positive predictive values (PPV). We retrained the original model (OM) and assessed the performance via an external validation design. A multivariate linear regression model was used to identify independent variables associated with the model's performance. Agreements in volumetric measurements and segmentation were evaluated using Pearson's correlation coefficients (r) and intraclass correlation coefficients (ICC), respectively. With 1040 patients, the OM had a median DSC, sensitivity, and PPV of 0.84, 0.79, and 0.93, compared to thoseo f 0.83, 0.80, and 0.91 in the retrained model (RM). However, the median DSC for infratentorial ICH was relatively low and improved significantly after retraining, at $p < 0.001$. ICH volume and location were significantly associated with the DSC, at $p < 0.05$. The agreement between volumetric measurements (r > 0.90, $p > 0.05$) and segmentations (ICC ≥ 0.9, $p < 0.001$) was excellent. Conclusion: The model demonstrated good generalization in an external validation cohort. Location-specific variances improved significantly after retraining. External validation and retraining are important steps to consider before applying deep learning models in new clinical settings.

Keywords: intracerebral hemorrhage; automated segmentation; deep learning; multicenter; external validation

1. Introduction

Spontaneous intracerebral hemorrhage (ICH) is a major cause of morbidity and mortality worldwide despite the relatively small contribution to all stroke types of up to 27% [1–3]. The prognosis after ICH is particularly affected by the ICH volume in addition to its location, the presence of intraventricular hemorrhage (IVH), and acute hematoma expansion (HE) [4]. Thus, instruments for accurate ICH and IVH quantification upon neuroimaging are crucial to guide further patient management and to inform future clinic trials [5–8]. The ABC/2 method has remained a clinically well-established formula to manually estimate the ICH volume [9] despite the consistent reports of underestimation or overestimation in large and irregular bleedings. The semiautomatic measurements of ICH and IVH are equally limited as they are labor-intensive and time-consuming [10]. Novel deep learning-based models have the potential to quantify ICH and IVH volumes rapidly and accurately in a fully automated approach and are therefore in high demand [11]. The DeepBleed network presented by Sharrock et al. is the first publicly available 3D neural network for the segmentation of ICH and IVH [12]. Despite being trained and internally validated on a large dataset from the MISTIE II and III trial series [13,14], its performance in an independent cohort has not been described yet. This step is of particular importance in imaging-based segmentation networks as they have increasingly shown inconsistent performance results on external datasets [15]. Therefore, it is important that clinicians are aware of the quality assessment steps that need to be taken before the local implementation of these models. In particular, the performance of DeepBleed in the detection and segmentation of infratentorial and small ICH remains undetermined as these two subsets were excluded from the MISTIE trials [13,14,16]. The aim of this study was to evaluate the generalizability and further improve the robustness of the proposed DeepBleed network. The objective of this study was to assess the performance of the existing DeepBleed network before and after retraining. Therefore, we hypothesized that the DeepBleed network would accurately detect and segment ICH and IVH regardless of its location and size. To test and evaluate this, the following threefold steps were performed. First, we externally validated the original DeepBleed model (OM) in an independent multicenter cohort. Secondly, we retrained the model (RM) to test the effect on the validation accuracy through an internal validation design. Third, we compared the interrater reliabilities between the OM and RM network and independent human raters. This study serves as a use case illustrating how the generalizability of deep learning models may be addressed via local retraining.

2. Materials and Methods

2.1. Study Population

This retrospective study was approved by the local ethics committee (Charité Berlin, Germany (protocol number EA1/035/20), University Medical-Center Hamburg, Germany (protocol number WF-054/19), and IRCCS Mondino Foundation, Pavia, Italy (protocol number 20190099462]). Written informed consent was waived by the institutional review boards. All study protocols and procedures were conducted in accordance with the Declaration of Helsinki. The study included patients of ≥ 18 years who were diagnosed with primary spontaneous ICH upon noncontrast computed tomography (NECT) between January 2017 and June 2020. Patients with multiple ICH, artifacts, external ventricular drain (EVD) or any other type of surgical procedure, and secondary hemorrhage following head trauma, ischemic infarction, neoplastic mass lesions, ruptured cerebral aneurysms, or vascular malformations were excluded from the study as presented in Figure 1.

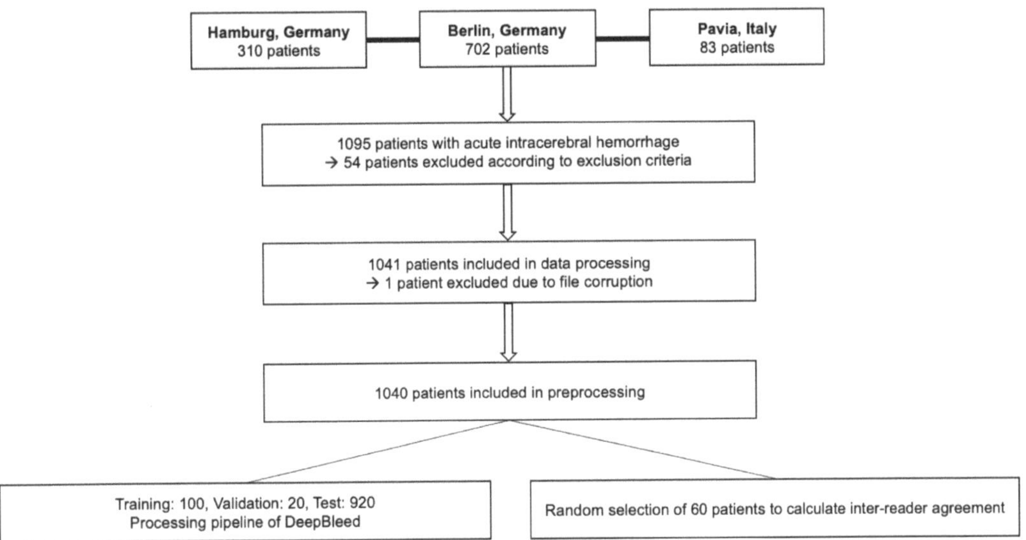

Figure 1. Flow diagram of patients included from three European participating sites and final patient cohort after further exclusion.

2.2. Image Acquisition and Manual Segmentation

Participating sites acquired NECT images according to their local imaging protocols. De-identified and pseudomyzed imaging data were retrieved from the local picture archiving and communication system (PACS) servers and converted into a Digital Imaging and Communications in Medicine (DICOM) format according to local guidelines. DICOM data were then transformed in Neuroimaging Informatics Technology Initiative (NifTI) for further imaging analysis. Images were analyzed for the presence of IVH and the ICH location by one experienced neuroradiologist (J.N., who has 5 years of experience in ICH imaging research). Supratentorial bleedings in cortical and subcortical locations were classified as lobar and hemorrhages involving the thalamus, basal ganglia, internal capsule and deep periventricular white matter [17]. Infratentorial bleedings were classified within the brainstem, pons and cerebellum [18]. Ground truth (GT) masks of ICH and IVH were manually segmented on CT scans by two experienced raters (both with 3 and 5 years of experience in ICH imaging research) who were supervised by one neuroradiologist (J.N.) who inspected each ICH and IVH mask for the quality of segmentations and corrected them if necessary. Segmentation of the ICH and IVH was performed using ITK-SNAP software version 3.8.0 (Penn Image Computing and Science Laboratory, Philadelphia, PA, USA) [19]. Two expert raters segmented the test set six months apart for 60 patients to calculate inter-reader agreement (J.N. and FM, with 5 years of experience in ICH imaging research). The function shuffle from Python library NumPy random was applied on the subject ID list [20]. All readers independently analyzed and segmented images in a random order while blind to all demographic data and were not involved in the clinical care of assessment of the enrolled patients.

2.3. Preprocessing and Postprocessing

The preprocessing comprised two steps as described in the original study (Figure 2): brain extraction and the coregistration. Briefly, after setting to zero the CT scan intensities lower than 0 or higher than 100, brain extraction was performed with FSL Brain Extraction Tool (BET) [21], setting the fractional intensity parameter (-f flag) to 0.01. Rigid coregistration was performed with ANTs [22] using a 1.5 mm^3 isotropic CT template [23]. The resulting transformation was applied to the GT masks as well. During postprocessing, a threshold of 0.6 was applied to the resulting probability maps, setting the higher values to 1 and the others to 0. Finally, inverse coregistration of the resulting mask was performed using the inverse transformation matrix.

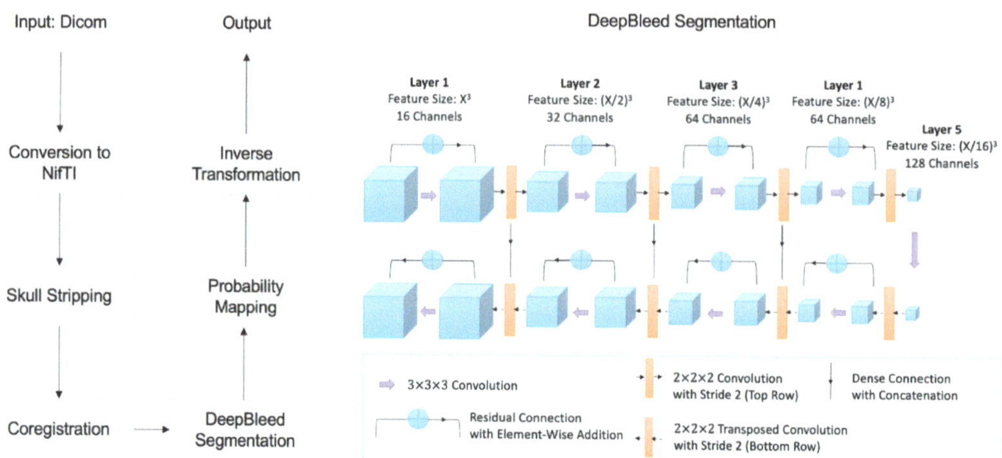

Figure 2. Processing pipeline of the 3D DeepBleed network.

After gantry tilt and unequal slices were corrected, DICOM data were converted using NIfTI. For brain extraction and coregistration, Python preprocessing pipelines were used. DeepBleed was then used to predict intracerebral hemorrhage (ICH) and intraventricular hemorrhage (IVH). In the final step, the predictions from the previous template registration were inversely transformed in the native space.

2.4. Model Retraining

In addition to the general exclusion criteria described in the above, the following additional exclusion criteria were applied to the training cohort in accordance with those of MISTIE studies [13,14]: symptom onset > 24 h prior to the admission CT or an unknown time of symptom onset as well as an admission ICH volume of >30 mL. As described in the original study, adaptive moment estimation (Adam) was used as the optimization function [24], the dice similarity coefficient (DSC) was used as a loss function [25], and 100, randomly selected subjects were used as a training cohort, whereas 20 were used as a validation cohort. After testing various combinations, the optimal learning rate was 1×10^{-4} and a batch size of 3 was chosen. Initially, the training dataset was shuffled; we stopped the training if, after the first 10th epoch, the current epoch did not improve, and validation was performed every five epochs.

2.5. Model Testing

For the model testing, preprocessing and postprocessing as described above were performed on the OM and RM test dataset. Contrary to the training dataset, the test dataset was defined according to the original criteria of our study with no additional exclusion criteria applied. This decision was made because we were interested in evaluating the performance of DeepBleed on small hemorrhages.

2.6. Code Availability

The code is publicly available in Jupiter Notebooks on GitHub (Microsoft, San Fransisco, CA, USA) with the following link: https://github.com/Orangepepermint/retrain deepbleed accessed on 9 May 2023. It is written in Python v3.9 [26] using the following libraries: NumPy v1.23.3 [20], FSLpy v3.9.0 [27], and ANTsPy v0.3.1 [22].

2.7. Statistical Analysis

Statistics were conducted using GraphPad Prism v9.0.2 (GraphPad Software, Inc., San Diego, CA, USA) [28] and R (the R project for statistical computing, Vienna, AT) using tidyverse [29]. Various metrics were used to evaluate the DeepBleed performance and compare the segmentations from our RM with the OM. These included DSC, sensitivity, positive predictive value (PPV), and volume measurements. *t*-tests were used to compare the DSC, sensitivity, and PPV distributions between the OM and RM. Based on the central limit theorem, the *t*-test assumptions were fulfilled. To determine factors influencing segmentation performance, a linear regression model with the following formula was used:

$$DSC \sim volume + hemorrhage\ location + IVH\ presence + participating\ center$$

Pairwise correlations among volumes measured from each of the three segmentation methods (GT masks, OM and RM DeepBleed network) were assessed using the Pearson correlation coefficient (r). Agreements between two raters and the OM and RM DeepBleed network were assessed using the intraclass correlation coefficient (ICC) in the DSC. Moreover, a repeated measures ANOVA was performed with a pairwise *t*-test as a post hoc. The homogeneity of variance assumption for ANOVA was evaluated using the Levene test. Cohen's d effect size was determined as well. A *p*-value of < 0.05 was considered significant. Bonferroni adjustment was applied where necessary. Adjusted *p*-values are indicated as p_{adj}-values.

3. Results

3.1. Demographics and Characteristics of the Study Cohort

The manual review of the images led to an elimination of 54 patients due to exclusion criteria and manual segmentation errors. The final dataset was composed of 1040 patients, NECT scans and respective masks, where the numbers of patients in training, validation and test were $n = 100$, $n = 20$ and $n = 920$. The mean age was 69.6 (SD 14.2) years. The median NIHSS and GCS scores were 7.5 (IQR 12) and 13 (IQR 7), respectively. Imaging was performed within a median symptom onset time of 4.3 h (IQR 13.6). In total, 519 patients (49.9%) presented with ICH and IVH with a mean volume of 74.9 (SD 41.6) ml. A total of 521 patients (50.1%) presented with ICH only with a mean volume of 41.5 (SD 32.8) mL. No significant differences in demographic characteristics were found between training, validation, and test subjects as presented in Table 1.

Table 1. Comparison of the demographics and volumetry.

Variable		Training Cohort (*n* = 100)	Validation Cohort (*n* = 20)	Test Cohort (*n* = 920)	*p*-Value
Age (years), mean ± SD		70.5 ± 13.1	68 ± 13.4	69.6 ± 14.2	0.84 [1]
Sex, *n* (%)					
	Male	41 (41)	11 (55)	516 (56)	
	Female	59 (59)	9 (45)	404 (44)	
NIHSS score, median (IQR)		7.5 (10)	10 (10)	7 (5)	0.70 [1]
GCS score, median (IQR)		13 (7)	14 (4)	13 (8)	0.11 [2]
Symptom onset to imaging (hours), median (IQR)		4.3 (13.6)	3.9 (7.6)	4.23 (12.8)	0.89 [2]
ICH location, *n* (%)					
	Lobar	44 (44)	6 (30)	338 (36.7)	
	Deep	46 (46)	7 (35)	455 (49.5)	
	Brainstem	3 (3)	3 (15)	41 (4.5)	
	Cerebellum	7 (7)	4 (20)	86 (9.3)	
ICH + IVH volume (ml), mean ± SD		83.8 ± 47.2	56.4 ± 89.7	76.5 ± 44.9	0.67 [2]
ICH volume (ml), mean ± SD		27.7 ± 30.2	34.2 ± 30.9	44.1 ± 14.2	0.30 [2]
IVH volume (ml), mean ± SD		61.1 ± 49.1	22.2 ± 77.1	34.5 ± 42.5	0.64 [2]

Demographics and descriptive characteristics compared between the training, validation, and test datasets. NIHSS, National Institutes of Health Stroke score; GCS, Glasgow Coma Scale; ICH, intracerebral hemorrhage; IQR, interquartile range; IVH, intraventricular hemorrhage; SD, standard deviation. [1] one-way-ANOVA test; [2] Kruskal–Wallis test, if the data do not fulfill the normal distribution.

3.2. Model Retraining and Testing

With Nvidia RTX 3090 GPU, the model was trained for 810 epochs in 16 h. Illustrative examples of segmentations are displayed in Figure 3. Model performance metrics of the OM and RM derived in the test set are presented in detail in Table 2 with additional DSC metrics illustrated in Figure 4A. The overall DSC values for the segmentations across all locations were relatively similar in both DeepBleed models with a median DSC of 0.84 (95% CI, 0.73–0.88) in the OM and 0.83 (95% CI, 0.74–0.88) in the RM. DSC values given separately for each location were also almost equally high in supratentorial ICH with a median DSC of 0.86 (95% CI, 0.80–0.89) in deep ICH and 0.84 (95% CI, 0.78–0.89) in lobar ICH in the OM compared to 0.87 (95% CI, 0.81–0.90) and 0.83 (95% CI, 0.72–0.88) in the RM, respectively. In comparison, performance metrics in infratentorial locations demonstrated an overall lower DSC with a median DSC of 0.71 (95% CI, 0.46–0.78) in cerebellar ICH and 0.48 (95% CI, 0.23–0.64) in brainstem ICH in the OM. DSC metrics improved in the RM, especially in cerebellar ICH, with a median DSC of 0.79 (95% CI, 0.65–0.84) and 0.77 (95% CI, 0.57–0.83) in brainstem ICH. OM and RM performance metrics were significantly different for the DSC and PPV, at p_{adj}-values of < 0.001, compared to the sensitivity, p_{adj}-value of >0.05. DSC, and sensitivity showed a significant improvement in the cerebellum ($p < 0.001$) after retraining.

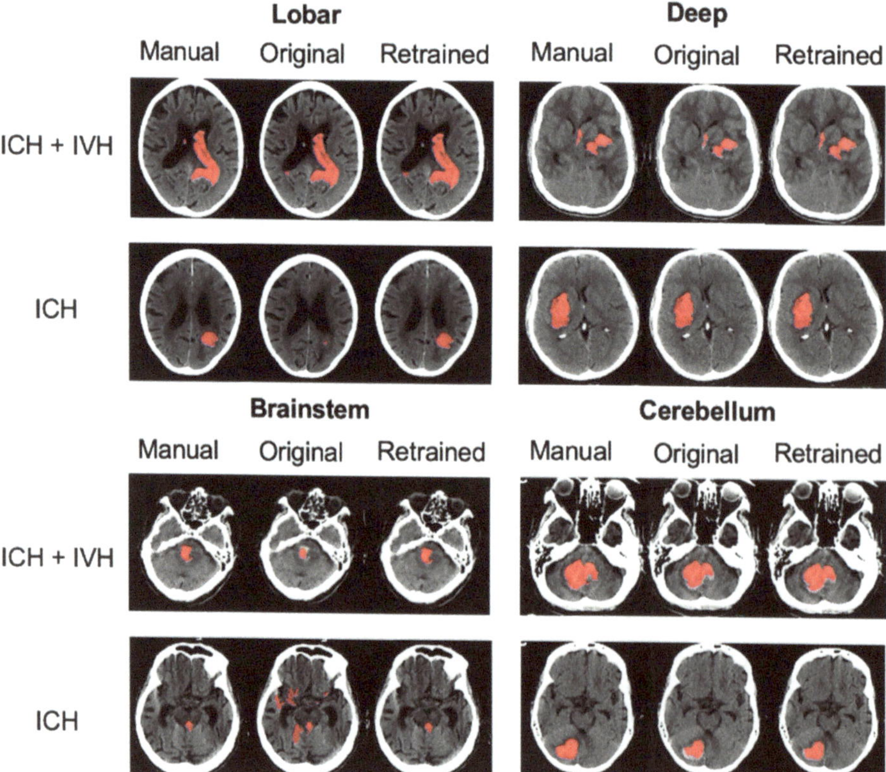

Figure 3. Segmentations of the original and retrained model across different locations. Illustrative examples of ground truth segmentations (red) for intracerebral hemorrhage (ICH) with intraventricular hemorrhage (IVH; upper rows) and ICH only (lower rows) given for deep, lobar, brainstem, and cerebellar ICH. Segmentations given are displayed for manual segmentations (left column), and DeepBleed segmentations with the original model (OM, middle column), and the retrained model (RM, right column).

Table 2. Original and retrained model performance metrics.

Metric	All Locations	Deep	Lobar	Brainstem	Cerebellum
OM					
DSC	0.84 (0.73, 0.88)	0.86 (0.80, 0.89)	0.84 (0.78, 0.89)	0.71 (0.46, 0.78)	0.48 (0.23, 0.64)
Sensitivity	0.79 (0.65, 0.86)	0.85 (0.79, 0.91)	0.80, (0.70, 0.87)	0.58 (0.38, 0.74)	0.34 (0.13, 0.49)
PPV	0.93 (0.85, 0.97)	0.91 (0.85, 0.95)	0.99 (0.85, 0.97)	0.88 (0.76, 0.94)	0.94 (0.76, 0.99)
RM					
DSC	0.83 (0.74, 0.88)	0.87 (0.81, 0.90)	0.83 (0.72, 0.88)	0.77 (0.57, 0.83)	0.79 (0.65, 0.84)
Sensitivity	0.80 (0.69, 0.87)	0.85 (0.79, 0.91)	0.79 (0.63, 0.88)	0.72 (0.57, 0.79)	0.75 (0.59, 0.84)
PPV	0.91 (0.84, 0.95)	0.91 (0.85, 0.95)	0.92 (0.63, 0.88)	0.87 (0.77, 0.94)	0.88 (0.79, 0.94)
t^1 OM vs. RM (p_{adj}-value)					
DSC	−5.9 (0.001)	1.64 (ns)	4.57 (0.001)	1.90 (ns)	12.94 (0.001)
Sensitivity	1.45 (ns)	3.05 (0.036)	4.03 (0.001)	3.33 (0.03)	16.49 (0.001)
PPV	−7.23 (0.001)	0.12 (ns)	2.33 (ns)	0.02 (ns)	0.30 (ns)

Model performance across the original (OM) and retrained (RM) DeepBleed models. All datasets and hemorrhage locations were evaluated for dice scores (DSCs), sensitivity, and positive predictive values (PPVs) and are given as medians with 95% confidence intervals. The metrics of the original (OM) and retrained weights (RM) were compared using *t*-tests with adjusted *p*-values. Briefly, 95% CI = 95% confidence interval; p_{adj}-value = adjusted *p*-value; t^1 = paired *t*-test between OM and RM for the specified metric; ns = not significant.

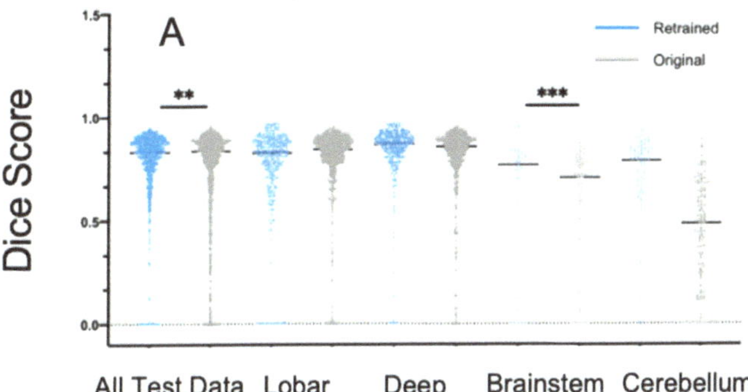

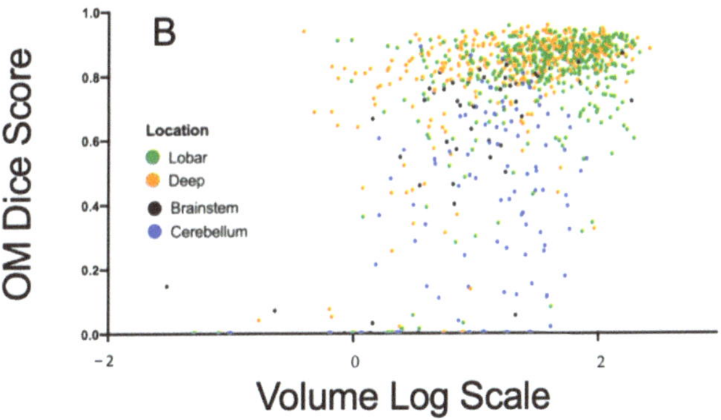

Figure 4. *Cont.*

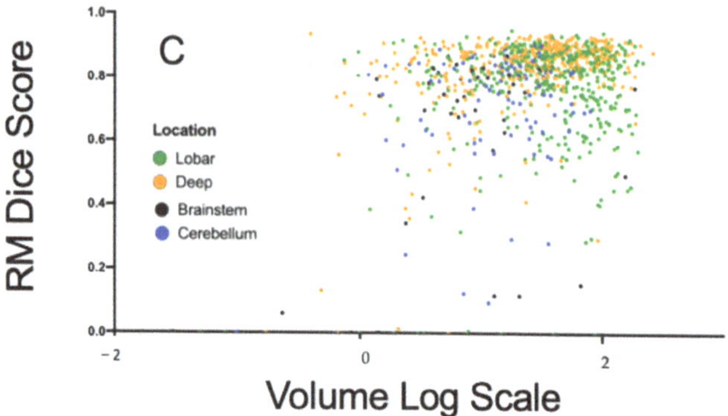

Figure 4. Model performance of the original and retrained model across different locations. (**A**) Comparison of dice scores across lobar, deep, brainstem and cerebellar hemorrhage with original weights (OM, grey) and retrained weights (RM, blue). (**B,C**) Relationship between hemorrhage volume and dice scores across different locations. (**B**) Automatic segmentation with OM. (**C**) Automatic segmentation with RM. **: p-value < 0.01, ***: p-value < 0.001.

3.3. Analysis of Factors Influencing the Model Performance

The results from the multivariate linear models are summarized in Table 3. Results for the univariate analysis are presented in the Supplementary Table S2. Overall model performance was negatively influenced by the ICH location. However, deep ICH was the only location that was not significantly associated with a DSC loss performance neither in the OM nor RM network, at p-values of > 0.05. While the slope coefficients for lobar ICH remained relatively similar in the OM and RM (-0.04, SD 0.01 and -0.06, SD 0.01), the negative effect of brainstem and cerebellar ICH decreased from -0.20 (SD 0.03) and -0.32 (SD 0.02) in the OM to -0.18 (SD 0.03) and -0.08 (SD 0.02) in the RM, at p-values of < 0.001. ICH volume increase had a strong positive effect on the DSC in the OM and RM network. The presence of IVH and the data's originating site had no significant effect on the DSC in the OM or RM. The correlation between ICH location and volume in the DSC are illustrated in Figure 4B,C.

Table 3. Factors influencing the original and retrained model performance.

Parameter	OM			RM		
	Slope	SD	p-Value	Slope	SD	p-Value
	0.75	0.01	<0.001	0.78	0.01	<0.001
Location (in respect to deep location)						
Lobar	−0.04	0.01	<0.01	−0.06	0.01	<0.001
Brainstem	−0.20	0.03	<0.001	−0.18	0.03	<0.001
Cerebellum	−0.32	0.02	<0.001	−0.08	0.02	<0.001
Volume (mm^3)	0.00	0.00	<0.001	0.00	0.00	<0.001
IVH Presence	0.02	0.01	0.17	0.02	0.01	0.15
Center (in respect to Berlin, DE)						
Hamburg, DE	0.003	0.01	0.81	−0.02	0.013	0.09
Pavia, IT	0.008	0.02	0.73	−0.01	0.023	0.66

Multivariate linear regression analysis of variables influencing the model performance in the original (OM) and retrained (RM) DeepBleed model. DE, Germany; IT, Italy; IVH, intraventricular hemorrhage; SD, standard deviation.

3.4. Volume and Segmentation Agreement Analysis

Figure 5A shows the correlation between the GT masks and DeepBleed's automatic volume prediction with the OM and RM. Overall strong correlations were observed among the three segmentation methods ($r > 0.9$, p-value < 0.001), however, correlations among both DeepBleed models, OM and RM, were highest, whereas their correlation with GT volumes was lowest ($r = 0.92$ for OM and $r = 0.94$ for RM). The mean volumes of GT, OM and RM ($\pm$SD) were 43.2 ($\pm$42.6), 36.0 ($\pm$35.9) and 36.2 ($\pm$35.9). The median volumes of the three methods are displayed in Figure 5B. The repeated-measure ANOVA showed no significant effect between the volume estimation of GT masks and automatic volume estimations in the DeepBleed OM and RM (F = 0.45, *p*-value > 0.05).

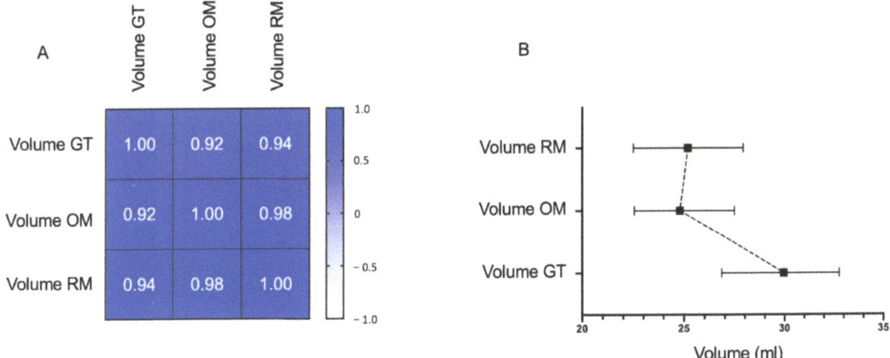

Figure 5. Volume agreement analysis of the original and retrained model with the ground truth. (**A**) Pearson's correlation matrix of agreement for ground truth volume and automatic segmentations of DeepBleed with original weights (original model, OM) and retrained weights (retrained model, RM). (**B**) Median intracerebral hemorrhage and intraventricular hemorrhage volumes of ground truth and automatic segmentation with OM and RM given with a 95% confidence interval. The mean volumes of GT, OM and RM ($\pm$ SD) were 43.2 ($\pm$ 42.6), 36.0 ($\pm$ 35.9) and 36.2 ($\pm$ 35.9). It is possible to see that DeepBleed normally underestimates volume, probably due to the probability map threshold.

Significant agreements were found between the DSCs of manual segmentations by two expert raters and those of the OM and RM DeepBleed network with the GT masks (ICC = 0.90 and ICC = 0.94, *p*-value < 0.0001) and are presented in Figure 6A,B. The repeated-measure ANOVA showed a significant effect of the rater and OM and RM on the DSC (F = 14.38, *p* < 0.0001). The post hoc test showed a significant effect between OM and RM (t = 2.78, p_{adj}-value < 0.05, d = 0.4, small), OM and both raters (t = -5.11 and t = -5.37, p_{adj}-value <0.001 and d = 0.7, moderate for both) and RM and both raters (t = -4.9 and t = -5.3, p_{adj}-value < 0.001 and d = 0.7, moderate for both). No significant difference was found between the two raters (t = -1.57, *p*-value > 0.05, d = 0.2, small).

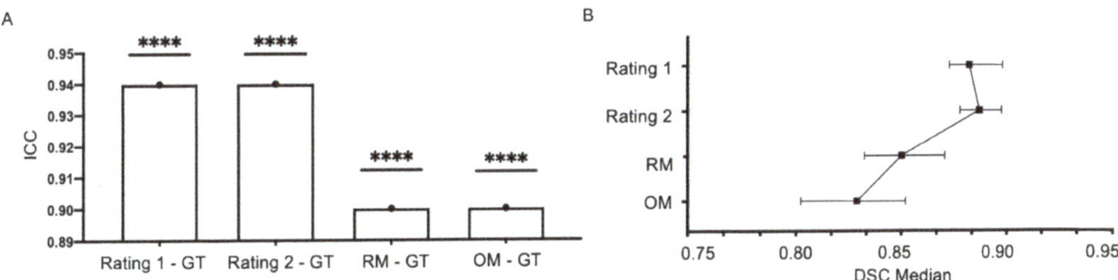

Figure 6. Segmentation agreement analysis of the original and retrained model and human expert raters. (**A**) Intraclass correlation coefficient (ICC) of dice score (DSC) from two experienced raters, automatic segmentations of DeepBleed with original weights (OM) and retrained weights (RM) compared to ground truth (GT) of a senior stroke imaging neuroradiologist. (**B**) Median with 95% CI of DSC from each group. DSC, dice score; ICC, intraclass correlation coefficient; OM, original model; RM, retrained model. ****: *p*-value < 0.0001.

4. Discussion

Our study confirmed the external validity of the first open-source 3D deep learning network for the automatic detection and segmentation of spontaneous ICH with the presence of IVH upon CT. Furthermore, we illustrated the importance of local retraining for a specific setting to increase the applicability of neural network models for ICH segmentation purposes.

In brief, the OM showed overall good results in our multicenter cohort during the validation process. However, location-specific performance metrics were comparatively low for infratentorial ICH lesions which improved significantly after retraining. In particular, performance metrics in cerebellar ICH improved with a 1.7-fold increase in the DSC. The negative effects on the DSC in our linear model improved with a slope increase of 25% for cerebellar ICH after retraining. In comparison, the model performance in supratentorial ICH was overall good for both the OM and RM, with deep ICH demonstrating the best and stable performance metrics. Nonetheless, even lobar ICH slightly improved after model retraining while initially demonstrating the second-best performance metrics in the OM. Additionally, the DSC performance was negatively associated with lobar ICH lesions in our linear regression model. Lobar ICH may demonstrate irregular margins and internal density heterogeneities upon imaging [30]. These imaging phenomena have especially been associated with the use of oral anticoagulants which in turn have been excluded from the MISTIE III trial (novel oral anticoagulants; NOAC) [14,31,32]. As DeepBleed adopts a binary prediction approach at the voxel level using a predetermined threshold, these nuanced voxel differences might be missed [12]. In comparison, another deep learning network, nnU-Net, utilizes the softmax output, as demonstrated by Zhao et al., for ICH segmentation [33–35]. However, we believe that these differences have only a minor impact on the general segmentation performance as shown by the overall good DSC during the validation process. A key strength of the network was that the DSC metrics were independent of the participating site's dataset as well as the presence of IVH. The level of heterogeneity in the developmental cohort of the DeepBleed network may directly relate to this high generalizability of the original model in our external validation cohort. In brief, the original DeepBleed network was multicenter-curated with data from the MISTIE II and III trial series that were conducted at 78 sites in North America, Europe, Australia, and Asia with over 500 patients included [13,14]. In comparison, most of the previous ICH segmentation models were single-center-curated and thus required even more generalization testing ahead of clinical implementation at other sites [35–39]. Secondly, the DeepBleed model employs a dice-based loss function and Adam optimizer, enabling the easy combination of various datasets and augmentation options making it easy to share the trained models as open-source models in order to adapt the network to specific settings [12].

We also observed some limitations in the DeepBleed network. We found that a volume increase had a significantly positive effect on the DSC. This finding is also consistent with that of previous studies showing a positive correlation between lesion size and the DSC in other segmentation networks [40,41]. Therefore, DeepBleed's limited performance in small ICH appears to be a general limitation of deep learning-based networks for segmentation purposes rather than a methodological limitation of DeepBleed—due to the inclusion of supratentorial ICH with an absolute volume greater 30 mL according to the inclusion criteria of the MISTIE III trial [14]. Considering other performance metrics, the absolute volume differences described by the DeepBleed authors were only small despite the variations in the DSC [12] and were thereby within the range with volume errors of 2 to 5 mL observed in similar studies [11,12,42]. In line with this, our post hoc analysis showed a high correlation between the manual and automatic volume predictions, with an underestimation of 5 mL in the automated approach, while the automatic segmentations had a statistically lower agreement in terms of the DSC compared to human expert raters [43,44]. The latter findings might have a stronger clinical implication than do the conclusions obtained from the DSC metrics, as in clinical practice the ICH volume is of great relevance [13,14,16,45–47].

The DSC evaluates the quality of the alignment, which denotes the overlap between the predicted and the GT segmentation [48]. Finally, our results are limited to an external validation cohort of ICH patients who presented within a symptom onset of 24 h. Hence, the performance upon follow-up CT scans beyond a time interval of 24 h may differ.

Our study has the following two main implications. First, our results illustrated that the generalizability of neural networks may be restricted, even when the development and validation cohorts have strong similarities in terms of patient population and healthcare context. Secondly, more extensive retraining may be required to improve the performance at a new site when generalizability is poor. From a clinical point of view, the open-source RM could support decision making for surgical interventions and aid outcome prediction [18,49] in both supra- and infra-tentorial ICH [17,18] with IVH [50,51].

To conclude, the DeepBleed network demonstrated good generalization in an external validation cohort of patients diagnosed with spontaneous ICH on CT and retraining improved location-specific variances significantly. Volumetric analysis showed strong agreement with the manual segmentations of expert raters. However, segmentation accuracy was statistically higher in ground truth masks. The code and RM weights have been made available online [52,53]. Our study illustrates the importance of local retraining for a specific setting to increase the applicability of neural network models for segmentation purposes in spontaneous ICH patients. ICH clinicians and decision makers may take this into account when considering applying externally designed neural network models to their local settings.

Supplementary Materials: The following supporting information can be downloaded at: https://www.mdpi.com/article/10.3390/jcm12124005/s1; Table S1: Original and retrained model performance metrics; Table S2: Factors influencing the original and retrained model performance using univariate analysis.

Author Contributions: Conceptualization, H.C., A.M., F.M., D.D., F.S., C.G., H.K., T.P., J.F., U.H., A.D. and J.N.; methodology, H.C., J.N. and A.D.; software, H.C. and A.D.; validation, H.C., J.N. and A.D.; formal analysis, H.C., A.M., F.M., D.D., F.S., C.G., H.K., T.P., J.F., U.H., A.D. and J.N.; investigation, H.C., J.N. and A.D.; resources, H.C., A.M., F.M., D.D., F.S., C.G., H.K., T.P., J.F., U.H., A.D. and J.N.; data curation, H.C., A.M., F.M., D.D., F.S., C.G., H.K., T.P., J.F., U.H., A.D. and J.N.; writing—original draft preparation, H.C., J.N. and A.D.; writing—review and editing, H.C., A.M., F.M., D.D., F.S., C.G., H.K., T.P., J.F., U.H., A.D. and J.N.; visualization, H.C., J.N. and A.D.; supervision, J.N., A.D., T.P. and U.H. All authors have read and agreed to the published version of the manuscript.

Funding: This research received no external funding.

Institutional Review Board Statement: The study was conducted in accordance with the Declaration of Helsinki, and approved by the Institutional Review Board of Charité Berlin, Germany (protocol number EA1/035/20), University Medical-Center Hamburg, Germany (protocol number WF-054/19), and IRCCS Mondino Foundation, Pavia, Italy (protocol number 20190099462).

Informed Consent Statement: Written informed consent was waived by the institutional review boards due to the retrospective nature of the study.

Data Availability Statement: The data that support the findings of this study are available from the corresponding authors upon reasonable request and in accordance with the institution's data security regulations.

Acknowledgments: J.N. is grateful for being supported by the Berlin Institute of Health (Digital Clinician Scientist Grant funded by Charité—Universitaetsmedizin Berlin, the Berlin Institute of Health and the German Research Foundation, DFG). F.S. was supported by the Berlin Institute of Health (Clinician Scientist Grant). T.P. was supported by the Berlin Institute of Health (Clinician Scientist Grant and Platform Grant), Ministry of Education and Research (BMBF, 01KX2021, and 68GX21001A), German Research Foundation (DFG, SFB 1340/2), and Horizon 2020 (952172). F.M. was supported by the Italian Ministry of Health (Ricerca Corrente 2022–2024).

Conflicts of Interest: Tobias Penzkofer reports research agreements (with no personal payments, outside of submitted work) with AGO, Aprea AB, ARCAGY-GINECO, Astellas Pharma Global Inc.

(APGD), Astra Zeneca, Clovis Oncology, Inc., Dohme Corp, Holaira, Incyte Corporation, Karyopharm, Lion Biotechnologies, Inc., MedImmune, Merck Sharp, Millennium Pharmaceuticals, Inc., Morphotec Inc., NovoCure Ltd., PharmaMar S.A. and PharmaMar USA, Inc., Roche, Siemens Healthineers, and TESARO Inc., and has received fees for a book translation (Elsevier). All other authors have no relevant financial or non-financial interests to disclose.

References

1. Feigin, V.L.; Stark, B.A.; Johnson, C.O.; Roth, G.A.; Bisignano, C.; Abady, G.G.; Abbasifard, M.; Abbasi-Kangevari, M.; Abd-Allah, F.; Abedi, V.; et al. Global, regional, and national burden of stroke and its risk factors, 1990–2019: A systematic analysis for the Global Burden of Disease Study 2019. *Lancet Neurol.* **2021**, *20*, 795–820. [CrossRef] [PubMed]
2. Caplan, L.R. Intracerebral haemorrhage. *Lancet* **1992**, *339*, 656–658. [CrossRef] [PubMed]
3. Qureshi, A.I.; Mendelow, A.D.; Hanley, D.F. Intracerebral haemorrhage. *Lancet* **2009**, *373*, 1632–1644. [CrossRef] [PubMed]
4. Pinho, J.; Costa, A.S.; Araújo, J.M.; Amorim, J.M.; Ferreira, C. Intracerebral hemorrhage outcome: A comprehensive update. *J. Neurol. Sci.* **2019**, *398*, 54–66. [CrossRef]
5. Hemphill, J.C., III; Greenberg, S.M.; Anderson, C.S.; Becker, K.; Bendok, B.R.; Cushman, M.; Fung, G.L.; Goldstein, J.N.; Macdonald, R.L.; Mitchell, P.H.; et al. Guidelines for the Management of Spontaneous Intracerebral Hemorrhage: A Guideline for Healthcare Professionals from the American Heart Association/American Stroke Association. *Stroke* **2015**, *46*, 2032–2060. [CrossRef]
6. Anderson, C.S.; Huang, Y.; Wang, J.G.; Arima, H.; Neal, B.; Peng, B.; Heeley, E.; Skulina, C.; Parsons, M.W.; Kim, J.S.; et al. Intensive blood pressure reduction in acute cerebral haemorrhage trial (INTERACT): A randomised pilot trial. *Lancet Neurol.* **2008**, *7*, 391–399. [CrossRef]
7. Garg, R.K.; Liebling, S.M.; Maas, M.B.; Nemeth, A.J.; Russell, E.J.; Naidech, A.M. Blood pressure reduction, decreased diffusion on MRI, and outcomes after intracerebral hemorrhage. *Stroke* **2012**, *43*, 67–71. [CrossRef]
8. Morgan, T.; Zuccarello, M.; Narayan, R.; Keyl, P.; Lane, K.; Hanley, D. Preliminary findings of the minimally-invasive surgery plus rtPA for intracerebral hemorrhage evacuation (MISTIE) clinical trial. *Acta Neurochir. Suppl.* **2008**, *105*, 147–151.
9. Webb, A.J.; Ullman, N.L.; Morgan, T.C.; Muschelli, J.; Kornbluth, J.; Awad, I.A.; Mayo, S.; Rosenblum, M.; Ziai, W.; Zuccarrello, M.; et al. Accuracy of the ABC/2 Score for Intracerebral Hemorrhage: Systematic Review and Analysis of MISTIE, CLEAR-IVH, and CLEAR III. *Stroke* **2015**, *46*, 2470–2476. [CrossRef]
10. Delcourt, C.; Carcel, C.; Zheng, D.; Sato, S.; Arima, H.; Bhaskar, S.; Janin, P.; Salman, R.A.-S.; Cao, Y.; Zhang, S.; et al. Comparison of ABC methods with computerized estimates of intracerebral hemorrhage volume: The INTERACT2 study. *Cerebrovasc. Dis. Extra* **2019**, *9*, 148–154. [CrossRef]
11. Wang, T.; Song, N.; Liu, L.; Zhu, Z.; Chen, B.; Yang, W.; Chen, Z. Efficiency of a deep learning-based artificial intelligence diagnostic system in spontaneous intracerebral hemorrhage volume measurement. *BMC Med. Imaging* **2021**, *21*, 125. [CrossRef]
12. Sharrock, M.F.; Mould, W.A.; Ali, H.; Hildreth, M.; Awad, I.A.; Hanley, D.F.; Muschelli, J. 3D Deep Neural Network Segmentation of Intracerebral Hemorrhage: Development and Validation for Clinical Trials. *Neuroinformatics* **2021**, *19*, 403–415. [CrossRef]
13. Hanley, D.F.; Thompson, R.E.; Morgan, T.C.; Ullman, N.; Mould, W.A.; Carhuapoma, J.; Kase, C.; Ziai, W.; Thompson, C.B.; Yenokyan, G.; et al. Safety and efficacy of minimally invasive surgery plus alteplase in intracerebral haemorrhage evacuation (MISTIE): A randomised, controlled, open-label, phase 2 trial. *Lancet Neurol.* **2016**, *15*, 1228–1237. [CrossRef]
14. Hanley, D.F.; Thompson, R.E.; Rosenblum, M.; Yenokyan, G.; Lane, K.; McBee, N.; Mayo, S.W.; Bistran-Hall, A.J.; Gandhi, D.; Mould, W.A.; et al. Efficacy and safety of minimally invasive surgery with thrombolysis in intracerebral haemorrhage evacuation (MISTIE III): A randomised, controlled, open-label, blinded endpoint phase 3 trial. *Lancet* **2019**, *393*, 1021–1032. [CrossRef]
15. Yu, A.C.; Mohajer, B.; Eng, J. External validation of deep learning algorithms for radiologic diagnosis: A systematic review. *Radiol. Artif. Intell.* **2022**, *4*, e210064. [CrossRef]
16. Hanley, D. Minimally Invasive Surgery Plus Rt-PA for ICH Evacuation Phase III (MISTIE III). Available online: https://clinicaltrials.gov/ct2/show/study/NCT01827046 (accessed on 28 January 2022).
17. Falcone, G.J.; Biffi, A.; Brouwers, H.B.; Anderson, C.D.; Battey, T.W.K.; Ayres, A.; Vashkevich, A.; Schwab, K.; Rost, N.S.; Goldstein, J.N.; et al. Predictors of hematoma volume in deep and lobar supratentorial intracerebral hemorrhage. *JAMA Neurol.* **2013**, *70*, 988–994. [CrossRef]
18. Chen, R.; Wang, X.; Anderson, C.S.; Ronbinson, T.; Lavados, P.M.; Lindley, R.I.; Chalmers, J.; Delcourt, C.; The INTERACT Investigators. Infratentorial intracerebral hemorrhage: Relation of location to outcome. *Stroke* **2019**, *50*, 1257–1259. [CrossRef]
19. Yushkevich, P.A.; Piven, J.; Hazlett, H.C.; Smith, R.G.; Ho, S.; Gee, J.C.; Gerig, G. User-guided 3D active contour segmentation of anatomical structures: Significantly improved efficiency and reliability. *Neuroimage* **2006**, *31*, 1116–1128. [CrossRef]
20. Harris, C.; Millman, K.; van der Walt, S.; Gommers, R.; Virtanen, P.; Cournapeau, D.; Smith Kern, R.; Picus, M.; Hoyer, S.; van Kerkwijk, M.H.; et al. Array programming with NumPy. *Nature* **2020**, *585*, 357–362. [CrossRef]
21. Smith, S.M. Fast robust automated brain extraction. *Hum. Brain Mapp.* **2002**, *17*, 143–155. [CrossRef]
22. Avants, B.B.; Tustison, N.; Song, G. Advanced normalization tools (ANTS). *Insight J.* **2009**, *2*, 1–35.
23. Rorden, C.; Bonilha, L.; Fridriksson, J.; Bender, B.; Karnath, H.O. Age-specific CT and MRI templates for spatial normalization. *Neuroimage* **2012**, *61*, 957–965. [CrossRef] [PubMed]
24. Kingma, D.P.; Ba, J. Adam: A method for stochastic optimization. *arXiv* **2014**, arXiv:14126980.

25. Dice, L.R. Measures of the amount of ecologic association between species. *Ecology* **1945**, *26*, 297–302. [CrossRef]
26. Van Rossum, G.; Drake, F.L. *Python 3 Reference Manual*; CreateSpace: Scotts Valley, CA, USA, 2009.
27. McCarthy, P.; Cottaar, M.; Webster, M.; Fitzgibbon, S.; Craig, M. fslpy (3.10.0). 2022. Available online: https://git.fmrib.ox.ac.uk/fsl/fslpy/ (accessed on 9 May 2023).
28. GraphPad. Available online: www.graphpad.com (accessed on 9 May 2023).
29. Wickham, H.; Averick, M.; Bryan, J.; Chang, W.; McGowan, L.D.A.; François, R.; Grolemund, G.; Hayes, A.; Henry, L.; Hester, J.; et al. Welcome to the tidyverse. *J. Open Source Softw.* **2019**, *4*, 1686. [CrossRef]
30. Muschelli, J.; Sweeney, E.M.; Ullman, N.L.; Vespa, P.; Hanley, D.F.; Crainiceanu, C.M. PItcHPERFeCT: Primary intracranial hemorrhage probability estimation using random forests on CT. *NeuroImage Clin.* **2017**, *14*, 379–390. [CrossRef]
31. Morotti, A.; Goldstein, J.N. Anticoagulant-associated intracerebral hemorrhage. *Brain Hemorrhages* **2020**, *1*, 89–94. [CrossRef]
32. Gerner, S.T.; Kuramatsu, J.B.; Sembill, J.A.; Sprügel, M.I.; Hagen, M.; Knappe, R.U.; Endres, M.; Haeusler, K.G.; Sobesky, J.; Schurig, J.; et al. Characteristics in Non–Vitamin K Antagonist Oral Anticoagulant–Related Intracerebral Hemorrhage. *Stroke* **2019**, *50*, 1392–1402. [CrossRef]
33. Isensee, F.; Jaeger, P.F.; Kohl, S.A.; Petersen, J.; Maier-Hein, K.H. nnU-Net: A self-configuring method for deep learning-based biomedical image segmentation. *Nat. Methods* **2021**, *18*, 203–211. [CrossRef]
34. Isensee, F.; Jäger, P.F.; Kohl, S.A.; Petersen, J.; Maier-Hein, K.H. Automated design of deep learning methods for biomedical image segmentation. *arXiv* **2019**, arXiv:190408128.
35. Zhao, X.; Chen, K.; Wu, G.; Zhang, G.; Zhou, X.; Lv, C.; Wu, S.; Chen, Y.; Xie, G.; Yao, Z. Deep learning shows good reliability for automatic segmentation and volume measurement of brain hemorrhage, intraventricular extension, and peripheral edema. *Eur. Radiol.* **2021**, *31*, 5012–5020. [CrossRef]
36. Patel, A.; Leemput SCvd Prokop, M.; Ginneken, B.V.; Manniesing, R. Image Level Training and Prediction: Intracranial Hemorrhage Identification in 3D Non-Contrast CT. *IEEE Access* **2019**, *7*, 92355–92364. [CrossRef]
37. Ironside, N.; Chen, C.-J.; Ding, D.; Mayer, S.A.; Connolly, E.S., Jr. Perihematomal edema after spontaneous intracerebral hemorrhage. *Stroke* **2019**, *50*, 1626–1633. [CrossRef]
38. Dhar, R.; Falcone, G.J.; Chen, Y.; Hamzehloo, A.; Kirsch, E.P.; Noche, R.B.; Roth, K.; Acosta, J.; Ruiz, A.; Phuah, C.-L.; et al. Deep Learning for Automated Measurement of Hemorrhage and Perihematomal Edema in Supratentorial Intracerebral Hemorrhage. *Stroke* **2020**, *51*, 648–651. [CrossRef]
39. Yu, N.; Yu, H.; Li, H.; Ma, N.; Hu, C.; Wang, J. A robust deep learning segmentation method for hematoma volumetric detection in intracerebral hemorrhage. *Stroke* **2022**, *53*, 167–176. [CrossRef]
40. Lin, L.; Dou, Q.; Jin, Y.-M.; Zhou, G.-Q.; Tang, Y.-Q.; Chen, W.-L.; Su, B.-A.; Liu, F.; Tao, C.-J.; Jiang, N.; et al. Deep learning for automated contouring of primary tumor volumes by MRI for nasopharyngeal carcinoma. *Radiology* **2019**, *291*, 677–686. [CrossRef]
41. Rudie, J.D.; Weiss, D.A.; Saluja, R.; Rauschecker, A.M.; Wang, J.; Sugrue, L.; Bakas, S.; Colby, J.B. Multi-disease segmentation of gliomas and white matter hyperintensities in the BraTS data using a 3D convolutional neural network. *Front. Comput. Neurosci.* **2019**, *13*, 84. [CrossRef]
42. Scherer, M.; Cordes, J.; Younsi, A.; Sahin, Y.-A.; Götz, M.; Möhlenbrunch, M.; Stock, C.; Bösel, J.; Unterberg, A.; Maier-Hein, K.; et al. Development and Validation of an Automatic Segmentation Algorithm for Quantification of Intracerebral Hemorrhage. *Stroke* **2016**, *47*, 2776–2782. [CrossRef]
43. Koo, T.K.; Li, M.Y. A Guideline of Selecting and Reporting Intraclass Correlation Coefficients for Reliability Research. *J. Chiropr. Med.* **2016**, *15*, 155–163. [CrossRef]
44. Cicchetti, D.V. Guidelines, Criteria, and Rules of Thumb for Evaluating Normed and Standardized Assessment Instruments in Psychology. *Psychol. Assess.* **1994**, *6*, 284–290. [CrossRef]
45. Hanley, D. Clot Lysis: Evaluating Accelerated Resolution of Intraventricular Hemorrhage Phase III (CLEAR-III). 2013. Available online: https://clinicaltrials.gov/ct2/show/NCT00784134 (accessed on 28 January 2022).
46. Hanley, D.F.; Lane, K.; McBee, N.; Ziai, W.; Tuhrim, S.; Lees, K.R.; Dawson, J.; Gandhi, D.; Ullman, N.; Mould, W.A.; et al. Thrombolytic removal of intraventricular haemorrhage in treatment of severe stroke: Results of the randomised, multicentre, multiregion, placebo-controlled CLEAR III trial. *Lancet* **2017**, *389*, 603–611. [CrossRef] [PubMed]
47. MIND. Artemis in the Removal of Intracerebral Hemoorrhage. Available online: https://clinicaltrials.gov/ct2/show/NCT03342664 (accessed on 28 January 2022).
48. Taha, A.A.; Hanbury, A. Metrics for evaluating 3D medical image segmentation: Analysis, selection, and tool. *BMC Med. Imaging* **2015**, *15*, 29. [CrossRef] [PubMed]
49. Hemphill, J.C., III; Bonovich, D.C.; Besmertis, L.; Manley, G.T.; Johnston, S.C. The ICH score: A simple, reliable grading scale for intracerebral hemorrhage. *Stroke* **2001**, *32*, 891–897. [CrossRef] [PubMed]
50. Hill, M.D.; Silver, F.L.; Austin, P.C.; Tu, J.V. Rate of stroke recurrence in patients with primary intracerebral hemorrhage. *Stroke* **2000**, *31*, 123–127. [CrossRef]
51. Hallevi, H.; Albright, K.C.; Aronowski, J.; Barreto, A.D.; Martin-Schild, S.; Khaja, A.M.; Gonzales, N.R.; Illoh, K.; Noser, E.A.; Grotta, J.C. Intraventricular hemorrhage: Anatomic relationships and clinical implications. *Neurology* **2008**, *70*, 848–852. [CrossRef]